In accordance with the latest syllabus prescribed by the
Council for the Indian Certificate of Secondary Education Examination, New Delhi.

ICSE BIOLOGY

Most Likely Question Bank

Class IX

By

A PANEL OF AUTHORS

OSWAL PUBLISHERS

1/12, Sahitya Kunj, M. G. Road, Agra-282 002

Edition : 2021

Price : ₹ 299·00

ISBN : 978-93-91184-60-5

OSWAL PUBLISHERS

Head office : 1/12, Sahitya Kunj, M.G. Road, Agra-282 002
Phone : (0562) 2527771– 4, + 91 75340 77222
E-mail : contact@oswalpublishers.com
Website : www.oswalpublishers.com
Printed at : Upkar Printing Unit, Agra

The cover of this book has been designed using resources from Freepik.com

PREFACE

It is a matter of immense pride for us to present the 'ICSE MOST LIKELY QUESTION BANK' series, especially prepared for students appearing for Board examinations in the oncoming year.

This book has been created with the specific purpose of making the students' journey of learning, understanding and revising the concepts, effortless and simple. The topical approach with ample questions for every category is adopted to reinforce the students' understanding of each chapter. The category-wise division also allows them to peruse their progress as well as keep a check on their grasp of the theory.

Meticulous care has been taken in writing the book in simple, student-friendly language without compromising with the clarity of style.

We are confident that the book will enable the candidates to develop a better understanding of the curriculum and help them organize their learning process. This book shall definitely prove to be a fruitful tool for the students and encourage them towards scholastic excellence.

Constructive suggestions for further improvement of the book are always welcome.

—Publisher

IMPORTANT NOTE

The global outbreak of the Novel Coronavirus (COVID-19) has impacted all aspects of life including the educational life at schools. Schools across the country have been shut since March, 2020 due to the pandemic. While numbers of CISCE affiliated schools have tried to adapt to this changed scenario and have tried to keep alive the teaching learning process through online classes, there has been a significant shortening of the academic year and loss of the instructional hours.

To make up for the loss in instructional hours during the current session 2020-2021, the CISCE has worked with its subject experts, to reduce the syllabi for all major subjects at the ICSE and ISC levels. Syllabus reduction has been done, keeping in mind the linear progression across classes while ensuring that the core concepts related to the subject are retained.

The following reduced syllabi, for the current Academic Year 2020-2021 have been made available on the CISCE website www.cisce.org under 'Publications':

- ICSE Reduced Syllabus for Class IX
- ICSE Reduced Syllabus for Class X
- ISC Reduced Syllabus for Class XI
- ISC Reduced Syllabus for Class XII

Heads of CISCE affiliated schools have been asked to ensure that the concerned subject teachers at the ICSE and ISC levels transact the syllabus strictly according to sequence of topics, so as to facilitate further reduction in syllabus, if required, depending on the situation of the pandemic in the country.

We at Oswal Publishers, have developed all our books on the basis of the original syllabi with complete subjective knowledge of all the subjects, so that the students have an access to the entire syllabus. However, for the examination purpose, the students are advised to structure their preparations considering the latest alterations by the Council.

Scan this to know the recent changes in Syllabus

CONTENTS

How to choose a

GREAT CAREER

1. Know where your heart is

Whatever inspires, motivates and comes naturally to you should be the basis for making a career choice.

2. Explore related courses

It could be science, marine biology, astrophysics or anything, search what courses are there in your area of interest.

3. Reality check

Follow your heart but only after a reality check. It is advisable to explore other fields if your area of interest offers a limited scope.

4. Eligibility criteria

Check the eligibility criteria and know the requirements for the course.

5. Assess yourself

Weigh the pros and cons to see where can you apply.

6. Share and seek advice

Don't hesitate. Speak to your parents or seek professional advice for the course selected.

7. The final decision

Once you are through with all the steps, you will know where you stand and will be better placed to make the right choice.

Go where your strength is, not where your friends are.

HOW TO MANAGE YOUR TIME BETTER

This is the time when you are going to be promoted to higher classes. Going a level up also means there will be more books, more syllabus, more tests & exams. We have pieced together some of the most effective tips that will surely help you stay ahead of your peers by staying organised and managing your time as well as energy in an improved manner.

- **Make a list:** Every morning, jot down all the things you have to do for the day. Arrange all your tasks according to their importance and urgency. Then, before going to bed, strike off the tasks you managed to finish, giving yourself a sense of accomplishment and helping you stay on track.

- **Segregate your time:** You should divide your day into hourly-chunks, based on your routine. For example, separate about 6-7 compulsory hours for school; then, 7-8 hours for sleep, 4-5 hours for self-study, 1 hour each for leisure and meals. You can keep the remaining hours free as they get used up in chores and other mundane tasks.

- **Improve your focus:** You must strive to finish your tasks in a decided time limit. Learn to eliminate distractions. You should give your undivided attention to their completion. This will improve your efficiency and help you finish your work on time.

- **Take little breaks:** In your study schedule, make sure to assign small breaks between long study sessions to give your brain some rest and restore your energy for another round of rigorous learning. You can have a snack or simply close your eyes and quietly meditate.

- **End procrastination:** Procrastination is the biggest hurdle in your path to success. If you have a daunting task at hand, it's better to break it into smaller chunks and work on them than just postponing it for later. Slow and steady wins the race, after all!

Follow these tips and watch your productivity increase with time. Share your experience with us at contact@oswalpublishers.com .

WEEKLY SCHEDULE

MONDAY	TUESDAY	WEDNESDAY	THURSDAY	FRIDAY	SATURDAY	SUNDAY

GOALS

DON'T FORGET!

NOTES

Correct Statement and Order

Set 1

Chapter 1. Introducing Biology

Rewrite correct statments by changing the underlined words:

1. Aristotle is said to be the father of botany.
2. Cultivation of wheat was introduced in India.
3. Horticulture is the practice of bee keeping for the manufacturing of honey and beeswax.
4. Silk is produced by rearing honeybee.
5. Taxonomy (also called eugenics) is science of naming, grouping, and classifying plants and animals.

Ans. 1. Theophrastus is said to be the father of botany.
2. Cultivation of rice was introduced in India.
3. Apiculture is the practice of bee keeping for the manufacturing of honey and beeswax.
4. Silk is produced by rearing silkworm.
5. Taxonomy (also called systematics) is science of naming, grouping, and classifying plants and animals.

Chapter 2. Cell: The Unit of Life

Given below are some set of terms. Rewrite the terms in logical sequence beginning with the underlined term:

1. Cytoplasm, Plasma membrane, Capsule, Cell wall, Nucleoid
2. Chromatin network, Nucleoplasm, Nuclear membrane, Nucleolus, Protein synthesis
3. Chromosomes, Nucleoplasm, Nuclear membrane, Cytoplasm, Cell membrane **[November, 2019]**

Ans. 1. Capsule, Cell wall, Plasma membrane, Cytoplasm, Nucleoid
2. Nuclear membrane, Nucleoplasm, Chromatin network, Nucleolus, Protein synthesis
3. Nuclear membrane, Nucleoplasm, Chromatin network, Nucleolus, Protein synthesis

Rewrite the correct statements by changing the underlined words:

1. Prokaryotic cells have welldeveloped nucleus separated by a nuclear membrane.
2. Robert Hooke was the first scientist who discovered the living cells in pond water in 1960.
3. Ribosomes are the factories for the synthesis of lipids.
4. In plant cells vacuoles are usually quite small and the liquid which they contain is called cell sap.
5. Every cell in the human body has 23 chromosomes. **[November, 2019]**

Ans. 1. Eukaryotic cells have well developed nucleus separated by a nuclear membrane.
2. Anton von Leeuwenhoek was the first scientist who discovered the living cells in pond water in 1960.
3. Ribosomes are the factories for the synthesis of proteins.
4. In plant cells vacuoles are usually quite large and the liquid which they contain is called cell sap.
5. Every cell in the human body has 46 chromosomes.

Chapter 3. Tissues: Plants and Animal tissue

Rewrite the correct statements by changing the underlined words:

1. Cartilage is semi-transparent and fluid muscular tissue.
2. Phloem fibres mainly concern with storage of starch, fat and other organic food materials.
3. Apical meristem is responsible for increase in diameter of the stem.
4. Epithelial tissue helps in maintaining the shape of the body and its organs.
5. Voluntary muscles are unstriated. **[November, 2019]**

Ans. 1. Cartilage is semi-transparent and elastic supportive fluid connective tissue.
2. Phloem parenchyma mainly concern with storage of starch, fat and other organic food materials.
3. Cambial or lateral meristem is responsible for increase in diameter of the stem.
4. Connective tissue helps in maintaining the shape of the body and its organs.
5. Voluntary muscles are striated.

Given below are some of terms. Rewrite the terms in logical sequence beginning with the underlined terms.

1. Dendron, Dendrite, Cell body, Nissl granules, Nucleus
2. Companion cell, Sieve cell, Phloem parenchyma cell, Sieve plate,
3. Cork cambium, Ray, Cork, Phloem, Cambium
4. Permanent tissue, Collenchyma, Phloem, Supportive tissue, Plant tissue

Ans. 1. Dendrite, Dendron, Cell body, Nissl granules, Nucleus
2. Phloem parenchyma cell, Companion cell, Sieve cell, Sieve plate
3. Cambium, Ray, Phloem, Cork cambium, Cork
4. Plant tissue, Supportive tissue, Permanent tissue, Collenchyma, Phloem

Chapter 4. The Flower

Given below are some sets of terms. Rewrite the terms in logical sequence beginning with the underlined terms:

1. Ovary, Style, Stigma, Stalk, Receptacle, Ovule.
2. Petal, Sepal, Stamen, Pistil.
3. Ovary, Pollen grain, Ovule, Stigma, Style **[November, 2019]**

Ans. 1. Stalk, Receptacle, Ovule, Ovary, Style, Stigma
2. Sepal, Petal, Stamen, Pistil.
3. Pollen grain, Stigma, Style, Ovary, Ovule.

Rewrite the correct statements by changing the underlined words:

1. Gynoecium represents the male reproductive organ.
2. An inflorescence is categorised on the basis of the arrangement of leaveson a main axis (peduncle).
3. Sepals are brightly coloured which attract insects for pollination.

Ans. 1. Androecium represents the male reproductive organ.
2. An inflorescence is categorised on the basis of the arrangement of flowers on a main axis (peduncle).
3. Petals are brightly coloured which attract insects for pollination.

Chapter 5. Pollination and Fertilisation

Rewrite the correct statements by changing the underlined words:

1. The ovary enlarges to form the seeds and the ovarian wall forms the fruit wall.
2. Self-pollination produces healthier offspring and new varieties.

3. In heterostyly, the pollen of a flower cannot reach the stigma of the same flower.
4. Pollination of flowers by insects is called ornithophily.
5. In double fertilisation, from the two sperm cells, one sperm fertilises the egg cell, forming a triploid zygote and the other sperm fuses with the two polar nuclei, forming a diploid cell that develops into the endosperm.

Ans. 1. The ovary enlarges to form the fruit and the ovarian wall forms the fruit wall.
2. Cross-pollination produces healthier offspring and new varieties.
3. In herkogamy, the pollen of a flower cannot reach the stigma of the same flower .
4. Pollination of flowers by birds is called ornithophily.
5. In double fertilisation, from the two sperm cells, one sperm fertilises the egg cell, forming a diploid zygote and the other sperm fuses with the two polar nuclei, forming a triploid cell that develops into the endosperm.

Given below are some sets of terms. Rewrite the terms in logical sequence beginning with the underlined terms.

1. Anther, Stigma, Style, Petal, Ovary.
2. Central cell, Nucellus, Ovule, Egg cell, Micropyle, Embryo sac, Synergids.

Ans. 1. Petal, Stigma, Anther, Style, Ovary.
2. Ovule, Nucellus, Embryo sac, Central cell, Synergids, Egg cell, Micropyle.

Chapter 6. Seed-Structure and germination

Given below are some sets of terms. Rewrite the terms in logical sequence beginning with the underlined terms:

1. Cotyledon, Testa, Radicle, Tegmen, Hypocotyl.
2. Epithelium, Aleurone layer, Scutellum, Endosperm, Radicle.
3. Seed coat bursts, Hypocotyl elongates, Radicle grows downward, Hypocotyl forms loop above the soil, Epicotyl elongates.
4. Embryo, 1st male gamete, Zygote, Egg cell, Micropyle.
5. Zygote, Embryo, Seed, Allogamy, Fusion of gametes.
6. Seed, Seedling, Pollination, Germination, Fertilization. **[February, 2020]**

Ans. 1. Testa, Tegmen, Cotyledon, Hypocotyl, Radicle.
2. Aleurone Layer, Endosperm, Epithelium, Scutellum, Radicle.
3. Seed coat bursts, Radicle grows downward, Hypocotyl forms loop above the soil, Epicotyl elongates.
4. Micropyle, 1st male gamete, Egg cell, Zygote, Embryo.
5. Allogamy, Fusion of gametes, Zygote, Embryo, Seed.
6. Pollination, Fertilisation, Seed, Germination, Seedling.

Rewrite the correct statements by changing the underlined words:

1. Fruit is the enlarged ripened ovule.
2. Pollination is the process of formation of a seedling developed from the embryo.
3. Endosperm is the embryonic leaf which possess plumule and radicle of the embryonic plant.
4. In dry seeds the embryo is in active state.
5. The embryo of monocot consists of one large, shield-shaped cotyledon known as aleurone layer.

Ans. 1. Fruit is the enlarged ripened ovary.
2. Germination is the process of formation of a seedling developed from the embryo.
3. Cotyledon is the embryonic leaf which possess plumule and radicle of the embryonic plant.

4. In dry seeds the embryo is in dormant state.
5. The embryo of monocot consists of one large, shield-shaped cotyledon known as scutellum.

Chapter 7. Respiration in Plants

Given below are some sets of terms. Rewrite the terms in logical sequence beginning with the underlined terms:

1. Pyurvate, glucose, cytoplasm, mitochondria, 38 ATP
2. Ethanol, anaerobic respiration, glucose, fungi, carbon dioxide
3. Glucose, glycolysis, pyruvate, cytoplasm, ATP
4. Carbon dioxide, germinating seeds, respiration, oxygen, glucose
5. Muscle cells, pyruvate, anaerobic respiration, lactic acid, heat

Ans. 1. Cytoplasm, glucose, pyruvate, mitochondria, 38 ATP
2. Fungi, anaerobic respiration, glucose, ethanol, carbon dioxide
3. Glycolysis, cytoplasm, glucose, pyruvate, ATP
4. Germinating seeds, respiration, oxygen, glucose, carbon dioxide
5. Anaerobic respiration, muscle cells, pyruvate, lactic acid, heat

Rewrite the correct statements by changing the underlined words:

1. Kreb's cycle is the first phase of breaking down of glucose.
2. Respiration is an anabolic process.
3. Carbon dioxide turns soda-lime milky.
4. Pyurvate is the product formed during Kreb's cycle.
5. End products of anaerobic respiration in plants is lactic acid.

Ans. 1. Glycolysis is the first phase of breaking down of glucose.
2. Photosynthesis is an anabolic process.
3. Carbon dioxide turns lime water milky.
4. Pyurvate is the product formed during glycolysis.
5. End products of anaerobic respiration in plants is ethyl alcohol.

Chapter 8. Five Kingdom Classification

Given below are some sets of terms. Rewrite the terms in logical sequence beginning with the underlined terms:

1. Class, phylum, family, order, genus **[November, 2019]**
2. Multicellular, kingdom animalia, kingdom plantae, eukaryotes, organisms
3. Invetebrates, cnidaria, animal kingdom, porifera, nematode
4. Arthropoda, mollusca, annelida, coelomata, invertebrates
5. Gymnosperms, algae, ferns, mosses, dicots
6. Mollusca, cnidaria, chordata, flatworms, Annelids
7. Amphibians, pisces, aves, reptilies, mammals
8. Cuttlefish, jellyfish, silverfish, dogfish, starfish

Ans. 1. Phylum, class, order, family, genus
2. Organisms, eukaryotes, multicellular, kingdom animalia, kingdom plantae
3. Animal kingdom, invetebrates, porifera, cnidaria, nematoda
4. Mollusca, arthropoda, annelida, coelomata, invertebrates
5. Dicots, gymnosperms, ferns, mosses, algae

6. Cnidaria, flatworms, annelids, mollusca, chordates
7. Mammals, aves, reptiles, amphibians, pisces
8. Jelly fish, silverfish, cuttlefish, starfish, dogfish

Rewrite the correct statements by changing the underlined words:

1. In a genus, members can interbreed among themselves to produce fertile young ones.
2. Two-kingdom classification was proposed by Whittaker.
3. The plants belonging to Pteridophytes bear naked seeds.
4. Osculum is present in phylum Echinodermata.
5. Class Reptilia have smooth, non-scaly, moist skin.
6. Presence of pinna is a characteristic of birds. **[November, 2019]**

Ans. 1. In a species, members can interbreed among themselves to produce fertile young ones.
2. Two-kingdom classification was proposed by Linnaeus.
3. The plants belonging to Gymnosperms bear naked seeds.
4. Water vascular system is present in phylum Echinodermata.
5. Class Amphibia have smooth, non-scaly, moist skin.
6. Presence of pinna is a characteristic of mammals.

Chapter 9. Economic Importance of Bacteria and Fungi

Given below are some sets of terms. Rewrite the terms in logical sequence beginning with the underlined terms:

1. Cell membrane, capsule, cytoplasm, cell wall, nucleus
2. Spores, mycelium, sporangium, hyphae, sporangiophores
3. Casing, preservation, composting, cropping and harvesting, spawning
4. Ripening, milk curdling, salting, curd processing, cheese
5. Nitrites, Ammonia, nitrates, ammonium compounds, nitrogen

Ans. 1. Capsule, cell wall, cell membrane, cytoplasm, nucleus
2. Mycelium, hyphae, sporangiophores, sporangium, spores
3. Composting, spawning, casing, cropping and harvesting, preservation
4. Milk curdling, curd processing, salting, ripening, cheese
5. Nitrogen, ammonia, ammonium compounds, nitrites, nitrates

Rewrite the correct statements by changing the underlined words:

1. Free living bacteria fix nitrogen by forming close associations with other organisms.
2. Spore formation is a method of sexual production in bacteria.
3. Curdling of milk is done by Rhizopus.
4. Mushroom seed consists of buds of the selected type of mushrooms.
5. Sterilisation process is employed to preserve milk.

Ans. 1. Symbiotic bacteria fix nitrogen by forming close associations with other organisms.
2. Conjugation is a method of sexual production in bacteria.
3. Curdling of milk is done by *Lactobacillus.*
4. Mushroom seed consists of mycelium of the selected type of mushrooms.
5. Pasteurisation process is employed to preserve milk.

Chapter 10. Nutrition

1. All individuals need same kind of food.
2. Fruit peels also contain more nutrient and they are not easily digestible.

3. Calcium is present in haoemoglobin.
4. We should take tasty food when we are sick.
5. Iodine controls nerve and muscle activity.
6. Iron is the mineral required for proper functioning of the thyroid gland. **[February, 2020]**

Ans. 1. All individuals need different kind of food.
2. Fruit peels also contain less or no nutrients and they are not easily digestive.
3. Iron is present in haemoglobin.
4. We should take tasty food when we are healthy.
5. Potassium controls nerve and muscle cavity.
6. Iodine is the mineral required for proper functioning of the thyroid gland.

Chapter 11. Digestive System

Rewrite the correct statements by changing the underlined words:

1. Food is rolled into small balls called pylorus and pushed into food pipe.
2. Bile is produced by bile duct.
3. Maximum absorption of nutrients takes place in large intestine.
4. Chymotrypsin converts casein to paracasein.
5. Large intestine is large in length.

Ans. 1. Food is rolled into small balls called bolus and pushed into food pipe.
2. Bile is produced by liver.
3. Maximum absorption of nutrients takes place in small intestine.
4. Renin converts to paracasein.
5. Lumen is large in length.

Given below are some sets of terms. Rewrite the terms in logical sequence beginning with the underlined terms:

1. Large intestine, stomach, mouth, small intestine, oesophagus. **[November, 2019]**
2. Stomach, oesophagus, large intestien, small intestine, pharynx.
3. Caecum, rectum, colon, jejunum, ileum. **[February, 2020]**

Ans. 1. Mouth, oesophagus, stomach, small intestine,large intestine.
2. Pharynx, oesophagus, stomach, small intestine, large intestine.
3. Jejunum, ileum, caecum, colon, rectum.

Chapter 12. Skeleton-Movement and Locomotion

Rewrite the correct statements by changing the underlined words:

1. Human skull has 23 bones.
2. Hinge joint is fixed joint.
3. There are eight pairs of true ribs.
4. Skull bones have moveable joints.
5. Backbone is also called vertebrae.
6. Humerus is a bone of leg.
7. Pelvic girdle articulates with humerus.
8. The vertebral column has 12 neck vertebrae. **[November, 2019]**

Ans. 1. Human skull has 22 bones.
2. Hinge joint is slightly moveable.

3. There are seven pairs of true ribs.
4. Skull bones have fixed joints.
5. Backbone is also called vertebral column.
6. Humerus is bone of arm.
7. Pelvic girdle articulates with femur.
8. The vertebral column has 7 neck vertebrae.

Given below are some sets of terms. Rewrite the terms in logical sequence beginning with the underlined terms:

1. Atlas, coccyx, Lumbar, Axis, Sacrum **[November, 2019]**
2. Axis is the first cervical vertetrae. **[February, 2020]**
3. Lumbar, Cervical, Coccyx, Sacrum, Thoracic. **[February, 2020]**
4. Metacarpals, Radius, Humerus, Carpals, Pholanges. **[February, 2020]**

Ans. 1. Atlas, axis, lumber, sacrum, coccyx.
2. Atlas is the first cervical vertebrae.
3. Cervical, thoracic, lumbar, sacrum, coccyx
4. Humerus, radius, carpals, metacorpals, phalanges.

Chapter 13. Skin-"The jack of all Trades"

Rewrite the correct statements by changing the underlined words:

1. Skin protects us from infra red radiations.
2. Dermis is very thick on finger tips.
3. Sweat contains traces of keratin.
4. All mammals are ectothermal.
5. Aestivism is also called autumn sleep.
6. Vasoconstriction occurs during summer. **[February, 2020]**

Ans. 1. Skin protects us from ultra-violet radiations
2. Dermis is very thick on palms.
3. Sweat contains traces of urea.
4. All mammals are endothermal.
5. Aestivism is also called summer sleep.
6. Vasodilation occurs during summer.

Chapter 14. The Respiratory System

Rewrite the correct statements by changing the underlined words:

1. In mitochondria glucose is broken down to produce energy.
2. C shaped bony rings protect the bronchi from collapsing.
3. Air expelled from lungs contains almost 21% oxygen.
4. Pharynx is called voice box.
5. CO_2 and water are produced in anaerobic respiration.
6. Respiration takes place in the chloroplast. **[February, 2020]**
7. Larynx, Alveolus, Trachea, Bronchiole, Nose. **[February, 2020]**

Ans. 1. In mitochondria glucose is broken down to pyruvic acid.
2. Cartilaginous protects the bronchi from collapsing.
3. Air expelled from lungs contains almost 16.40% oxygen.
4. Larynx is called voice box.

5. CO_2 and water are produced in aerobic respiration.
6. Respiration takes place in the mitochondria.
7. Nose, Larynx, Alvelus, Trachea, Bronchiole, Alveolus.

Chapter 15. Hygiene

Rewrite the correct statements by changing the underlined words:

1. Cholera is marked by intestinal pain and mild fever.
2. Blood circulation improves by rest and sleep.
3. Infection of intestinal tract is by bacillary dysentery.
4. Dysentery causes severe diarrhea and vomiting.
5. The infectious hepatitis A is transmitted through blood and other body fluids.
6. Aedes mosquito spreads malaria due to the malaria germs.
7. Presence of sulphides in drinking water causes tooth decay and bone disorders.

Ans. 1. Cholera is marked by severe diarrhea and no urination.
2. Blood circulation improves by physical exercise.
3. Infection of intestinal tract is by Bacterium (*Vibrio cholera).*
4. Cholera causes severe diarrhea and vomiting.
5. The infectious hepatitis A is transmitted through contaminated water.
6. *Anopheles* mosquito spreads malaria due to the malaria germs.
7. Presence of fluorides in drinking water causes tooth decay and bone disorders.

Chapter 16. Disease : Cause and Control

Rewrite the correct statements by changing the underlined words:

1. BCG vaccine is used to prevent typhoid.
2. Ascariasis is caused by *Wuchereria bancrofti*.
3. The incubation period of Hepatitis A is few hours.
4. March 21 is observed as anti-tuberculosis day.
5. *Vibrio cholerae* causes malaria.
6. Salk's vaccine is used to prevent Tuberculosis. **[November, 2019]**

Ans. 1. BCG vaccine is used to prevent tuberculosis.
2. Ascariasis is caused by *Ascaris lumbricoides.*
3. The incubation period of Hepatitis A is 14 to 45 days.
4. March 24 is observed as anti-tuberculosis day.
5. *Vibrio cholerae* causes cholera.
6. BCG vaccine is used to prevent Tuberculosis.

Chapter 17. Aids to Health

Rewrite the correct statements by changing the underlined words:

1. The term immunity means body's defence against its own cell.
2. The prevention of formation of antibody is called local defence system.
3. The blood clotting occurs when any cut in the skin to allow the germs the cell.
4. Keratin is the slimy secretion of the epithelial lining of various organs.
5. The immunity by virtue of genetic constitutional make up is called acquired immunity.
6. DDT is an antiseptic.

7. Penicillin is a disinfectant.
8. Disinfectants are applied on the body.
9. Deodorants are both antiseptic as well as disinfectants.
10. Alexander Fleming discovered sulphonamide.
11. Antibiotics can not be made synthetically.

Ans. 1. The term immunity means body's defence against foreign bodies.
2. The prevention of entry of germs is called local defence system.
3. The blood clotting occurs during any cut in the skin to prevent the germs from entering the cell.
4. Mucus is the slimy secretion of the epithelial lining of various organs.
5. The immunity by virtue of genetic constitutional make up is called innate immunity.
6. DDT is a disinfectant.
7. Penicillin is an antibiotic.
8. Disinfectants are applied on the surface or floor.
9. Deodorants are neither antiseptics nor disinfectants.
10. Alexander Fleming discovered penicillin.
11. Antibiotics are made synthetically.

Chapter 18. Health Organisation

Rewrite the correct statements by changing the underlined words:

1. Small towns are type of habitations where there are multi-storeyed buildings and too many vehicles are found.
2. The two main health organisations are UNO and WHO.
3. Typhoid is an insect-borne disease.
4. International bodies are one of the categories of health problem.
5. Red Cross is the specialised agency of UNO.

Ans. 1. Big cities are type of habitations where there are multi-storeyed buildings and too many vehicles are found.
2. The two main health organisations are Red Cross and WHO.
3. Malaria is an insect-borne disease.
4. International bodies are one of the Categories of health organisation.
5. WHO is the specialised agency of UNO.

Chapter 19. Waste Generation and Management

Rewrite the correct statements by changing the underlined words:

1. Garbage should be disposed off in one bin.
2. Fruit peels, vegetables, leftover food is agricultural waste.
3. Re-useable articles are collected by rag pickers.
4. We should cultivate habit of making manure from commercial waste.
5. Recycled paper can damage trees.
6. Segregation is the disposal of waste by burning. **[February, 2020]**

Ans. 1. Garbage should be disposed off in two bin.
2. Fruit peels, vegetables, leftover food is domestic waste.
3. Recycleable articles are collected by rag pickers.
4. We should cultivate habit of making manure from domestic waste.
5. Recycled paper can save trees.
6. Incineration is the disposal of waste by burning.

Differentiation Based Questions

Set 2

Differentiate between the following terms:

Chapter 1. Introducing Biology

1. Botany and Zoology (Type of study)
2. Histology and Cytology (Type of study)
3. Physiology and Morphology (Type of study)
4. Apiculture and Sericulture (Application in biology)
5. Biotechnology and Bioinformatics (Applied biology)

Ans.

1.

S. No.	Botany	Zoology
(i)	It deals with the study of plants.	It deals with the study of animals.
(ii)	Theophrastus is known as the father of botany.	Aristotle is considered as the father of zoology.

2.

S. No.	Histology	Cytology
(i)	It refers to the study of the microscopic structure of tissues.	It refers to the study of structure and function of plant and animal cells.
(ii)	It studies the architecture of tissues.	It only studies the cellular areas of cells.

3.

S. No.	Physiology	Morphology
(i)	It deals with the study of normal function and activities of organisms and their parts.	It deals with the study of form and structure of plants and animals.
(ii)	For example, study of human body.	For example, morphology of leaf, morphology of cockroach, etc.

4.

S. No	Apiculture	Sericulture
(i)	It is also called bee keeping.	It is also called silk faming.
(ii)	It is a process of making honey, beeswax and other products of the beehive.	It is the process of making silk fibres.
(iii)	The species involved in this process is honey bee.	The commercial species involves in sericulture is silkworm.

5.

S. No	Biotechnology	Bioinformatics
(i)	It is process of using living processes, organisms or systems in industries and technology.	It is the collection, classification, storage, and analysis of biochemical and biological information using computers.
(ii)	It includes area like production of vaccine, enzyme, antibodies, tissue culture, genetic engineering, etc.	It increases the understanding of biological processes.

Chapter 2. Cell: The Unit of Life

1. Cell wall and Cell membrane
2. RER and SER
3. Plant cell and Animal cell (Type of cell)
4. Eukaryotic cell and Prokaryotic cell **[February, 2020]**
5. Nucleolus and Nucleus (Type of organelle)

Ans.

1.

S. No.	Cell wall	Cell membrane
(i)	It is present in plants, bacteria, fungi and algae.	It is present in all type of cells.
(ii)	It is fully permeable.	It is semipermeable.
(iii)	It is made up of different components like chitin (fungi), cellulose (plants), etc.	It is made up of protein, lipid and carbohydrates.
(iv)	It provides protection, strength, and rigidity to the cell.	It controls passage of materials in and out of the cell.

2.

S. No.	RER	SER
(i)	RER is rough endoplasmic reticulum.	SER is smooth endoplasmic reticulum.
(ii)	It appears rough due to presence of ribosomes on their membrane.	It appears smooth due to absence of ribosomes on their membrane.
(iii)	It is involved in the synthesis of proteins.	It is involved in the synthesis of fats.

3.

S. No.	Plant cell	Animal cell
(i)	Cell wall is present.	Cell wall is absent.
(ii)	Centrosome is absent.	Centrosome is present.
(iii)	Vacuoles are large and prominent.	Vacuoles are small and temporary.
(iv)	Plastids are present.	Plastids are absent.

4.

S. No.	Eukaryotic cell	Prokaryotic cell
(i)	Organisms with cells containing a well-defined nucleus with a nuclear membrane are called eukaryotes.	Organisms with cells containing a primitive, undefined nucleus are called prokaryotes.
(ii)	They contain larger ribosomes.	They contain smaller ribosomes.
(iii)	They contain other cell organelles.	They lack in other cell organelles.
(iv)	Examples: *Euglena*, Human beings	Examples: Bacteria, blue-green algae.

5.

S. No.	Nucleolus	Nucleus
(i)	It is the sub-organelle in the nucleus.	It is a membrane-enclosed, large spherical organelle found in the eukaryotic cell.
(ii)	It is rich in RNA.	It is rich in DNA.
(iii)	Responsible for the production of ribosomes.	Responsible for the transmission of genetic information and regulation of cell function.

Chapter 3. Tissues: Plants and Animal Tissue

1. Skeletal muscle and Smooth muscle
2. Cartilage and Bone (Nature of tissue)
3. Meristematic plant tissue and Permanent plant tissue
4. Xylem and Phloem
5. Blood and Lymph
6. Tendon and Ligament (function) **[November, 2019]**

Ans.

1.

S. No.	Skeletal muscle	Smooth muscle
(i)	These are voluntary muscles.	These are involuntary muscles.
(ii)	These muscles are under the control of their will.	These muscles are not under the control of their will
(iii)	Cells are nucleated and striated.	Cells are uninucleated and spindle shaped.
(iv)	These muscles are found in the arms, legs, face, neck, etc.	These muscles are found in the walls of intestine, lining of blood vessels, uterus, urinary bladder skin, etc

2.

S. No.	Cartilage	Bone
(i)	It is non-porous tissue.	It is porous tissue.
(ii)	It is a soft and flexible tissue.	It is the hardest part of the body.
(iii)	It is made up of protein salt and connective tissue.	It is made up of calcium and potassium.
(iv)	Cartilage does not have rich blood supply.	Bones have rich blood supply.

3.

S. No.	Meristematic plant tissue	Permanent plant tissue
(i)	These tissues have capacity to divide.	These tissues have lost the capacity to divide.
(ii)	These tissues do not have intercellular spaces.	These tissues have intercellular spaces.
(iii)	They are responsible for primary and secondary growth of plants.	These tissues add to various growths and take a permanent shape to perform some functions.

4.

S. No.	Xylem	Phloem
(i)	It is composed of four types of elements: xylem vessels, tracheids, xylem parenchyma and xylem fibres.	It is composed of four types of elements: sieve tubes, companion cells, phloem parenchyma and phloem fibres
(ii)	It consists of mainly dead cells (except xylem parenchyma).	It consists of mainly living cells(except phloem fibres).
(iii)	It conducts water and minerals.	It conducts food.
(iv)	Conduction is mainly unidirectional *i.e.,* from roots to apical part of the plants.	Conduction is mainly bidirectional *i.e.,* from leaves to various parts or from storage organ to growing new leaves of the plants.

5.

S. No.	Blood	Lymph
(i)	It is reddish colour fluid due to presence of red blood cells (haemoglobin).	It is a colourless (pale yellow) fluid due to absence of red blood cells.

(ii)	It consists of plasma, red blood cells, white blood cells and platelets.	It consists of plasma and white blood cells (most abundant lymphocytes)
(iii)	It circulates oxygen, carbon dioxide, nutrients, hormones and waste products.	It helps in body defence and a part of immune system.

6.

S. No.	Tendons	Ligaments
(i)	Connect bones to mucles.	Connect bones to bones.

Chapter 4. The Flower

1. Gynoecium and Androecium (Type of floral part)
2. Calyx and Corolla (Part of flower and function)
3. Petals and Petaloid (Function)
4. Inflorescence and Flower
5. Polyandrous androecium and Polyadelphous androecium

Ans.

1.

S. No.	Gynoecium	Androecium
(i)	It is a female reproductive organ.	It is a male reproductive organ.
(ii)	It is also called pistil or carpel.	It is also called stamen.
(iii)	It is composed of style, stigma and ovary.	It is composed of filament and anther.
(iv)	Flowers without gynoecium are called staminate.	Flowers without androecium are called pistillate or carpellate.

2.

S. No.	Calyx	Corolla
(i)	It is whorl of sepals of a flower.	It is a whorl of petals of flowers.
(ii)	They are green in colour.	They are bright in colours.
(iii)	It protects the inner structure of the flower.	It attracts pollinators to the flowers.

3.

S. No.	Petal	Petaloid
(i)	It is a non-essential part of the flower.	It contains modified sepals resembling petals.
(ii)	It helps in protection of reproductive parts and makes the attractive for pollination.	It attract pollinators to the flower.

4.

S. No.	Inflorescence	Flower
(i)	Inflorescence is the mode of arrangement of flower on the axis of plant.	Flower is defined as modified dwarf shoot which is meant for sexual reproduction.
(ii)	Peduncle is the stalk of the inflorescence.	Pedicle is the stalk of the inflorescence.
(iii)	It increases the chance of cross pollination.	Produces gametes, undergoes fertilisation, and form seeds.

5.

S. No.	Polyandrous androecium	Polyadelphous androecium
(i)	In polyandrous androecium, filaments of the stamens are many and free.	In polyadelphous androecium, filaments of the stamens are united in several groups-with their anthers remaining free.

Chapter 5. Pollination and Fertilisation

1. Self-pollination and Cross-pollination (Type of pollination)
2. Pollination and Fertilisation (Definition)
3. Herkogamy and Heterostyly (Nature)
4. Anemophily and Entomophily (Agents) **[November, 2019]**
5. Anther and Stigma (Reproductive parts)

Ans.

1.

S. No.	Self-pollination	Cross-pollination
(i)	It occurs within the same flower or between two flowers of the same plant.	It occurs between two flowers on different plants of the same species.
(ii)	Flowers do not depend on other agencies for pollination.	Flowers depend on external agents (such as insects, water and wind) for pollination.
(iii)	Pollen grains are produced in small number.	Pollen grains are produced in large number.
(iv)	It does not introduce any variation in daughter flower.	It introduces variation within the characteristics and appearances.
(v)	It can occur even when the flowers is closed.	It occurs only when the flower is open.

2.

S. No.	Pollination	Fertilisation
(i)	It is the deposition of pollen grains on the stigma of a flower.	It is the fusion of haploid male and female gametes forming a diploid zygote.
(ii)	It is achieved by pollinating agents like wind, water, insects, etc.	It does not require any external agents.
(iii)	It is of two types: self- pollination and cross - pollination.	It involves double fertilisation in flowering plants.

3.

S. No.	Herkogamy	Heterostyly
(i)	It is a condition in which the pollen of a flower cannot reach the stigma of same flower due to barrier such as pansy flower.	It is the condition (e.g. in primroses) of having styles of different lengths relative to the stamens in the flowers of different individual plants.
(ii)	It reduces sexual interference between male and female function.	It reduces self-pollination.

4.

S. No.	Anemophily	Entomophily
(ii)	Flowers are small and inconspicuous.	Flowers are large and showy.
(ii)	They do not produce nectar.	They produce nectar as a food for the insects.

(iii)	The stigma is feathery and hangsout of the flower to trap the pollen grains.	The stigma is sticky and does not hang out from the flower.
(iv)	Pollen grains are light, dry and smooth so that they can easily carry away by wind.	Pollen grains are sticky and spiny to enable them to be carried by the insects easily.

5.

S. No.	Anther	Stigma
(i)	It is a part of stamen.	It is a part of pistil.
(ii)	It produces pollen grain.	It receives pollen grain.

Chapter 6. Seeds-Structure and Germination

1. Bean seed and maize grain (Structure)
2. Hypogeal and epigeal germination (Type of germination)
3. Albuminous and ex-albuminous seed (Classification based on endosperm)
4. Monocot and dicot seed (Classification based on cotyledons)
5. Cotyledon and endosperm (Define)

Ans.

1.

S. No.	Bean seed	Maize grain
(i)	Two cotyledons and no endosperm.	One cotyledon and endosperm is present.
(ii)	Hilum and micropyle can be seen.	Hilum and micropyle are not clearly visible.
(iii)	Seed coat is thick.	Seed coat is fused with pericarp.
(iv)	Embryo is large.	Embryo is small.
(v)	Plumule leaves folded.	Plumule leaves rolled.

2.

S. No.	Hypogeal germination	Epigeal germination
(i)	Cotyledons remain underground.	Cotyledons pushed above the ground.
(ii)	Epicotyl elongates.	Hypocotyl elongates.
(iii)	This type of germination is seen in pea, gram, etc.	This type of germination is seen in castor, bean, etc.
(iv)	Cotyledons remain non-green, dries up gradually and falls.	Cotyledons turns green and starts photo-synthesis process.

3.

S. No.	Albuminous seed	Ex-albuminous seed
(i)	Food is stored in endosperm.	Food is stored in cotyledons.
(ii)	Cotyledons are thin.	Cotyledons are thick.
(iii)	Endosperm present.	Endosperm absent.
(iv)	Examples: Cereals, poppy, millets, etc.	Example: Gram, pea, orchids, etc.

4.

S. No.	Monocot seed	Dicot seed
(i)	Only one cotyledon is present.	Two cotyledons are present.
(ii)	They mostly contain fibrous root and parallel venation.	They contain tap root and reticulate venation.
(iii)	Example: Maize, grass, etc.	Example: pea, gram, etc.

5.

S. No.	Cotyledon	Endosperm
(i)	It possess plumule and radicle of the embryonic plant.	It is the nutritive tissue in the seed which contains stored starch, proteins and oils.
(ii)	It becomes the first leaves of the plant during growth of seedling.	It is the nutritive tissue that provides nutrients to the development of the embryo.

Chapter 7. Respiration in Plants

1. Respiration and Burning
2. Aerobic and Anaerobic respiration
3. Photosynthesis and respiration (gas given out) **[February, 2020]**
4. Respiration in plants and respiration in animals
5. Glycolysis and Kreb's cycle (Amount of energy released) **[February, 2020]**

Ans.

1.

S. No.	Respiration	Burning
(i)	It is a biochemical process which occurs in a series of steps.	It is a physico-chemical process which occurs in a single step.
(ii)	It is a cellular process carried out by enzymes.	It is a non-cellular process carried out by heat.
(iii)	It occurs at body temperature.	It occurs at very high temperature.
(iv)	Energy is liberated as ATP and heat.	Energy is liberated as heat and light.

2.

S. No.	Aerobic respiration	Anaerobic respiration
(i)	It occurs in the presence of oxygen.	It occurs in the absence of oxygen.
(ii)	Complete breakdown of glucose occurs.	Glucose is incompletely broken down.
(iii)	38 ATP molecules are released from 1 mole of glucose.	2 ATP molecules are released from one mole of glucose.
(iv)	End products are carbon dioxide, water.	End products are carbon dioxide, ethyl alcohol [plants], lactic acid [animals]

3.

S. No.	Photosynthesis	Respiration
(i)	It occurs in chlorophyll containing cells in the presence of sunlight.	It occurs all the time in all living cells.
(ii)	It uses carbon dioxide and water. Glucose and oxygen are released as end product.	It uses oxygen and glucose and carbon dioxide is released as end product.
(iii)	It results in gain of weight.	It results in loss of weight.
(iv)	It is an anabolic process where light energy is converted into chemical energy.	It is a catabolic process where chemical energy is converted into ATP and heat energy.

4.

S. No.	Respiration in plants	Respiration in animals
(i)	Respiratory gases diffuses from cell to cell. There is no transport system.	There is a transport system and blood plays an important role in diffusion of respiratory gases.

(ii)	End product of anaerobic respiration is ethyl alcohol.	End product of anaerobic respiration is lactic acid.
(iii)	Less amount of heat is produced.	Large amount of heat is produced.
(iv)	Carbon dioxide is released in anaerobic respiration.	No carbon dioxide is released during anaerobic respiration.

5.

S. No.	Glycolysis	Kreb's cycle
(i)	It occurs in cytoplasm. It is also known as EMP pathway.	It occurs in mitochondria. It is also known as TCA cycle.
(ii)	Here glucose is converted into pyruvate.	Here pyruvate is converted into carbon dioxide and water.
(iii)	It is the common phase of both aerobic and anaerobic respiration.	It occurs only aerobically.
(iv)	8 ATP molecules are produced.	24 ATP molecules are produced.

Chapter 8. Five Kingdom Classification

1. Vertebrates and Invertebrates
2. Flatworms and Segmented worms
3. Aves and Mammals
4. Gymnosperms and Angiosperms
5. Arthropods and Molluscs

Ans.

1.

S. No.	Vertebrates	Invertebrates
(i)	Backbone present.	Backbone absent.
(ii)	Heart ventral in position.	Heart dorsal in position.
(iii)	Endoskeleton is present.	Endoskeleton is absent.
(iv)	Nerve cord is dorsal and hollow.	Nerve cord is vental and solid.
(v)	Haemoglobin is present in RBCs. Examples: Phylum Porifera, Mollusca, Cnidaria, etc.	Haemoglobin is almost in dissolved form. Examples: Class pisces, Reptilia, Mammals, etc.

2.

S. No.	Flatworms	Segmented worms
(i)	They belong to phylum *Platyhelminthes.*	They belong to phylum Annelida.
(ii)	Flame cells are their excretory organs.	Nephridia are their excretory organs.
(iii)	They are acoelomate.	They possess true coelom.
(iv)	Body is flat, unsegmented. Examples: *Planaria,* liverfluke.	Body is metamerically or ringlike segmented. Examples: Earthworms, leech.

3.

S. No.	Aves	Mammals
(i)	They are oviparous.	They are viviparous.
(ii)	Body covered with feathers.	They have hairy skin.
(iii)	They excrete uric acid.	They excrete urea as nitrogenous waste.
(iv)	They are mostly aerial in habitat. Examples: Pigeon, parrot, peacock, crow, sparrow, etc.	They live on land, water as well as in air. Examples: Bat, monkey, man, whales, etc.

4.

S. No.	Gymnosperms	Angiosperms
(i)	They are plants with naked seeds.	They are flowering plants with covered seeds.
(ii)	They are perennial.	They are annual, biennial or perennial.
(iii)	Cones and unisexual.	Present in flowersand can be unisexual or bisexual.
(iv)	Leaves are flat in shape.	Leaves are scale-like and needle-like in shape.
(v)	They are found mostly in cold climate where snowfall occurs. Examples: Cycas, pine, etc.	They grow in almost every kind of habitat. Examples: Bamboo, mango, neem, sugar-cane, cereals, pulses, etc.

5.

S. No.	Arthropoda	Mollusca
(i)	Body is all segmented.	Body is unsegmented.
(ii)	Body has an exoskeleton made up of chitin.	They have a hard calcareous shell, made up of calcium carbonate over their soft body.
(iii)	They have jointed legs or appendages for locomotion.	They have a muscular foot for locomotion.
(iv)	Malpighian tubules are their excretory organs.	Meta-nephridia are their excretory organs.
(v)	Compound eyes with mosaic vision present. Body is divided into head, thorax and abdomen.	Body is divided into head, visceral mass and ventral foot.
(iv)	Respiration by gills or trachea or book lungs. Examples: Spider, cockroach, prawn, fly, etc.	Respiration by ctenidia or mantle or lungs. Examples: Snail, octopus, unio, etc.

Chapter 9. Economic Importance of Bacteria and Fungi

1. Bacteria and Fungi
2. Saprophyte and Parasite
3. Decay and Putrefaction
4. Vaccines and Toxoids
5. Sterilisation and Pasteurisation

Ans.

1.

S. No.	Bacteria	Fungi
(i)	They are prokaryotes, microscopic, single-celled.	They are eukaryotes and multicellular. Most of them are visible by naked eye.
(ii)	They reproduce mostly asexually.	They reproduce both by sexual and asexual methods.
(iii)	Some bacteria use flagella as locomotory organ.	They are immobile.
(iv)	Cell wall is made up of peptidoglycan.	Cell wall is made up of chitin.

2.

S. No.	Saprophyte	Parasite
(i)	They feed on dead and decay organisms.	They take nourishment from host.

(ii)	They use extracellular digestion.	They use intracellular digestion.
(iii)	They depends on dead and decay organisms and are cleansing agents.	They cause harm to their hosts.
(iv)	They are not specialised to certain food source.	Most of the parasites are host specific.

3.

S. No.	Decay	Putrefaction
(i)	Complete breakdown of organic matter by bacteria.	Incomplete break-down of organic matter by bacteria.
(ii)	Does not emit foul smell.	Produces a foul smell.

4.

S. No.	Vaccine	Toxoid
(i)	Vaccine can be bacteria or foreign body cells.	Toxoid is only a bacterial toxin.
(ii)	Vaccine is preparation of dead or weakened germs.	Toxoids are inactivated toxins of bacteria.
(iii)	Examples: Vaccines for cholera, typhoid, measles, etc.	Examples: Toxoids for diphtheria, tetanus, etc.

5.

S. No.	Sterilisation	Pasteurisation
(i)	It is the process of heating at very high temperature to kill microbes as well as their spores.	It is the process of heating liquids mainly milk at 60°C for 30 minutes followed by sudden cooling.
(ii)	It kills bacteria their vegetative and spore forms.	It kills only the vegetative forms of bacteria.
(iii)	Sterilised products have a longer shelf life.	Pasteurised products have shorter shelf life.

Chapter 10. Nutrition

1. Protein and Carbohydrate (Main function)
2. Minerals and Proteins
3. Cellulose and Fats (Function)
4. Perinicious Anaemia and Anaemia (Lack of nutrient)
5. Disaccharides and Polysaccharides.
6. Carbohydrates and Proteins.
7. Kwashiorkor and Marasmus
8. Cellulose and Glycogen (Occurance in organism) **[February, 2020]**

Ans.

1.

S. No.	Protein	Carbohydrate
(i)	These are required for growth, repair of body tissues and building cells.	These are source of energy.

2.

S. No.	Minerals	Proteins
(i)	Minerals are some micro grams in amount.	Proteins are some macro grams in amount.
(ii)	These are needed in our diet for proper strengthening, working of thyroid, forming haemoglobin.	Proteins are needed for growth and repair of cells in body.

3.

S. No.	Cellulose	Fats
(i)	Movement of food in food pipe.	It is important storage form of food.

4.

S. No.	Perinicious anaemia	Anaemia
(i)	This disease occurs due to lack of Vitamin B_{12}	This disease occurs due to lack of Iron.

5.

S. No.	Disaccharides	Polysacchariddes
(i)	These have two sugar molecules.	These have more than two sugar molecules.
(ii)	General formula is $C_{12}H_{22}O_{11}$.	General formula is $(C_6H_{10}O_5)_n$
(iii)	These are first digested and then aborbed in the body.	Sources include meat, egg, soyabean, etc.

6.

S. No.	Carbhydrates	Proteins
(i)	The elements present are carbon, hydrogen and oxygen.	They contain carbon, hydrogen , oxygen and nitrogen.
(ii)	These provide energy.	These are required for growth and repair of the body tissues.
(iii)	Sources are potato, cereals, etc.	Sources include meat, egg, soyabean, etc.

7.

S. No.	Kwashiorkor	Marasmus
(i)	Affect children (1-5 years of age).	Affect infants.
(ii)	Darkening of the skin.	No darkening of the skin.
(iii)	Caused due to deficiency of proteins.	Caused due to the deficiency of carbohydrate fats and proteins.

8.

S. No.	Cellulose	Glycogen
(i)	Present in plant cell.	Present in animal cells.
(ii)	In cell wall.	In liver and muscles.

Chapter 11. Digestive System

1. Amylase and Maltase (Main substrate)
2. Proteins and Fats (End product)
3. Chyme and Chyle (Composition)
4. Small intestine and large intestine (Main function)
5. Gastro intestinal tract and accessory organs (Definition)
6. Pepsin and Trysin
7. Hormone and Enzyme

Ans.

1.

S. No.	Amylase	Maltase
(i)	The enzyme which act on starch to form out maltage	The enzyme which act on maltose to form glucose.

2.

S. No.	Proteins	Fats
(i)	The end product of proteins are amino acids in the presence of enzyme trypsin.	The end product of fats are fatty acids in the presence of lipase.

4.

S. No.	Small intestine	Large intestine
(i)	The functions of small intestine are digestion and absorption.	The functions of large intestine are absorption of water.

5.

S. No.	Gastro intestinal tract	Accessory organs
(i)	Alimentary canal is the long tube through which the food that we eat is passed. It begins at the mouth (buccal or oral cavity), passes through the pharynx, esophagus or food pipe, stomach, small intestines, large intestines, rectum and finally ends at the anus. The food particles get digested gradually as they travel through various compartments of the alimentary canal.	Accessory organs are organs which participate in the digestion process but are not actually a part of GIT.

6.

S. No.	Pepsin	Trypsin
(i)	Acts in acidic medium.	Acts in alkaline medium.
(ii)	Converts proteins into proteoses and peptones (peptides).	Converts proteins into peptones (peptides) and polypeptides.

7.

S. No.	Hormone	Enzyme
(i)	It is secreted by the endocrine gland.	It is secreted by the exocrine gland.
(ii)	It helps in chemical coordination.	It catalyses a biochemical reaction.
(iii)	Its action is irreversible	Its action is reversible.
	It is used up in the action.	It is not used up in the action. It only catalyse the reaction.

Chapter 12. Skeleton: Movement and Locomotion

1. Femur and Enamel (Location)
2. Ear ossicles and Femur (Size)
3. True ribs and Floating ribs (Number) **[November, 2019]**
4. Pectoral girdle and Pelvic girdle (Location)
5. Movement and Locomotion (Function)

Ans.

1.

S. No.	Femur	Enamel
(i)	Femur is located in leg.	Enamel is located in tooth.

2.

S. No.	Ear ossicles	Femur
(i)	Shortest bone in human body.	Longest bone of human body .

3.

S. No.	True ribs	Floating ribs
(i)	True ribs are attached to sternum.	False ribs are attached to the other ribs.
(ii)	The first seven pairs of ribs.	The last two pairs of ribs.

4.

S. No.	Pectoral girdle	Pelvic girdle
(i)	Pectoral girdle is located in shoulder.	Pelvic girdle is located in hips.

5.

S. No.	Movement	Locomotion
(i)	Orientation or the displacement of a body part from its orginal position to new position.	The displacement of entire body from change of place.

Chapter 13. Skin—"The jack of all Trades"

1. Hair follicle and Hair shaft (Location)
2. Vasodilation and Vasoconstriction.
3. Mammary gland and Ceruminous gland (Modification)
4. Ceruminous gland and Sebaceous gland (Location)
5. Sweat and Sebum
6. Leucoderma and Albinism (Type of disorder) **[November, 2019]**
7. Sweat gland and Sebaceous gland

Ans.

1.

S. No.	Hair follicle	Hair shaft
(i)	These are located in dermis.	These are located on epidermis.

2.

S. No.	Vasodilation	Vasoconstriction
(i)	It is a secretion of sweat gland.	It is a secretion of sebaceous gland.
(ii)	It keeps the body cool.	It makes the skin soft and waterproof.
(iii)	It is a watery fluid.	It is an oily secretion.

3.

S. No.	Mammary gland	Ceruminous gland
(i)	These are modified sweat gland.	These are modified sebaceous gland.

4.

S. No.	Ceruminous gland	Sebaceous gland
(i)	These glands are present in ear.	These glands are present in skin.

5.

S. No.	Sweat	Sebum
(i)	It is a secretion of sweat gland.	It is a secretion of sebaceous gland.
(ii)	It keeps the body cool.	It makes the skin soft and waterproof.
(iii)	It is a watery fluid.	It is an oily secretion.

6.

S. No.	Leucoderma	Albinism
(i)	It is a condition in which the skin pigment is lost in patches.	It is the complete loss of skin pigmentation.
(ii)	The cause is not yet known.	It is an inherited disease.

7.

S. No.	Sweat gland	Sebaceous gland
(i)	These are duct glands.	These are ductless glands.
(ii)	These secrete sweat.	These secrete sebum.

Chapter 14. The Respiratory System

1. Trachea and Food pipe (Function)
2. Bronchi and Alveoli (Structure)
3. Pleura and Pericardium (Function)
4. Ciliated epithelium and Glandular epithelium (Location)
5. Intercostal muscles and Diaphragm (Function)
6. Inspiration and Expiration
7. Inspired air and Expired air
8. External respiration and Internal respiration
9. Aerobic respiration and Anaerobic respiration
10. Pleural fluid and Pericardial fluid

Ans.

1.

S. No.	Trachea	Food pipe
(i)	Trachea is the windpipe for the passage for air.	Food pipe is for the passage for food.

2.

S. No.	Bronchi	Alveoli
(i)	Inverted Y shaped tube.	Balloon like structure at end of bronchioles.

3.

S. No.	Pleura	Pericardium
(i)	Protect lungs	Protect heart

4.

S. No.	Ciliated epithelium	Glandular epithelium
(i)	Located in respiratory passage	Located in stomach

5.

S. No.	Intercostal muscles	Diaphragm
(i)	These muscles move ribs in and out.	It makes chest cavity larger by moving up and down.

6.

S. No.	Inspiration	Expiration
(i)	It is an active process.	It is a passive process.
(ii)	It is the taking in of oxygen.	It is the expelling of CO_2.
(iii)	It increases the volume of thoracic cavity.	It decreases the volume of thoracic cavity.

7.

S. No.	Inspired Air	Expired Air
(i)	In this O_2 content is high.	In this O_2 content is low.
(ii)	CO_2 content is low.	CO_2 content is high.
(iii)	The temperature is variable.	The temperature is same as that of the body (37°C).

8.

S. No.	External Respiration	Internal Respiration
(i)	It is physical process.	It is a chemical process.
(ii)	It includes exhange of gases between, the blood and inhaled air.	It is exchange of gases between the tissues celles and the blood.

9.

S. No.	Aerobic respiration	Anaerobic respiration
(i)	Respiration in which breakdown of glucose occurring in the presence of oxygen.	Respiration in which breakdown of glucose occurring in the absence of oxygen.

10.

S. No.	Pleural fluid	Pericardial fluid
(i)	The fluid formed in the pleural sac surrounding the lungs.	The fluid formed in the pericardial sac surrounding the heart.

Chapter 15. Hygiene

1. Bacillary dysentery and Amoebic dysentery (Agent)
2. Cholera and Dysentery (Symptoms)
3. Hepatitis and Dysentery (Prevention)
4. Vector and Water borne disease (Definition)
5. Housefly and Mosquitoes (Transmission)
6. Culex and Anopheles mosquitoes

Ans.

1.

S. No.	Bacillary dysentery	Amoebic dysentery
(i)	The causative agent of bacillary dysentery is a bacterium (*Shigella*).	The causative agent of amoebic dysentery is a protozoan *Entamoeba histolytica*.

2.

S. No.	Cholera	Dysentery
(i)	Cholera causes severe diarrhea, vomiting and no urination.	Dysentery causes loose motion, intestinal pain, and mild fever.

3.

S. No.	Hepatitis	Dysentery
(i)	Hepatitis is prevented by hepatitis vaccine, proper hygiene, avoid taking fat rich substances.	Dysentery is prevented by antibiotics, drinking safe (boiled) water, and preventing flies from contaminating food.

4.

S. No.	Vector	Water borne disease
(i)	Vector is any agent that acts as an intermediate carrier of the pathogen.	Water borne diseases are caused by drinking contaminated water or dirty water.

5.

S. No.	Mosquito	Housefly
(i)	The germs of the diseases caused by mosquitoes are spread when the mosquitoes bite to suck blood. With this blood mosquitoes receive germs of the particular disease, which grow and multiply within their body. This is how mosquito diseases are transmitted when a such infected mosquito bite a healthy person.	The houseflies easily pick up filth containing germs from dirty places. Such flies contaminate our food by sitting on it. This is how housefly transmits diseases.

6.

S. No.	Culex mosquito	Anopheles mosquito
(i)	Sits parallel while resting.	Sits obliquely while resting.

(ii)	Wings are unspotted	Wings have dark sports.
(iii)	Eggs are laid in rafts and erect.	Eggs are laid singly and horizontally.
(iv)	These mosquitoes spread elephantiasis.	These type of mosquitoes spread malaria.

Chapter 16. Disease: Cause and Control

1. Viruses and bacteria (Size)
2. Typhoid and AIDS (Incubation period)
3. *Vibrio cholera* and *Plasmodium falciparum* (Type of pathogen)
4. Cancer and AIDS (Types of disease)
5. Tuberculosis and Cholera (Body part)
6. Infectious and non-infectious disease
7. Ascariasis and Taeniasis (Pathogen) **[February, 2020]**

Ans.

1.

S. No.	Virus	Bacteria
(i)	Viruses are very small that it can be visible only by electron microscope.	Bacteria are larger that it can be seen by light microscope.

2.

S. No.	Typhoid	AIDS
(i)	The incubation period of typhoid is 7-12 days.	The incubation period of AIDS is up to 12 or more years.

3.

S. No.	Vibrio cholerae	Plasmodium falciparum
(i)	*Vibrio cholerae* is the bacterium which causes bacterial disease called Cholera.	*Plasmodium falciparum* is the protozoan which causes malaria.

4.

S. No.	Cancer	AIDS
(i)	Cancer is the non-infectious disease.	AIDS is an infectious disease.

6.

S. No.	Infectious disease	Non infectious disease
(i)	Infectious diseases are caused by some germs and it can be spread from a patient to healthy person.	In these diseases, they are not caused by germs and they cannot spread from a patient to another person.

7.

S. No.	Ascariasis	Taeniasis
(i)	Caused by common Round-warm (ascaris).	Cause by type-warm (Taeniasolium).

Chapter 17. Aids of Health

1. Vaccination and Immunisation (Function)
2. Active immunity and Passive immunity
3. Antiseptics and Disinfectants (Definition) **[November, 2019]**
4. Non-specific innate immunity and specific innate immunity (Definition)
5. Antiseptic and Antibiotic

6. Disinfectant and Deodorant

Ans.

1.

S. No.	Vaccination	Immunisation
(i)	The vaccines are inactive form of bacteria or virus that is injected into the body to stimulate an actual infection. As the injected microorganism is dead they do not cause harm to the body and immune the body against some specific diseases.	The immunisation is the process which occurs inside the body after vaccination. The vaccine stimulates the immune system so that it can recognise the disease and protect from infection in future.

2.

S.No.	Active immunity	Passive immunity
(i)	The active immunity generally involves cellular responses, serum antibodies or combination acting against one or more antigens on the infecting organism.	The passive immunity occurs naturally, when maternal antibodies are transferred to the fetus through the placenta and it can also be induced artificially, when high levels of antibodies specific to a pathogen or toxin.
(ii)	Produced by one's own body.	Recieved from outside, *i.e.,* not developed by one's own body.
(iii)	Induced by injections or by contact with immunogens.	Provided by ready-made antibodies.
(iv)	Provides effective and long lasting protection.	Protection is less efficient and does not ensure about infections.
(v)	Immunity effective after certain time period. Example: Lymphocytes	Immediately effective. Example: Antivenin for treating snake-bite.

3.

S. No.	Antiseptics	Disinfectant
(i)	Antiseptics are designed for use on skin and wounds *i.e.,* in the living tissue.	The difference is that the disinfectants kill bacteria on non-living surfaces such as ceramic, wood, stone, or metals antiseptics are designed for use on skin and wounds *i.e.,* in the living tissue.

4.

S. No.	Non-specific innate immunity	Specific innate immunity
(i)	Non-specific innate immunity is a degree of natural resistance to all infections in general. For example, human do not suffer from infection of plants or certain diseases o animals for example crown goal disease is a highly infectious disease in plants but it do not harm humans.	Specific innate immunity is a natural resistance to a particular kind of germs only. Some races or some individuals do not suffer from certain infection diseases. For example; distemper disease which causes on dogs does not affect humans.

5.

S. No.	Antiseptic	Antibiotics
(i)	Antiseptics are designed for use on skin and wounds *i.e.,* in the living tissue.	Antibiotics are chemical substances that introduced to kill the germs inside the body.

6.

S. No.	Disinfectants	Deodorants
(i)	Disinfectants kill bacteria on non-living surfaces such as ceramic, wood, stone, or metals antiseptics are designed for use on skin and wounds *i.e.,* in the living tissue.	Deodorants are used to prevent bad smell from the body. It is neither antiseptic nor disinfectants.

Chapter 18. Health Organisation

1. Diarrhea and Malaria (Types of Disease)
2. Village and Remote areas (Types of habitations)
3. Red Cross and WHO (Year of establishment)
4. National and International bodies (Functions)
5. Food borne diseases and Insect borne disease (Causes)

Ans.

1.

S. No.	Diarrhea	Malaria
(i)	Diarrhea is a food borne disease.	Malaria is an insect borne disease.

2.

S. No.	Village	Remote area
(i)	Village consists of small population, agriculture based; diaries, poultry farming are the main sources of livelihood.	A remote area consists of tribal people, mostly thriving on forests and forests products.

3.

S. No.	Red Cross	WHO
(i)	Red Cross was established in 1864.	WHO was established in 1984.

4.

S. No.	National bodies	International bodies
(i)	National bodies are engaged in studying the health problems of the country.	International bodies look after the health problems of people of the world.

5.

S. No.	Food borne diseases	Insect borne diseases
(i)	Food borne diseases are caused due to improper water supply. The water from hand- pumps and other sources is contaminated in many places leading to diseases.	Insect borne diseases are caused due to the lack of cleanliness leading to the breeding of houseflies, mosquitoes and other insects.

Chapter 19. Waste Generation and Management

1. Paper and Plastic (Type of waste)
2. Dead leaves and Hide of animals (Source of production)
3. Mineral dust and Glass dust (Place of production)
4. Kitchen waste and Hospital waste (Degree of harmlessness)
5. Biodegradable and Non-biodegradable waste

Ans.

1.

S. No.	Paper	Plastic
(i)	Biodegradable material	Non-biodegradable material

2.

S. No.	Dead leaves	Hide of animals
(i)	Agricultural waste	Processing remains

3.

S. No.	Mineral dust	Glass dust
(i)	Mines	Glass cutting

4.

S. No.	Kitchen waste	Hospital waste
(i)	Easily decomposable make it harmless	Hospital waste is more harmful

5.

S. No.	Biodegradable was]te	Non- biodegradable waste
(i)	These waste can be broken down by micro-organisms.	These wastes cannot be broken down by micro-organisms.
(ii)	Non-toxic waste	Toxic waste
(iii)	Example: Vegetable peels, dry leaves, food wastes, etc.	Example: Glass, polythene bags, plastics, radioactive materials, toxic industrial waste, etc.

❑

Name the Following

Set 3

Q. Name the following:

Chapter 1. Introducing Biology

1. Study of the structure and function of a cell.
2. Science which deals with the study of resistance to diseases or defence against diseases.
3. Study of reptiles, snakes and lizards.
4. Use of living cells or microbes in industry and technology.
5. Study of fish.
6. The branch of science that deal with feeding, shelter, health and breeding of domestic animals.
7. Cultivation of ornamental flower plants.
8. Application of scientific knowledge to question of civil and criminal law.
9. Study of virus.
10. Study of science which improves the human race through controlled heredity.
11. Animals who gave birth to young ones.
12. Study of disease of plants and animals.
13. Study of chemical and reactions that takes place inside the living things.
14. Study of prehistoric form of life.

Ans. 1. Cytology 2. Immunology 3. Herpetology 4. Biotechnology
5. Pisciculture 6. Animal husbandry 7. Floriculture 8. Forensic science
9. Virology 10. Eugenics 11. Viviparous animal 12. Pathology
13. DNA fingerprinting 14. Palaeontology

Chapter 2. Cell: The Unit of Life

1. It contains various hydrolytic enzymes which help in digestion of lipids, proteins and carbohydrates.
2. It is called the power house of the cell.
3. The cell concerned with transmission of hereditary characters from parents to offspring.
4. The cell concerned with initiation of cell division.
5. Organisms concerned with the secretion of the cell including enzymes, hormones, etc.
6. Colourless plastid which store starch.
7. The cell concerned with liberation of energy.
8. The cell concerned with synthesis of proteins.
9. Longest cell of the body which starts from the fingertip up to the spinal cord inside the backbone.
10. The cell concerned with hydrolytic in function.
11. The cell concerned with entry of only certain substances into and out of the cell.
12. Outermost covering of plant cell. **[November, 2019]**

Ans. 1. Lysosome 2. Mitochondria 3. Chromosomes 4. Centrosomes
5. Golgi apparatus 6. Leucoplast 7. Mitochondria 8. Ribosomes
9. Nerve cell 10. Lysosomes 11. Cell membrane 12. Cell wall

Chapter 3. Tissues: Plants and Animal tissue

1. Epithelium which secretes hormones, sweat, mucus, enzymes and form tubular glands.
2. Tissues which form padding under the skin and around kidneys.
3. Strong water proof material present in the bark.
4. Supportive plant tissue which consists of dead cells and develops thick walls due to deposition of lignin.
5. Epithelium containing thin, flat cell with prominent nuclei found in the lining of mouth and nasal cavities, blood vessels, etc.
6. Part of neuron which contain nucleus and cytoplasm.
7. Cells are flat, cuboidal or columnar forms protective layer.
8. Cells can contract and relax.
9. Cells can conduct impulses.
10. Cells produce and pour out chemical substances.
11. The tissue which conducts water in a plant. **[February, 2020]**

Ans. 1. Glandular epithelium 2. Adipose tissue
3. Cork cells 4. Sclerenchyma
5. Squamous epithelium 6. Cyton or cell body
7. Epithelial tissue 8. Muscular tissue
9. Nervous tissue 10. Grandular epithelium tissue
11. Xylem.

Chapter 4. The Flower

1. The arrangement of ovules within the ovary.
2. The way in which filaments are united in two bundles.
3. A flower which contains only the carpels.
4. When flower arises in the axil of a leaf-like structure.
5. Sweet fragrant liquid produced by the flower.
6. A condition in which the sepals are free.
7. A condition in which the petals are united.
8. A monocot plant which contains female and male flowers on the same plant.
9. The expanded tip of the pedical bearing floral parts.

Ans. 1. Placentation 2. Diadelphous 3. Pistillate flower 4. Bract
5. Nectar 6. Polysepalous 7. Gamopetalous 8. Maize
9. Thalamus

Chapter 5. Pollination and Fertilisation

1. Every ovule has one or two protective coverings.
2. Flowers having affinity to water for pollination.
3. Manually transfer of pollen to the stigma.
4. The pollen of another flower of the same plant may fall on the stigma.
5. The condition of flowers whose male parts mature before the female ones.

Ans. 1. Integuments 2. Hydrophilous 3. Artificial pollination
4. Geitonogamy 5. Protandry

Chapter 6. Seeds: Structure and Germination

1. The opening through which water enters into a seed.
2. The protective sheath that encloses the radicle.
3. Type of germination in which seed germinates inside the fruit while it is attached to the plant.
4. The layer of endosperm of maize rich in protein.
5. A seed with folded plumule leaves.
6. A monocotyledonous endospermic seed.
7. Part of the plumule above the embryonic axis of the seed.
8. Part of flower which forms seed.
9. Protective covering of radicle in maize grain.
10. A plant which shows viviparous germination.

Ans. 1. Micropyle 2. Coleorhiza 3. Viviparous germination 4. Aleurone layer 5. Bean seed 6. Cereals 7. Epicotyl 8. Ovule 9. Coleorhiza 10. Rhizophora

Chapter 7. Respiration in Plants

1. Structures responsible for gaseous exchange in leaves of plants.
2. The energy currency of the cell.
3. A respiratory substrate.
4. The cycle in which pyruvic acid undergoes oxidation.
5. A chemical mostly used to test the presence of carbon dioxide in laboratory.
6. Oxidative breakdown of carbohydrates to release energy.
7. An organism which respires throughout life anaerobically.
8. A common phase in both aerobic and anaerobic respiration.
9. A chemical which removes carbon dioxide from the air.
10. The process in which sugar is used to form alcohol with the help of yeast.

Ans. 1. Stomata 2. ATP 3. Glucose 4. Kreb's cycle 5. Lime water 6. Respiration 7. Yeast 8. Glycolysis 9. Soda-lime 10. Fermentation

Chapter 8. Five Kingdom Classification

1. Common name of *Rana tigrina*.
2. A single large opening present in Poriferans.
3. An animal possessing stinging cells.
4. An animal with water vascular system.
5. Simplest group of Kingdom Plantae.
6. Excretory organs of tapeworms.
7. Body can be divided into two equal halves by one plane only.
8. Body can be divided into two equal halves by any plane.
9. Amphibians of plant kingdom.
10. A group in which organisms do not possess cell wall.

Name the Following

11. The phylum which consists of animals with unsegmented body with a protective calcareous shell.

Ans. 1. Frog 2. Osculum 3. Hydra 4. Starfish 5. Thallophytes/algae 6. Flame cells 7. Bilateral symmetry 8. Radial symmetry 9. Bryophyta 10. Animalia 11. Mollusca

Chapter 9. Economic Importance of Bacteria and Fungi

1. A substance produced in animal bodies which reacts with poison produced by the invading germs.
2. Introduction of weakened or dead germ substances into the body for developing resistance to a particular disease.
3. Blood plasma from which fibrinogen has been removed.
4. The mould observed by Alexender Fleming and lead to the discovery of antibiotics.
5. Scientific name of the most commonly used yeast in fermentation process.
6. The symbiotic association of fungi with plants.
7. Three common shapes of bacteria.
8. Kind of nutrition in bacteria.
9. Inactivated toxins of the bacteria which can stimulate the production of antibodies.
10. The whole mass of hyphae of bread mould.

Ans. 1. Antitoxin 2. Vaccination 3. Serum 4. *Penicillium notatum* 5. Saccharomyces cerevisae 6. Mycorrhiza 7. Cocci, Bacilli, Spirilla 8. Saprotrophic or Parasitic 9. Toxoids 10. Mycellium

Chapter 10. Nutrition

1. Diet which contains all the required nutrients.
2. State of being under nourished.
3. State of being unwell due to lack of nutrient or pathological reason.
4. Nutrient which is used as energy resource.
5. Nutrient present in table salt.
6. A mineral required for clotting of blood.
7. A rich source of carbohydrates.
8. A fat soluble vitamin.
9. A mineral present in bones and teeth.
10. Deficiency disease of vitamin C and D.
11. The structural unit of proteins.
12. A wate soluble vitamin.
13. A disease caused by the deficiency of iron.
14. Sunshine vitamin.
15. An essential element of proteins.
16. An essential element of proteins. **[November, 2019]**

Ans. 1. Balanced Diet 2. Malnutrition 3. Disease. 4. Carbohydrate

5. Sodium
6. Calcium
7. Potato
8. Vitamin-A
9. Calcium
10. Vitamin C-Scurvy and Vitamin D-Rickets
11. Amino Acid
12. Vitamin-C
13. Anaemia
14. Vitamin-D
15. Nitrogen
16. Calcium.

Chapter 11. Digestive System

1. The passage through which food passes during process of digestion.
2. Glands associated with breakdown of food.
3. Breaking down of complex food molecule into simpler form.
4. End product of fats.
5. Biocatalyst which help in digestion.
6. The enzyme present in saliva.
7. An enzyme which converts fats into fatty acids.
8. An enzyme which functions in acidic medium.
9. The finger-like structures present in the small intestine which helps in the absorption of food.
10. The reserved food (glucose) in animals.
11. The end product of digestion of fats.
12. Thee ring-like muscles present in the pylorus of the stomach.
13. The hardest substance in the human body.
14. The living part of the tooth.
15. The pulp like form of food in the stomach. **[November, 2019]**
16. The three subdivisions of the small intestine.
17. The three subdivisions of the large intestine.
18. The three salivary glands.
19. The formula which represents the number and kind of teeth in man. **[February, 2020]**

Ans. 1. Alimentary canal
2. Digestive glands
3. Digestion
4. Fatty acids
5. Digestive enzymes
6. Salivary amylase/Ptyalin
7. Lipase
8. Pepsin
9. Villi
10. Glycogen
11. Fatty acids and glycerol
12. Sphincter
13. Enamel
14. Pulp cavity
15. Chyme
16. Dudenum, jejunum, ileum
17. Caecum, colon, rectum
18. Parotid, submandibular, sublingual
19. Dental formula.

Chapter 12. Skeleton-Movement and Locomotion

1. Hard framework of bones which provide support and protection to internal organs.
2. Bones of arms and legs.
3. Flat bone of hip.
4. Cage like structure formed by bones of ribs and sternum.

5. Bones of back bone.
6. A joint which allows movement in all directions.
7. The longest bone in human body.
8. The protective covering of spinal cord.
9. The number of ribs.
10. The bones of pectoral girdle.
11. Tthe regions of vertabral column.
12. The bone present in the upper arm.
13. Total number of muscles present in our body.
14. The membrane covering the external surjace of the bone. **[November, 2019]**

Ans. 1. Skeleton system 2. Apendicular skeleton
3. Pelvic girdle 4. Rib cage or thoracic basket
5. Vertebrae 6. Ball and socket joint
7. Femur 8. Vertebral column
9. Twelve pairs 10. Scapula and clavicle
11. Cervical, thoracic, lumber, sacral and coccygal.
12. Humerus 13. 650 (approx.)
14. Periosteum

Chapter 13. Skin: "The jack of all Trades"

1. Animal body system related to skin.
2. Outermost layer of epidermis.
3. Nails are formed due to modification in this layer of skin.
4. Skin colour is fair or dark due to it.
5. Layer of skin which lies below the epidermis.
6. Any two glands found in the human skin.
7. The layer of epidermis which divides repeatedly and gives rise to new cells.
8. The process of dilation of blood vessels due to high tempreraturе.
9. The dead protein present in the stratum corneum.
10. Location of the heat-regulating centre of the body.
11. The tissue in the skin where the fat is stored.
12. The pigment present in the skin of man.
13. The branch of science dealing with the study of skin disorders.
14. The inflammed skin glands, which cause acne.
15. A modified sweat gland. **[February, 2020]**

Ans. 1. Integumentary system 2. Stratum Corneum 3. Epidermis
4. Melanin. 5. Dermis 6. Sweat gland and sebaceous gland
7. Malpighian layer 8. Vasodilation 9. Keratin
10. Hypothalamus 11. Adipose tissue 12. Melanin
13. Dermatology 14. Sebaceous glands 15. Mammary gland

Chapter 14. The Respiratory System

1. Process of using oxygen and removing carbon dioxide.
2. Ultimate end part of respiratory system for exchange of gases.

3. Muscle which help in respiration.
4. Bones which protect lungs from injury.
5. Taking in and taking out of air.
6. Cells that remove dust particle in the trachea.
7. The condition in which the oxygen supply to the respiratory system is cut off.
8. The small amount of air inhaled that remains in the respiratory tract but does not take part in respiration.
9. An unstable bright red compound formed between haemoglobin and oxygen to carry the oxygen to the body cells.
10. The ultimate end parts of the resspiratory passages in mammalian lungs.
11. The technical term for air breathed in and out during normal breathing.
12. The kind of animal tissue that lines the respiratory organs.
13. The structure responsible for preventing food from entering the trachea during swallowing.
14. The structrue which helps to incrase the volume of the chest cavity lengthwise.
15. The cavity in the body in which the human lungs are located.
16. Prevents food from entering the trachea during swallowing.
17. Transports oxygen to the body cells.
18. Help to increase the volume of the chest cavity lengthwise.
19. Combines with the oxygen in the lungs.
20. Protects the lungs from mechanical injuries.
21. Provides actual diffusion of respiratory gases in lungs.
22. The physical process by which air is taken in and forced out to the lungs. **[February, 2020]**

Ans.

1. Respiration	2. Alveoli	3. Diaphragm
4. Ribs	5. Breathing	6. Ciliated epithelial cells.
7. Asphyxiation	8. Dead air space	9. Oxyhaemoglobin
10. Alveoli	11. Tidal volume	12. Ciliated epithelium
13. Epiglottis	14. Diaphragm	15. Thoracic cavity
16. Epiglottis	17. RBCs	18. Diaphragm
19. Haemoglobin	20. Ribs	21. Aveoli
22. Breathing		

Chapter 15. Hygiene

1. Mosquito that spreads elephantiasis.
2. The science and practice of maintaining good health.
3. Taking proper care of surroundings.
4. Substance injurious to health or may even lead to cancer.
5. Any agent that acts as an intermediate carrier of the pathogen.
6. A disease spread by rats.
7. The scientific name of housefly.
8. The kind of natural scavenger for consuming left-overs of the food.
9. The fish used for biological control of mosquito.
10. The mouth part of housefly.
11. The period between entry of the germs and the appearance of the first symptom of the disease. **[November, 2019]**

12. Organisms that transmit microbes that cause disease.
13. Organisms that cause disease
14. Science of maintaing good health.

Ans. 1. Culex mosquito 2. Hygiene 3. Social hygiene 4. Tobacco
5. Vector 6. Plague 7. *Musca nebulo* 8. Housefly
9. *Gambusia* 10. Proboscis 11. Incubation period
12. Vectors (disease carriers) 13. Pathogens 14. Hygiene

Chapter 16. Disease: Cause and Control

1. Disease that is widely distributed worldwide.
2. The germ that cause infectious disease.
3. The bacteria that cause tuberculosis.
4. Amoebic dysentery is caused by the protozoan.
5. The causative virus of AIDS.
6. Category of pathogen that causes diseases, like common cold and mumps.
7. The vaccine for preventing tuberculosis.
8. An organ usually affected by tuberculosis.
9. Germ of germ-substance introduced into the body to prevent occurrence of an infectious disease.
10. The vector responsible for transmission of sleeping sickness.
11. The microorganism which need a host to produce.
12. The popular name of filariasis.

Ans. 1. Pandemic disease 2. Pathogen
3. Mycobacterium tuberculosis 4. Entamoeba histolytica
5. Human Immunodeficiency Virus 6. Viral diseases
7. BCG 8. Lungs
9. Vaccine
10. Trypanosoma bruceigambiense and Trypanosoma bruceirhodesiense
11. Plasmodium 12. Elephantiasis

Chapter 17. Aids of Health

1. Substance formed by lymphocytes on exposure to antigen.
2. Barrier that prevents the entry of germs inside stomach.
3. The person who discovered Penicillin in 1955.
4. Chemical substances produced during entry of any poisonous substance.
5. Prevention of diseases by studying the biological behaviour, transmission, etc.
6. The immunity developed after vaccination.
7. A foreign body that introduce the formation of antibodies in the body.
8. The category of immunity acquired in the treatment of snake bite.
9. The viral disease caused by the bite of a mad dog.
10. The process by which WBCs engulf bacteria.
11. The drug based on arsenic compound, produced in 1910, which killed germs of syphilis.

12. The antibiotic that was discovered first.
13. The category of immunity required in the treatment of snake-bite.
14. The vaccine that helps to produce immunity against polio.

Ans. 1. Antibody 2. Hydrochloric acid 3. Alexander Fleming 4. Antitoxins
5. Prophylaxis 6. Acquired immunity 7. Antigen
8. Passive immunity 9. Rabies 10. Phagocytosis
11. Arsphenamine of Salvarsan 12. Penicillin
13. Passive acquired immunity 14. Oral polio vaccine (OPV)

Chapter 18. Health organisation

1. The disease that occurs due to contaminants in water supply.
2. The emblem particularly used in hospitals.
3. The category of health organization under which WHO falls.
4. The organization concerned with the campaigning for control of epidemic disease.
5. The body that looks after the health problems of the people of the world.
6. An organisation that extends relief during floods, earthquakes, etc. **[February, 2020]**

Ans. 1. Water borne disease 2. Red Cross
3. International bodies 4. WHO
5. International bodies 6. Red cross

Chapter 19. Waste Generation and Management

1. Things which are no longer of use or wanted.
2. Waste water coming out of a factory.
3. End result of composting.
4. Burning of organic waste.
5. Discarded electronic devices.
6. The solid precipitated material produced during secondary treatment of the effluent carried out in the effluent treatment plants.
7. The two types of devices commonly used for removing the participate air pollutants.

Ans. 1. waste 2. Effluent 3. Humus 4. Incineration
5. waste 6. Sludge
7. Scrubber, plate type electrostatic precipitator.

Location Based

Set 4

Chapter 2. Cell: The Unit of Life

1. Cristae

Ans. The finger like projections present in the inner membrane of mitochondria

2. Cellulose

Ans. Present in the plant cell wall.

3. Centrioles

Ans. Short bundles of microfilaments resent in centrosome.

4. Nucleolus

Ans. A small dense spherical structure present in the nucleus.

5. Chlorophyll

Ans. A green coloured photosynthetic pigment present in the chloroplast.

6. Ribosomes **[November, 2019]**

Ans. Small granules found attached to outside of endoplasmic reticulum or ffound scattered in cytoplasm.

Chapter 3. Tissues: Plants and Animal Tissue

1. Platelets

Ans. A component of blood t hat flows in blood vessels.

2. Columnar epithelium

Ans. Found in the inner lining of stomach and intestine.

3. Cartilage

Ans. Present in the tip of nose, external ear, trachea, bronchial tubes.

4. Apical tissue

Ans. Found near the tips of roots and stem and tips of axillary buds.

5. Sclerenchyma

Ans. Found in the stems and veins of the leaves.

6. Epidermis

Ans. Found at the lower surface of the leaf.

7. Intestinal meristem

Ans. Present in the inner lining of the intestine.

8. Ligament

Ans. The tissue present at the joints between two long bones is ligament.

9. Vascular cambium

Ans. Present in the walls of the veins of leaves.

10. Sclerids

Ans. The tissue found as gritty masses in the skin of pears is sclerids.

11. Ciliated epithelium

Ans. Found in the lining of wind pipe.

12. Squamous epithelium

Ans. Present in the linings of nasal and mouth cavities, blood vessels.

13. Lenticels **[November, 2019]**

Ans. Found in the bark of old and woody stems.

14. Sieve Tubes **[February, 2020]**

Ans. Present in phloem tissue which are elongated cells placed end to end forming long tube like structures.

Chapter 4. The Flower

1. Placenta

Ans. The swollen part in ovary which attaches ovules to the walls of ovary.

2. Pollen sac **[November, 2019]**

Ans. Present inside anther

3. Receptacle or thalamus

Ans. Tip of flower stalk

4. Nectaries **[February, 2020]**

Ans. Nectaries are situated at the base of pistil.

5. Sepals or calyx

Ans. Outermost whorl of the flower

Chapter 5. Pollination and Fertilisation

1. Ovule

Ans. Inside the ovary of gynoecium.

2. Embryo sac

Ans. Nucellus contains an embryo sac with eight nuclei.

3. Seed

Ans. Seeds are the product of the ripened ovule.

4. Tube nucleus and generative nucleus

Ans. Present in pollen grain.

5. Polar nuclei

Ans. Present in central cell of embryo sac.

Chapter 6. Seeds: Structure and Germination

1. Hilum

Ans. A distinct whitish oval scar on the concave side of the seed.

2. Coleoptile

Ans. A protective sheath that encloses plumule located towards upper broader side of embryonic region.

3. Tegmen

Ans. Inner layer of seed coat lying next to testa.

4. Hypocotyl

Ans. The region of axis below the cotyledons.

5. Epicotyl

Ans. The region of axis between the point of attachment of cotyledons and the plumule.

6. Micropyle

Ans. A tiny pore situated close to the hilum.

7. Aleurone layer

Ans. The outermost layer of endosperm present in maize grain.

8. Scutellum

Ans. Present near the endosperm of maize grain.

Chapter 7. Respiration in Plants

1. Stomata **[February, 2020]**

Ans. Found on epidermis of leaves.

2. Cytoplasm

Ans. Found within the cell membrane excluding the nucleus of a cell.

3. Mitochondria

Ans. A cell organelle which lies in cytoplasm of a cell.

4. Lenticels

Ans. Minute openings found on old, woody stems.

Chapter 8. Five Kingdom Classification

1. Notochord

Ans. Present in the mid-dorsal axis of the body.

2. Cnidoblasts

Ans. Present on the tentacles of the organisms of Phylum Cnidaria.

3. Clitellum

Ans. A short cylindrical band of thick glandular skin which surrounds the body in 14-16 segments about one-third distance of the body from front end.

4. Rhizoids

Ans. False root-like structures found in bryophytes.

5. Tube feet

Ans. Small tube-like projections on the underside of Echinoderms.

Chapter 9. Economic Importance of Bacteria and Fungi

1. Capsule

Ans. A slimy layer present outside the cell wall of a bacterial cell.

2. Peptidoglycan

Ans. Present in cell wall of bacteria.

3. Sporangiophore

Ans. The erect hyphae of bread mould.

4. Vacuole

Ans. Clear spaces present in cytoplasm of yeast cell.

5. Mesosome

Ans. The invasion or folding of cell membrane inwards towards the cytoplasm in bacterial cells.

6. Hyphae

Ans. A network of thread-like structures of Rhizopus.

7. Sporangium

Ans. The swollen tip of sporangiophore [erect hyphae] of bread mould.

Chapter 12. Skeleton: Movement and Locomotion

1. Vertebrae

Ans. A ring-like unit of backbone.

2. Sternum **[February, 2020]**

Ans. Long flat bone lying in middle of the front part of the chest.

3. Glinding joint

Ans. Occur between the bone of wrist and also between bones of ankle.

4. Pivot joint

Ans. Present in skull which rotate over axis vertebra.

5. Humerus

Ans. Present in upper arm of human.

6. Cranium

Ans. The protective covering of brain.

Chapter 13. Skin: "The jack of all Trades"

1. Malpighian layer

Ans. Between granular layer and dermis.

2. Sweat glands

Ans. They are located in the dermis.

3. Sebaceous glands

Ans. They are located in the dermis.

4. Subcutaneous fat layer

Ans. It is located below dermis.

5. Hair

Ans. They are located in the skin

6. Ceruminous gland **[November, 2019]**

Ans. Modified sebaceous gland found in auditony canal of ear.

7. Parotid gland **[November, 2019]**

Ans. Located just in front of and beneath each ear.

Chapter 14. The Respiratory System

1. Pharynx

Ans. The nasal chambers open at the back into a wide cavity, situated at the back of mouth.

2. Larynx

Ans. The voice-box located at the start of windpipe.

3. Bronchi

Ans. Closer to the lungs, the trachea divides into two tubes entering to the respective lungs.

4. Alvoeli

Ans. The repeating branches end up at air chambers inside the lungs as air sacs called alveoli.

Chapter 15. Hygiene

Mention the exact location of body, where following effects:

1. Trachoma

Ans. A germ that transfer to eye and causes infection.

2. Anthrax vacilli

Ans. A pathogen stays in tannery wastes as water pollutant.

3. Rat-flea

Ans. A parasite that lives on rat's skin.

4. Elephantiasis

Ans. Culex mosquito

5. Plasmodium

Ans. A germ received by anopheles mosquite of particular disease while biting to suck out blood from the body.

6. Pseudopodium

Ans. A temporary protrusion of the surface of an amoeboid cell for movement and feeding.

Chapter 16. Disease: Cause and Control

1. Cholera

Ans. Cholera is the disease that attacks the intestinal tract.

2. Malaria

Ans. Malaria is the disease that affects the red blood cells.

3. Ascariasis

Ans. Ascariasis is the disease cause in intestine.

4. Amoebic Dysentery

Ans. Amoebic dysentery is caused by an amoeba which causes destruction of the lining of the large intestine.

5. Filariasis

Ans. Filariasis attacks the lymphatic system and connective tissues of the human body mainly the limbs.

Chapter 17. Aids to Health

1. Moulds which produce penicillin.

Ans. Mould which produce penicillin is located in *Penicillium notatum.*

2. T-lymphocytes

Ans. T-lymphocyte is located in thymus.

3. Keratin

Ans. Keratin is present in hairs and nails.

4. Mucous membrane

Ans. Mucous membrane is found in nose, mouth, and lungs that trap germs.

5. Hydrochloric acid

Ans. Hydrochloric acid is present in the stomach.

6. Alvoeli

Ans. Alvoeli is located in the alveolar sacs in the lungs.

Chapter 18. Health Organisation

1. The headquarters of WHO.

Ans. The headquarters of WHO located in Geneva.

2. Headquarters of UNO.

Ans. The headquarters of UNO is in New York.

❑

Location Based

Multiple Choice Questions | Set 5 |

Chapter 1. Introducing Biology

1. Pathology is the study of:
   (a) Immune system
   (b) Bacteria and virus
   (c) Disease of plants and animals
   (d) Parasites live inside or outside the body
2. ______ is the study of chemicals and reactions that takes place inside the living organisms.
   (a) Parasitology
   (b) Embryology
   (c) Biochemistry
   (d) Biogeography
3. A person involved in the treatment and surgery of animals has a degree of ________.
   (a) genomics
   (b) veterinary science
   (c) marine biology
   (d) molecular biology
4. A person studies the science which deals with the life present elsewhere in the universe. Identify the branch.
   (a) Exobiology
   (b) Nuclear biology
   (c) Bioengineering
   (d) Space biology
5. Study of fossils is called ________.
   (a) Histology
   (b) Embryology
   (c) Morphology
   (d) Palaeontology
6. Palaeontology is the study of ________.
   (a) embryos
   (b) blood
   (c) Classification
   (d) fossils
7. Clones of animals are being produced through the practice of ________.
   (a) genetics
   (b) ecology
   (c) embryology
   (d) biotechnology

Ans. **1.** (c) **2.** (c) **3.** (b) **4.** (a) **5.** (d)
6. (d) **7.** (a)

Chapter 2. Cell: The Unit of Life

1. Which of the following would be found in an animal cell but not in a bacterial cell?
   (a) DNA
   (b) Cell wall
   (c) Ribosomes
   (d) Endoplasmic reticulum
2. Which of the following organelles plays an important role in the process of making food?
   (a) Lysosome
   (b) Ribosomes
   (c) Chloroplast
   (d) Mitochondria
3. In which part of the nucleus thread like structures (called chromatin fibres) is present?
   (a) Nucleolus
   (b) Nucleoplasm
   (c) Nuclear membrane
   (d) None of the above

4. Which one of the following is found both in the cells of a mango plant and a monkey?
   (a) Chloroplasts (b) Centrioles
   (c) Cell wall (d) Cell membrane
5. Lysosome is called as the suicidal bag of cell due to presence of:
   (a) ATP as an energy source of the cell.
   (b) Cell wall which gives rigidity and shape to the cell.
   (c) Hydrolytic enzymes which destroy and digest foreign substances around them.
   (d) Ribosomes on their membrane which plays an important role in the destruction of the damaged cell.
6. All life starts as:
   (a) an egg (b) a single cell
   (c) a gene (d) a chromosome
7. A plant cell can be identified from an animal cell by the:
   (a) absence of centrosome (b) presence of cell membrane
   (c) presence of vacuoles (d) none of these
8. Like the fingerprinting, pattern of which of the following will help in ascertaining the identity of a person?
   (a) ATP (b) RER
   (c) DNA (d) RNA
9. Which one of the following cell organelles is correctly matched with its function?
   (a) Ribosomes → Synthesis of proteins
   (b) Mitrochondria → Secretion of enzymes
   (c) Plasma membrane → Freely permeable
   (d) Centrosome → Carries genes

Ans. 1. (d) 2. (c) 3. (b) 4. (a) 5. (c)
6. (b) 7. (a) 8. (c) 9. (a)

Chapter 3. Tissues: Plants and Animal tissue

1. Cartilage is not found in:
   (a) ear (b) nose
   (c) larynx (d) kidney
2. Which of the following plant tissue carry out the functions like photosynthesis, storage, excretion?
   (a) Aerenchyma (b) Sclerenchyma
   (c) Parenchyma (d) Collenchyma
3. Classification of plant tissue into meristematic and permanent tissues is based on:
   (a) The position of the cells
   (b) whether they are living or dead
   (c) whether the cells being able to divide or not.
   (d) None of these
4. The kind of epithelium which forms the inner walls of blood vessels is:
   (a) cuboidal epithelium (b) columnar epithelium
   (c) squamous epithelium (d) ciliated columnar epithelium
5. A person met with an accident in which two long bones of his hands were dislocated. Which the following may be the possible reason?

(a) Tendon breaks
(b) Ligament breaks
(c) Break of skeletal muscle
(d)Areolar tissue breaks

6. In potato, starch is stored in:
(a) Sclerenchyma
(b) Collenchyma
(c) Parenchyma
(d) Chlorenchyma

7. Tendons and ligaments are examples of:
(a) Fibrous connective tissue
(b) Cartilage
(c) Muscular tissue
(d) Adipose tissue

8. Parenchyma containing chloroplasts is known as:
(a) Parenchyma
(b) Aerenchyma
(c) Collenchyma
(d) Chlorenchyma

9. Annual rings are the number of:
(a) Internodes in a stem
(b) Rings of vascular bundles in a monocot stem
(c) Barks layers in a woody stem
(d) Layers of xylem in a stem

10. Which of the following cells in plants are said to be nonliving?
(a) Meristem
(b) Parenchyma
(c) Collenchyma
(d) Schlerenchyma

11. Which of the following connects a muscle to a bone?
(a) Cartilage
(b) Ligament
(c) Tendon
(d) Interstitial fluid

12. Cardiac muscle is:
(a) Involuntary
(b) Smooth
(c) Striated
(d) Involuntary and striated

13. Which of the following is correctly matched?
(a) Meristem – Actively dividing cells
(b) Xylem – Transport of food
(c) Phloem – Transport of water
(d) Sclerenchyma – Storage of starch

Ans. **1.** (d) **2.** (c) **3.** (c) **4.** (c) **5.** (b)
6. (c) **7.** (a) **8.** (d) **9.** (d) **10.** (d)
11. (c) **12.** (d) **13.** (a)

Chapter 4. The Flower

1. Which one of the following characteristics is not related to gynoecium?
(a) It is composed of stamens.
(b) It is the female reproductive part of the flower.
(c) Stigma is usually at the tip of the style and is the receptive surface for pollen grains.
(d) Each ovary bears one or more ovules attached to a flattened, cushion-like placenta.

2. The term 'polyadelphous' is related to:
(a) calyx
(b) corolla
(c) gynoecium
(d) androecium

3. *Bougainviliea* flower is an example of:
(a) incomplete flower
(b) having a large nectary
(c) water pollination
(d) large colour bracts

4. The arrangement of ovules within the ovary is known as
(a) aestivation (b) placentation
(c) inflorescence (d) Both (a) and (c)

5. A flower is said to be complete when:
(a) It has the corolla and calyx (b) It has the corolla and gynoecium
(c) It has the androecium and gynoecium (d) It has all four whorls

6. Coloured, large, paper structures found in *Bougainvillea* are called: **[November, 2019]**
(a) Sepals (b) Petals
(c) Epcalyx (d) Bracts

7. The term neuter is given to a flower in which both: **[February, 2020]**
(a) Androecium and Gynoecium are present.
(b) Calyx and Corolla are present.
(c) Androcium and Gynoeccium are absent.
(d) Calyx and Corolla are absent.

8. A student was given a sample of flower to identify X and Y on the basis of the information given below.
'X is the outermost whorl of the flower and contains Y. Y is green, leaf-like and protect the other whorls of the 'flower'.
(a) X – Corolla; Y - Fruit (b) X - Calyx; Y - Sepals
(c) X - Sepals; Y - Corolla (d) X - Androecium; Y – Ovary

9. When the filaments of anther in a flower are fused into two groups, the condition is called:
(a) polyandrous (b) diadelphous
(c) polyadelphous (d) monoadelphous

10. The part of the flower that given rise to the fruit is:
(a) Sepals (b) Petals
(c) Ovary (d) Stamens

11. The part of the flower that gives rise to the seed is:
(a) Ovary (b) Placenta
(c) Ovule (d) Pollen grain

Ans. **1.** (a) **2.** (d) **3.** (d) **4.** (b) **5.** (d)
6. (d) **7.** (c) **8.** (b) **9.** (b) **10.** (c)
11. (c)

Chapter 5. Pollination and Fertilisation

1. Exine and intine are the wall layers of:
(a) ovule (b) ovary
(c) nucellus (d) pollen grain

2. The process of fusion of male gamete with polar nuclei is called:
(a) syngamy (b) triple fusion
(c) double fertilisation (d) All of these

3. Which of the following features are shown by the insect pollinated flowers?
(a) The flowers are small (b) Their stigmas are feathery
(c) Pollens are produced in large amount (d) They produce nectar to attract the insect

4. In wind pollinated maize:
(a) there is more chance of self-pollination
(b) anthers are tightly attached to the filament

(c) the female flowers mature earlier than male flowers
(d) the male flowers mature earlier than female flowers

5. Which of the following method help the pollen grain to reach the stigma?
(a) Allogamy (b) Autogamy
(c) Geitonogamy (d) All of these

6. Which one of the following is one of the characteristics of self-pollinated flowers?
(a) Flowers are large and showy
(b) Flowers remain closed and do not open
(c) Stigma and anthers mature at the same time
(d) Pollen is produced in very large quantities

7. In *Vallisneria*, pollination takes place by the agency of: **[November, 2019]**
(a) Water (b) Birds
(c) Wind (d) Insects

Ans. 1. (d) 2. (b) 3. (d) 4. (d) 5. (d) 6. (c) 7. (a)

Chapter 6. Seeds: Structure and Germination

1. The optimum temperature for seed germination is:
(a) 15°C (b) 40°C
(c) 25°C (d) 45°C

2. Scutellum is present in:
(a) Bean (b) Maize
(c) Pea (d) Gram

3. Seeds sown very deep in the soil fail to germinate because they:
(a) Cannot exert enough force to push the soil upward
(b) Do not get enough sunlight
(c) Get too much water
(d) Do not get enough oxygen

4. In pea, castor and maize the number of cotyledons are:
(a) 1,2 and 2 respectively (b) 2,2 and 1 respectively
(c) 2,1 and 2 respectively (d) 1, 2 and 1 respectively

5. In hypogeal seed germination, the structure help to push the cotyledon inside the soil is:
(a) Epicotyl (b) Hypocotyl
(c) Radicle (d) Plumule

6. Which one of the following plant parts is correctly matched with one of its stated characteristic?
(a) Mango seed ------------- Aleurone layer (b) Bean seed ------------- Endosperm
(c) Maize grain ------------- Coleoptile (d) Wheat grain ----------- exalbuminous

7. In the monocotyledon seeds, the endosperm is separated from the embryo by a distinct layer known as:
(a) Testa (b) Aleurone layer
(c) Tegmen (d) Epithelium

8. Endosperm is consumed by developing embryo in the seed of:
(a) Coconut (b) Maize
(c) Pea (d) Castor

9. Largest seed is:

(a) Double coconut (b) Coconut
(c) Walnut (d) Heartnut

10. An albuminous seed showing hypogeal germination is:
(a) Castor (b) Bean
(c) Gram (d) Maize

Ans. 1. (c) 2. (b) 3. (d) 4. (b) 5. (a) 6. (c) 7. (d) 8. (c) 9. (a) 10. (d)

Chapter 7. Respiration in Plants

1. The number of ATP molecules released during fermentation process by yeast from one mole of glucose is:
(a) 38 ATP (b) 2 ATP
(c) 12 ATP (d) 6 ATP

2. Which of the following statements comparing aerobic and anaerobic respiration is correct?
(a) Aerobic respiration releases a smaller amount of energy than anaerobic respiration
(b) Anaerobic respiration only occurs in microbes while aerobic respiration only occurs in multicellular organisms
(c) Aerobic respiration is the complete breakdown of glucose while anaerobic respiration is the incomplete break down of glucose
(d) Anaerobic respiration does not produce carbon dioxide while aerobic respiration does

3. The different steps of aerobic respiration are:
(a) Glycolysis, Kreb's cycle, Electron transport chain
(b) Glycolysis, fermentation
(c) Glycolysis, Kreb's cycle
(b) Fermentation, Electron transport chain, glycolysis

4. Site of Kreb's cycle is:
(a) Cytoplasm (b) Mitochondria
(c) Lysosome (d) Peroxisome

5. Glycolysis is a process:
(a) Which occurs in vacuole
(b) That occurs next to Kreb's cycle
(c) In which glycogen is broken down into glucose
(d) In which glucose is broken down into pyruvate

6. One same common function is performed by:
(a) Stomata and veins (b) Stomata and lenticels
(c) Lenticels and sepals (d) Lenticels and hydathodes

7. Anaerobic respiration normally occurs in:
(a) Grass (b) Cactus
(c) Coconut (d) Baker's yeast

8. Mitochondria is the storage house for:
(a) NADH (b) Pyurvic acid
(c) ATP (d) Citric acid

9. Less energy is liberated in anaerobic respiration in yeast in comparison with aerobic respiration because:
(a) Energy is not formed by oxygen (b) Less carbon dioxide is formed
(c) Energy is left in alcohol (d) Yeast requires less energy

10. The role of oxygen in cellular respiration:
(a) To form carbon dioxide
(b) To liberate energy from hydrogen storage
(c) To change pyruvate into acetyl coA
(d) To accept hydrogen and form water

Ans. 1. (b) 2. (c) 3. (a) 4. (b) 5. (d)
6. (b) 7. (d) 8. (c) 9. (c) 10. (d)

Chapter 8. Five Kingdom Classification

1. Identify the structure found in Protozoa for locomotion.
(a) Cilia
(b) Flagella
(c) Pseudopodia
(d) All of these
2. Pneumatic bones are present in:
(a) Mammals
(b) Pisces
(c) Aves
(d) Amphibians
3. The only one phylum with no terrestrial form is:
(a) Echinodermata
(b) Annelida
(c) Platyhelminthes
(d) Arthropoda
4. The basic taxonomic category is:
(a) Genus
(b) Species
(c) Order
(d) Family
5. Which of these are known for producing seeds?
(a) Pteridophyta and Gymnosperms
(b) Bryophyta and Thallophyta
(c) Angiosperms and bryophyta
(d) Angiosperms and Gymnosperms
6. The animals (Kingdom Animalia) are mainly grouped under two categories:
(a) Bilaterally symmetrical and radially symmetrical
(b) Coelomate and acoelomate
(c) Aquatic and terrestrial
(d) Vertebrates and invertebrates
7. Which one of the following examples does not come under the particular group?
(a) Mushroom, yeast, fern ------ Fungi
(b) Paramecium, Euglena, sponge ------ Protista
(c) Starfish, cuttlefish, dogfish ----- Pisces
(d) Bat, parrot, trout ---- Vertebrates
8. Which one of the following is an example of binomial scientific name?
(a) Green algae
(b) Snow leopard
(c) *Rana tigrina*
(d) China rose
9. Which one of the following is the correct statement about the respective animal?
(a) Donkey and horse should be considered one single species because they can successfully mate and produce an offspring.
(b) Donkey and horse are two different genuses.
(c) Mule is a separate species.
(d) Mule is neither a donkey nor a horse.
10. Which among the following is a feature that characterise vertebrates?
(a) Dorsal nerve cord and a tail behind the level of the anus.
(b) Haemoglobin dissolved
(c) Three or more pairs of limbs
(d) Dorsal heart and nerve cord ventral

Ans. 1. (d) 2. (c) 3. (a) 4. (b) 5. (d)
6. (d) 7. (d) 8. (c) 9. (d) 10. (a)

Chapter 9. Economic Importance of Bacteria and Fungi

1. The most common edible mushroom is:
   (a) Volvariella (b) Agaricus bisporous
   (c) Pleurotus (d) Aspergillus
2. Which one of the following do bacteria possess?
   (a) Ribosome (b) Golgi complex
   (c) Vacuole (d) Mitochondria
3. Yeast reproduces by:
   (a) Spores (b) Multiple fission
   (c) Budding (d) Binary fission
4. The bread dough rises during baking because of:
   (a) Increased heat (b) Growth of yeast cells
   (c) Kneading of dough (d) Grinding of dough
5. Which one of the following characteristics is found in all fungi but not in all bacteria?
   (a) Aerobic respiration
   (b) Cell wall
   (c) Spore formation
   (d) A long circular DNA lying loose in the cytoplasm
6. Production of ethanol occurs in one of the life processes of:
   (a) Bread mould (b) Yeast
   (c) Mushroom (d) Penicillium
7. Bacteria are referred to as prokaryotes because:
   (a) They have no chlorophyll (b) They are unicellular
   (c) They are free living (d) They do not have a true nucleus
8. Yeast is used in the production of:
   (a) Ethyl alcohol (b) Acetic acid
   (c) Cheese (d) Curd
9. Tick mark the correct statement.
   (a) All mushrooms are poisonous (b) All toadstools are poisonous
   (c) Some toadstools are poisonous (d) Some mushrooms are edible
10. Bacteria are no more classified as plants primarily because
    (a) These are unicellular (b) These are microscopic
    (c) Many of them are parasitic (d) They have no chlorophyll
11. Which bacteria is rod shaped?
    (a) Coccus (b) Spirillum
    (c) Bacillus (d) Vibrio
12. Which bacteria fixes nitrogen in the soil?
    (a) Nitrobacter (b) Nitrosomonas
    (c) Rhizobium (d) Clostridium
13. A particular species of which one of the following is the source bacterium of the antibiotic, discovered next to penicillin, for the treatment of tuberculosis?
    (a) Escherichia (b) Streptomyces
    (c) Rhizobium (d) Nitrobacter
14. Which one of the following diseases is caused by a fungus? **[February, 2020]**
    (a) Ring worm (b) Round worm
    (c) Tape worm (d) Typhoid

Ans. **1.** (b) **2.** (a) **3.** (c) **4.** (b) **5.** (c)
6. (b) **7.** (d) **8.** (a) **9.** (d) **10.** (d)
11. (c) **12.** (d) **13.** (b) **14.** (a)

Multiple Choice Questions

Chapter 10. Nutrition

1. Nutrient required for proper growth of bones is:
   (a) Protein (b) Carbohydrate
   (c) Calcium (d) Iron
2. Mineral present in fish oil:
   (a) Calcium (b) Iron
   (c) Iodine (d) Calciferol
3. Disease caused due to lack of iron is:
   (a) Anaemia (b) Bone deformation
   (c) Skin diseases (d) None of above
4. Vitamin which we get from sunlight is:
   (a) Vitamin B (b) Vitamin C
   (c) Vitamin D (d) Both A and C.
5. The hard material present on tooth:
   (a) Dentine (b) Enamel
   (c) Crown (d) Cement
6. Vitamin A is know as ....
   (a) Retinol (b) Ascorbic acid
   (c) Folic acid (d) Calciferol
7. An adult labourer need .... kcal daily.
   (a) 1400 (b) 3500
   (c) 2500 (d) 1100
8. .... is a double sugar.
   (a) Glucose (b) Sucrose
   (c) Fructose (d) Cellulose
9. Kwashiorkor in children is due to deficiency of ....
   (a) protein (b) Protein and Carbohydrate
   (c) Vitamin (d) Minerals
10. Vitamin B_1 deficiency causes .... disease.
    (a) Scurvy (b) Ricket
    (c) Beri-beri (d) Pellegra
11. For a balanced meal, which of the following should be in the highest quantity in a thali?
    (a) Rice (b) Dal
    (c) Vegetable curry (d) Eggs
12. Which of these is the best source of vitamin-A?
    (a) Apple (b) Carrot
    (c) Honey (d) Groundnut
13. Which of the following is good source of carbohydrates?
    (a) Meat (b) Almond
    (c) Groundnut (d) Rice
14. Marasmus is caused due to deficiency of which of these?
    (a) Vitamin-C (b) Carbohydrates
    (c) Fat (d) Iodine

15. Your grandmother always says you to enjoy the sunshine; as it is good for our bones. Which vitamin changes to active form on exposure to sunshine?
(a) Vitamin-A (b) Vitamin-B
(c) Vitamin-C (d) Vitamin-D

16. Which one of the following pairs of nutrients includes both as simple sugars (mono-accharides)?
(a) Sucrose and glucose (b) Glucose and maltose
(c) Fructose and glucose (d) Maltose and lactose

17. Which one of the following is the correct matching of a nutrient mineral element and its one rich source?
(a) Calcium-Potato (b) Sodium-Citrus fruits
(c) Iodine-Daiy milk (d) Potassium-Banana

18. Marasmus is due to the deficiency of:
(a) Vitamin C (b) Carbohydrates
(c) Proteins (d) Food

19. Fat soluble vitamins are:
(a) A, D and E (b) B, C and D
(c) B, D and E (d) A, B and C

20. Which of the following is the best sources of Vitamin A?
(a) Apple (b) Honey
(c) Carrot (d) Peanuts

21. Which one of the following is a vitamin deficiency disease? **[February, 2020]**
(a) Diabetes (b) Goitre
(c) Marasmus (d) Scurvy

Ans. **1.** (c) **2.** (d) **3.** (a) **4.** (a) **5.** (b)
6. (a) **7.** (b) **8.** (b) **9.** (a) **10.** (c)
11. (a) **12.** (b) **13.** (d) **14.** (b) **15.** (d)
16. (c) **17.** (d) **18.** (b) **19.** (a) **20.** (c)
21. (d)

Chapter 11. Digestive System

1. Starch turns:
(a) Blue litmus red (b) Red litmus blue
(c) Brown iodine blue (d) Blue iodine brown

2. ...... are directly absorbed from digestive tract.
(a) Fats (b) Carbohydrates
(c) proteins (d) Vitamins

3. Pylorus is opening from:
(a) Oesophagus into stomach (b) Mouth cavity into oesophagus
(c) Stomach into intestine (d) Intestine into rectum

4. Blind end of large intestine is:
(a) Vermiform appendix (b) Peristalsis
(c) Anus (d) Rectum.

5. The pyloric sphincter muscle present at opening of:
(a) stomach and the oesophagus (b) stomach and the duodenum
(c) oesophagus and the duodenum (d) duodenum and the ileum

6. Hepato-pancreatic duct opens into the duodenum and carries:
   (a) bile (b) pancreatic juice
   (c) both (a) and (b) (d) saliva
7. Villi are located in:
   (a) large intestine (b) small intestine
   (c) stomach (d) duodenum
8. Liver is the largest gland and is associated with various functions, choose one which is not correct?
   (a) Metabolism of carbohydrate (b) Digestion of fat
   (c) Formation of bile (d) Secretion of hormone called gastrin
9. Mark the right statement among the following
   (a) Trypsinogen is an inactive enzyme
   (b) Trypsinogen is secreted by intestinal mucosa
   (c) Enterokinase is secreted by pancreas
   (d) Bile contains trypsin
10. The water from the digested food is main absorbed by which of the following options:
    (a) Stomach (b) Duodenum
    (c) Colon (d) Rectum
11. The number of teeth in man:
    (a) 32 (b) 16
    (c) 20 (d) 30
12. Gastric juice is secreted by:
    (a) intestine (b) stomach
    (c) mouth (d) oesophagus
13. Protein is converted into amino acid by an enzyme called:
    (a) sucrase (b) lipase
    (c) trypsin (d) maltase
14. The enzyme present in saliva is:
    (a) Ptyalin (b) Lipase
    (c) Pepsin (d) Trypsin
15. An example of disaccharide is: **[November, 2019]**
    (a) Fructose (b) Glucose
    (c) Galactose (d) Maltose
16. Gastric Juice contains:
    (a) HCl and pepsin (b) Pepsin and trypsin
    (c) trypsin and HCl (d) Amylopsin and pepsin
17. Which one of the following pairs of types of teeth perform one common function as stated against it?
    (a) Incisors, canines .... Holding (b) Canines, premolars .... Tearing
    (c) Premolars, molars .... Grinding (d) Molars, incisors .... Tearing
18. The digestive juice which does not contain any enzyme: **[February, 2020]**
    (a) Saliva (b) Gastric juice
    (c) Bile (d) Pancreatic juice

Ans.					
	1. (c)	2. (d)	3. (a)	4. (a)	5. (b)
	6. (c)	7. (b)	8. (a)	9. (a)	10. (d)
	11. (a)	12. (c)	13. (b)	14. (a)	15. (d)
	16. (a)	17. (c)	18. (c)		

Chapter 12. Skeleton: Movement and Locomotion

1. Axial skeleton consists of:
(a) Skull (b) Backbone
(c) Ribs and sternum (d) All of these
2. Blood cells are formed in:
(a) Spleen (b) Liver
(c) Bone marrow (d) None of these
3. .... connects bone to muscles:
(a) Tendon (b) Ligament
(c) Skin (d) Fiber
4. Bone cells are called:
(a) Periosteum (b) Osteocytes
(c) Osteum (d) Both (a) and (c)
5. Collagen fibres are present in:
(a) Tendon (b) Ligament
(c) Bone (d) Muscles
6. Your external ear (pinna) is supported by:
(a) bone (b) cartilage
(c) tendon (d) capsule
7. The type of joint found at shoulder is also found at:
(a) elobw (b) knee
(c) ankle (d) hip
8. Which one of the following categories of vertabrae are correctly numbered:
(a) cervical-7 (b) thoracic-10
(c) lumbar-4 (d) sacral-4
9. Human skeleton altogether contains 213 bones. Which of these are the 6 bones?
(a) Neck vertebrae (b) Ear ossicles
(c) Carpals (d) Metacarpals
10. Axial skeleton is made up of:
(a) skull, and vertebral column
(b) skull, vertebral column
(c) skull, vertebral column, ribs and stermum
(d) skull
11. Pivot joint is present between:
(a) atlas and axis vertebrae (b) skull and axis vertebra
(b) skull and atlas vertebra (d) axis and first thoracic vertebra
12. Involuntary muscle is present in:
(a) leg (b) hand
(c) iris (d) head
13. Branched muscles are:
(a) skeletal (b) cardiac
(c) smooth (d) both (a) and (c)
14. Single nucleus is present in:
(a) skeletal muscle (b) smooth muscle
(c) cardiac muscle (d) both (a) and (c)

Ans. 1. (d) 2. (c) 3. (a) 4. (b) 5. (b)
6. (b) 7. (d) 8. (a) 9. (b) 10. (c)
11. (c) 12. (c) 13. (b) 14. (b)

Multiple Choice Questions

Chapter 13. Skin-"The jack of all trades"

1. Leucoderma consists of:
(a) White patches on skin (b) Black patches on skin
(c) Red patches on skin (d) Brown patches on skin
2. Bad smell of sweat is due to:
(a) Urea (b) Keratin
(c) Bacterial infection (d) None of these
3. Fair colour skin is due to:
(a) More of melanin (b) More of keratin
(c) Less of melanin (d) less of keratin
4. Excessive wax in ear is produced due to over activity of:
(a) Sebaceous gland (b) Sweat gland
(d) Ceruminous gland (d) None of these.
5. Sweat contains :
(a) Water (b) Urea
(c) Salt (d) All of these
6. The vitamin synthesised by the skin ....
(a) Vitamin-A (b) Vitamin-B
(c) Vitamin-D (d) Vitamin-C
7. The structure that covers the hair root ....
(a) follicle (b) bulb
(c) papilla (d) shaft
8. Human body temperature is:
(a) 40°C (b) 25°C
(c) 37°C (d) 28°C
9. The colouration of skin is due to the presence of: **[November, 2019]**
(a) Keratin (b) Melanin
(c) Haemoglobin (d) Anthocyanin
10. If for some reason the sebaceous glands fail to fuction:
(a) The body will not be able to regulate the body temperature
(b) The skin will turn darker with more melanin
(c) The hairs will fail to grow
(d) The skin will turn dry and rough
11. Which one of the following pairs includes the abnormal conditions of skin pigmentation?
(a) Leukoderma, Ringworm (b) Albinism, Leukoderma
(c) Baldness, Albinism (d) Rickets, baldness
12. Which one of the organs listed below, functions in regulating our body temperature most actively?
(a) Heart (b) Lungs
(c) Skin (d) Stomach
13. Sweat glands are situated in:
(a) Epiderms (b) Dermis
(c) Both (d) None of these
14. The epidermis is highly thickened in:
(a) Eyelid (b) Thigh
(c) Lip (d) Palm

Ans.	1. (a)	2. (c)	3. (c)	4. (c)	5. (d)
	6. (c)	7. (a)	8. (c)	9. (b)	10. (d)
	11. (b)	12. (c)	13. (b)	14. (d)	

Chapter 14. The Respiratory System

1. Small amount of air is left in lungs even after forceful exhalation. It is called:
   (a) Dead air (b) Tidal air
   (c) Residual air (d) Air capacity
2. Muscle(s) which help in respiration:
   (a) Intercostal muscle (b) Abdominal muscle
   (c) Diaphragm (d) All of these
3. On complete aerobic combustion of one molecule of glucose, ATP produced are:
   (a) 26 ATP (b) 38 ATP
   (c) 36 ATP (d) 6 ATP
4. End product produced in animals due to anaerobic respiration:
   (a) Lactic acid (b) ATP
   (c) Acetic acid (d) Urea
5. Glucose is converted to .... in liver.
   (a) Lactic acid (b) Starch
   (c) Glycogen (d) Ethanol
6. During inspiration, the diaphragm:
   (a) Relaxes (b) Contracts
   (c) Expands (d) Gets folded
7. The ultimate end parts of the respiratory system in humans is known as:
   (a) Alveoli (b) Bronchioles
   (c) Tracheoles (d) Bronchi
8. During respiration there is:
   (a) Gain in dry weight (b) Loss in dry weight
   (c) No change in dry weight (d) Increase in the overall weight
9. The sum of tidal volume, inspiratory reserve volume and expiratory reserve volume..........
   (a) total lungs capacity (b) vital capacity
   (c) residual capacity (d) inspiratory capacity
10. The ultimate end of trachea is:
    (a) Bronchi (b) Bronchioles
    (c) Alveoli (d) Pharynx
11. In which of the following, cartilaginous rings are present?
    (a) alveoli (b) trachea
    (c) bronchi (d) larynx
12. Where does glycolysis takes place?
    (a) Mitochondria (b) Cytoplasm
    (c) Chloroplast (d) Golgi body

Ans.	1. (c)	2. (d)	3. (b)	4. (a)	5. (c)
	6. (b)	7. (a)	8. (b)	9. (a)	10. (c)
	11. (b)	12. (b)			

Chapter 15. Hygiene

1. Which one of the following is a healthy habit?
   (a) Sharing towels with others
   (b) Going to bed at late hours
   (c) Put any sharp pointed object into the ear
   (d) Mouth should be washed after every meal
2. Urea in the blood is a typical symptom of:
   (a) Malaria (b) Cholera
   (c) Typhoid (d) Dysentery
3. Identify the mosquito which causes elephantiasis.
   (a) Anopheles (b) Aedes
   (c) Culex (d) All of these
4. Which of the following disease causes inflammation of liver?
   (a) Cholera (b) Dysentery
   (c) Hepatitis (d) Malaria
5. Which type of hepatitis is/are transmitted through blood and other body fluids?
   (a) A (b) B and C
   (c) A and C (d) D
6. No urination due to shortage of water in the body is a typical symptoms of:
   (a) Typhoid (b) Malaria
   (c) Dysentery (d) Cholera
7. Which one of the following is an unhealthy habit?:
   (a) Removing wax from inside the ears
   (b) Keeping the mouth closed while breathing
   (c) Sleeping for only 6-7 hours
   (d) Sharing towels with others
8. Protozoan that cause malaria is:
   (a) *Entamoeba histolytica* (b) *Planaria*
   (c) *Paramecium* (d) *Plasmodium*
9. The chemical substance mainly responsible for tooth disorders is:
   (a) Calcium (b) Chloride
   (c) Flourides (d) Iodides
10. Typhoid is caused due to:
    (a) Housefly (b) Bacteria
    (c) Virus (d) None of the these
11. The germ causing Hepatitis is transmitted through:
    (a) the bite of a mosquito (b) contamination of food by a housefly
    (c) personal contact with a patient (d) contaminated water

Ans. **1.** (d) **2.** (b) **3.** (c) **4.** (c) **5.** (b)
6. (d) **7.** (d) **8.** (d) **9.** (d) **10.** (a)
11. (d)

Chapter 16. Disease: Cause and control

1. The disease that found in a specific area attacking fewer people is:
   (a) Epidemic diseases (b) Endemic diseases
   (c) Pandemic diseases (d) Sporadic diseases

2. The disease which is non-communicable:
(a) AIDS (b) TB
(c) Cancer (d) Malaria
3. Chicken pox is caused by the virus called:
(a) Varicella Zoster (b) HIV
(c) Hepatitis A (d) Vibrio cholera
4. *Mycobacterium tuberculosis* cause infectious disease called:
(a) Cholera (b) Typhoid
(c) Tuberculosis (d) Malaria
5. BCG vaccination is useful for:
(a) Pneumonia (b) Tetanus
(c) Gonorrhoea (d) Tuberculosis
6. A disease widely spread worldwide is known as:
(a) Endemic (b) Epidemic
(c) Pandemic (d) Sporadic
7. The letter "B" in the name BCG vaccination stands for:
(a) Brief (b) Beri-beri
(c) Bacteria (c) Bacillus
8. Use of disposable syringes for injecting medicines, etc. is specially advised to prevent:
(a) Poliomyelitis (b) Mumps
(c) Rabies (d) AIDS
9. The vector that transmits the malarial pathogen is:
(a) Culex mosquito (b) Housefly
(c) Anopheles mosquito (d) Entamoeba
10. Amoebiasis is caused by the protozoan:
(a) Amoeba proteus (b) Euglena
(c) Plasmodium (d) Entamoeba
11. BCG vaccine gives immunity against:
(a) Tetanus (b) Cholera
(c) AIDS (d) Tuberculosis
12. The full form of AIDS is:
(a) Active Immunity Deficiency Syndrome
(b) Acquired Immuno Deficiency Syndrome
(c) Acquired Immuno Deficiency Status
(d) Active Immuno Deficiency Syndrome

Ans. 1. (b) 2. (c) 3. (a) 4. (c) 5. (d)
6. (c) 7. (d) 8. (d) 9. (c) 10. (d)
11. (d) 12. (b)

Chapter 17. Aids of Health

1. Vaccines are termed as:
(a) Extract of dead and weakened microbes
(b) Extract of toxins produced by bacteria
(c) Chemical applied on skin to defend against bacteria
(d) Chemical extracted from fungi

Multiple Choice Questions

2. The function of sulphonamides is:
   (a) It interferes in the metabolism of bacteria to kill it
   (b) It is involved in the killing of germs
   (c) It is a chemical that is produced when any toxin enter the body
   (d) It helps to cure sexually transmitted disease
3. A drug called "Pentidsulph" used to cure:
   (a) Common cold (b) Gonorrhoea
   (c) Sexually transmitted disease (d) Skin rashes
4. Antivenins are produce during:
   (a) Entering of toxin substance inside the body
   (b) Snake bite
   (c) Vaccination
   (d) Entry of bacteria
5. Toxoids are:
   (a) Inactive toxins produced from bacteria
   (b) Produced due to the entry of foreign substances
   (c) Produced during Acquired immunity
   (d) Produced from snake venom
6. Penicillin is:
   (a) An antiseptic (b) A disinfectant
   (c) An antibiotic (d) An antioxidant
7. 'T' in DTP vaccination stands for: **[November, 2019]**
   (a) Tuberculosis (b) Typhoid
   (c) Tetanus (d) Tonsillitis
8. World Health day is celebrated on:
   (a) April 7 (b) February 21
   (c) October 10 (d) January 15
9. DPT stands for:
   (a) Diphtheria, Polio and Tetanus (b) Diphtheria, Pertussis and Tetanus
   (c) Diphtheria, Polio and Typhoid (d) Dysentery, Polio and Typhoid
10. Salk's vaccine is used to build immunity against......
   (a) Tuberculosis (b) Malaria
   (c) Rabies (d) Poliomyelitis
11. Which one of the following is not an antibiotic? **[February, 2020]**
   (a) Penicillin (b) Steapsin
   (c) Streptomycin (d) Chloromycetin

Ans. **1.** (a) **2.** (b) **3.** (c) **4.** (b) **5.** (a)
6. (c) **7.** (c) **8.** (a) **9.** (b) **10.** (d)
11. (b)

Chapter 18. Health Organisation

1. The disease that cause air-borne disease:
   (a) Diarrhea (b) Dysentery
   (c) Malaria (d) Typhoid

2. The habitation in which small population, agriculture based, diaries, poultry farming are found:
(a) Small towns (b) Remote areas
(c) Big cities (d) Villages
3. The organization that has a national as well as an International agency.
(a) WHO (b) UNO
(c) Red Cross (d) None of these
4. The organization that is referred as an international body concerned with people's health.
(a) WHO (b) UNO
(c) International sanitation regulation (d) None of these
5. Red Cross day is celebrated on:
(a) 6th May (b) 7th May
(c) 8thMay (d) 9thMay
6. WHO head quarters are located in:
(a) Paris (b) Berlin
(c) Geneva (d) New York

Ans. 1. (c) 2. (d) 3. (c) 4. (a) 5. (c)
6. (c)

Chapter 19. Waste Generation and Management

1. All-most all human activities generate:
(a) Resource (b)Energy
(c) Waste (d) Product
2. Waste generated from living organisms and capable of being decomposed by bacteria are:
(a) Non biodegradable (b) Organic waste
(c) Biodegradable (d) Non of these
3. Waste generated from power plants is:
(a) Industrial waste (b) Radioactive waste
(c) Both (a) and (b) (d) none of these
4. The most rapidly increasing and much harmful waste today is:
(a) plastics (b) pesticides
(c) municipal sewage (d) electronic waste
5. Used syringes, cotton and surgical material is:
(a) Biomedical waste (b) Biodegradable waste
(c) Industrial waste (d) Commercial waste
6. The most convenint reusable wastes are:
(a) Old newspapers (b) Broken glass
(c) Flyash (d) Medical instruments
7. Domestic waste generated from our house consist of:
(a) Biodegradable waste (b) Commercial waste
(c) Non biodegradable waste (d) Both (a) and (c).

Ans. 1. (c) 2. (c) 3. (c) 4. (a) 5. (a)
6. (d) 7. (d)

True or False

| Set 6 |

Q. Mention wheather the following statements are True or False and also correct the false statement.

Chapter 3. Tissues: Plants and animal tissue

1. Axons of nerve cells are very long.
2. Dendrons bundles together form a nerve.
3. Cardiac muscles get tired soon.
4. Epithelial cells leave space in between.
5. Perikaryon is the nucleus of a nerve cell.
6. Muscles of the iris of the eye are of voluntary type.
7. Cartilage has no blood vessels or the nerves in it.

Ans. 1. True
2. False. Axon bundles together form a nerve.
3. False. Cardiac muscles can contract without exterior stimulation and yet not get fatigued.
4. False. Epithelial cells fit tight leaving no space.
5. True
6. False. Muscles of the iris of the eye are of involuntary type.
7. True

Chapter 4. The flower

1. Petaloid perianth is brightly coloured.
2. Anther produces ovule.
3. Nectaries produces mucus.
4. Androecium is the male part of flower.

Ans. 1. True
2. False. Anther produces pollen
3. False. Nectaries produces nector.
4. True

Chapter 5. Pollination and Fertilisation

1. Pollen tube enters ovule through micropyle.
2. Zygote gives rise to embryo.
3. *Vallisneria* is a wind pollinated flower.
4. Dry sepals may persist in some fruit.
5. Cross-pollination takes place in all dioecious plants.
6. Ornithophily is found in Rafflesia.

Ans. 1. True
2. True
3. False. *Vallisneria* is a water pollinated flower.
4. True
5. False. Cross-pollination takes place in all monoecious plants.
6. False. Elephophily is found in Rafflesia.

Chapter 6. Seeds: Structure and Germination

1. Single cotyledon present in maize grain is called scutellum.
2. Micropyle is the spot where ovule was attached to ovary.
3. Radicle forms the shoot and plumule forms the root.
4. Hypogeal germination is seen in bean seed.
5. Cereals are examples of monocot albuminous seeds.
6. Some seeds have no cotyledons at all.
7. Maize grain is fruit and not a seed.
8. Oxygen is necessary for the germination of seeds.
9. Seeds that fall in a flower bed from the previous crop usually do not germinate untill the next sowing season.
10. Hilum provides the diffusion of respiratory gases for the embryo.

Ans. 1. True
2. False. Hilum is the spot where ovule was attached to ovary.
3. False. Radicle forms the root and plumule forms the shoot.
4. False. Epigeal germination is seen in bean seed.
5. True
6. False. Monocotyledonous seeds have one cotyledon and dicotyledonous seeds have two cotyledons.
7. True
8. True
9. True
10. False. Micropyle provides the diffusion of respiratory gases for the embryo.

Chapter 7. Respiration in Plants

1. Aerobic respiration of one mole of glucose yields 138 ATP molecules.
2. Respiration is an anabolic process.
3. Glycolysis is the common phase of both aerobic and anaerobic respiration.
4. Aerobic respiration occurs in germinating seeds.
5. There is a gain in weight after respiration.
6. Carbon dioxide readily dissolves in lime water.
7. All leaves of a green plant normally respire anaerobically at night.
8. Respiration is faster than combustion.
9. Anaerobic respiration usually occurs in unicellular organisms.
10. Fermentation also occurs in muscles of animals.

Ans. 1. False. Aerobic respiration of one mole of glucose yields 38 ATP molecules.
2. False. Respiration is a catabolic process.
3. True
4. False. Anaerobic respiration occurs in germinating seeds.
5. False. There is a loss in weight after respiration.
6. True
7. False. All leaves of a green plant normally respire aerobically at night.
8. False. Respiration is slower than combustion.
9. True
10. True

Chapter 8. Five Kingdom Classification

1. Gymnosperms are said to be the first land plants having vascular tissues.
2. Fungi show heterotrophic mode of nutrition.
3. Dogfish is an invertebrate.
4. Echinoderms have tube feet for locomotion.
5. Exoskeleton of Arthropods is made up of calcium carbonate.
6. Mushroom is eukaryotic and heterotrophic organism.
7. Pila is an animal with spiny body and shows radial symmetry.
8. Hook worm belongs to phylum Annelida.
9. Planarian is a free living flat worm.
10. *Pheretimaposthuma* is scientific name of tapeworm.

Ans. 1. False. Pteridophytes are said to be the first land plants having vascular tissues.
2. True
3. False. Dogfish is a vertebrate.
4. True
5. False. Exoskeleton of Arthropods is made up of chitin.
6. True
7. False. Starfish is an animal with spiny body and shows radial symmetry.
8. False. Hook worm belongs to phylum Nematoda.
9. True
10. False. *Pheretimaposthuma* is scientific name of common Indian earthworm.

Chapter 9. Economic Importance of Bacteria and Fungi

1. Streptomycin was discovered by Fleming.
2. Streptomycin was the first antibiotic to be discovered.
3. Bacteria reproduces by fission.
4. Insulin was the first genetically engineered product produced from *E.coli*.
5. *Rhizopus* reproduce by budding.

Ans. 1. False. Streptomycin was discovered by Selman Waksman
2. False. Penicillin was the first antibiotic to be discovered.
3. True
4. True
5. False. *Rhizopus* reproduce by spore formation

Chapter 10. Nutrition

1. A rich source of iodine is sea food.
2. Cellulose connot be digested in human body.
3. Vegetables provide fats to the body.
4. B_{12} is a fat soluble vitamin.
5. Anaemia is caused by deficiency of calcium.
6. Cellulose in our food passes out undigested from body.
7. Kwashiorkor is a severe vitamin deficiency disease.
8. Iron is required for the proper working of thyroid.
9. Mineral salts contribute in regulating body processes.

Ans. 1. True
2. True
3. False. Vegetables provide vitamins and minerals to the body.
4. False. Vitamin B_{12} is water soluble.
5. False. Anaemia is caused by deficiency of iron.
6. True
7. False. Kwashiorkor is a protein deficiency disease.
8. False. Iodine is required for proper working of thyroid.
9. True

Chapter 11. Digestive System

1. Intestine is narrow for fast movement of food.
2. Intestine villi have a lymph vessel called lacteal.
3. Saliva moisten and lubricates food.
4. Saliva contains a protein-desting enzyme.
5. Saliva tends to destroy germs in the mouth.
6. Peristalsis occurs through all regions of the gut.
7. The fodd in stomach stays for about 10 hours.
8. Gastric juice is alkaline.
9. Gastric juice contains pepsinogen.

Ans. 1. False. Intestine is narrow for slow movement of food allowing absorption.
2. True
3. True
4. False. Saliva contains a starch-digesting enzyme.
5. True
6. True
7. False. The food in stomach stays for about 3 hours.
8. False. Gastric juice is highly acidic.
9. True

Chapter 12. Skeleton: Movement and Locomotion

1. Humerus of the upper arm is the longest bone.
2. Axial skeleton incudes shoulder and hip girdles
3. There are 10 vertebrae in the neck.
4. There are three kinds of ribs which are true, false and floating
5. The shoulder girdle is large and trough-shaped.
6. The flexor and externsor muscles are antagonistic muscles.
7. Muscles can contract as well as elongate.
8. Muscles are attached to bones by ligaments.
9. Cardiac muscle is a voluntary muscle.
10. Bending of arm is an example of first order of lever mechanism.

Ans. 1. False. Femur or the thigh bone is the longest bone.
2. False. Appendicular skeleton consists of shoulder and hip girdles.
3. False. The neck has 7 vertebrae
4. True
5. **False**. The hip pelvic girdle is a large trough- shaped part.

6. True
7. False. Muscles can contract and relax but cannot lengthen.
8. True
9. False. Cardiac muscles are involuntary muscles.
10. True

Chapter 13. Skin: "The Jack of all Trades"

1. All mammals are ectothermal.
2. The principal heat-regulating centre is located in thyroid.
3. Increased body heat causes vasodilation.
4. Fats are stored in the epidermis of human skin.
5. Stratum corneum is also called as Malpighian layer.
6. Cells in the cornified layer are dead.
7. Dermis is made up of epithelial tissue.
8. Sweat glands are present in dermis.
9. Temperature regulation centre in the brain is called thyroid.
10. Increased body heat causes dilation of blood vessels.
11. Stratum corneum is the middle layer of epidermis.
12. Skin is the largest organ.
13. Ceruminous glands are modified sebaceous glands.
14. Stratum corneum is made of dead cells having keratin.
15. The malpighian layer is hard and reisitant to bacterial invasion.
16. The pigment melanin which imparts coluration to the skin is found in the malpighian layer.
17. The dermis is very thick on palms and soles.

Ans. 1. False. Mammals are endothermal
2. False. Principal heat-regulating centre is located in brain.
3. True
4. False. Next to dermis fat is stored.
5. False. Stratum corneum is cornified layer.
6. True
7. False. Dermis is made of connective tissue.
8. True
9. False. Temperature regulating centre in brain is hypothalamus.
10. True
11. False. Stratum corneum is the outermost layer of epidermis.
12. True
13. True
14. True
15. False. Stratum corneum the cornified layer is hard and resistant to bacterial invasion.
16. True
17. True

Chapter 14. The Respiratory System

1. Strenuous physical exercise may cause fatigue due to accumulation of CO_2 in the blood.
2. No CO_2 is produced in anaerobic respiration in the human body.

3. Breathing and gaseous transport are and the same thing.
4. CO_2 is transported to the lungs by the blood in two forms: as bicarbonates and as carbamino-haemoglobin.
5. Tissue respiration means chemical changes occurring inside the cell.
6. Haemoglobin content decrease at high altitude.
7. Pharynx is the common passage for air and food.
8. During ingestion of food epiglottis close the entrance of trachea.
9. During respiration temperature rises.
10. In mammals, lactic acid is produced during anaerobic respiration.

Ans. 1. False. Strenuous physical exercise may cause fatigue due to accumulation of lactic acid in the blood.
2. True
3. False. Breathing and gaseous transport are two separate processes.
4. True
5. True
6. False. Haemoglobin content increases at high altitudes.
7. True
8. True
9. True
10. True

Chapter 15. Hygiene

1. Presence of fluorides in drinking water causes bone disorders.
2. Cockroaches are suspected to be carriers of certain bacterial disease.
3. Rat-flea is aparasite which lives on rat's skin.
4. Mercury causes muscular disorders.
5. Filariasis is caused by nematode worm *Wuchereria.*
6. Entamoeba has contractile vacuole.
7. Mosquitoes are the source of dysentery germs.
8. Spraying DDT is very dangerous for water pollution.
9. Sewage is the commonest source for the contamination of water in the cities.
10. Cholera is marked by acute diarrhoea and no urination.

Ans. 1. True
2. False. Cockroaches are suspected to be carriers of viral disease.
3. True
4. False. Mercury causes abnormalities in the nervous system.
5. True
6. False. Entamoeba does not have any contractile vacuole.
7. False. Dysentery is caused due to bacteria or protozoa.
8. True
9. True
10. True

Chapter 16. Disease: Cause and Control

1. The bacterium that infect intestine is *Salmonella typhi*.
2. Tsetse fly is the vector of Malaria.
3. The disease Taeniasis is caused due to consumption of infected pork.
4. World rabies day is celebrated on 8th September.
5. AIDS is transmitted by sexual intercourse.
6. Filiariasis is transmitted by the housefly.
7. Malaria is caused from protozoan.
8. BCG vaccine is used for chicken pox.
9. Louis Pasteur discovered remedy for malaria.
10. AIDS is caused by a bacterium.
11. HIV is a serious disease, usually fatal.
12. AIDS is not transmitted by contact with a patient's clothes.
13. Chicken pox and hepatitis are bacterial diseases.
14. Goitre is endemic in the sub-Himalayan regions of India.
15. Hay fever and asthma are allergies, not diseases.
16. Small pox still occurs in India.
17. The disease filariasis is caused by the bite of the female anopheles mosquito.

Ans. 1. True
2. False. Tsetse fly is the vector of Sleeping sickness.
3. True
4. True
5. True
6. False. Filiariasis is transmitted by the *Wuchereria bancrofti* worm.
7. True
8. False. BCG vaccine is used for tuberculosis.
9. False. Louis Pasteur discovered a cure for dog bite.
10. False. AIDS is caused by a virus.
11. True
12. True.
13. False. Chicken pox and hepatitis are viral diseases.
14. True
15. True
16. False. Small pox has been eradicated from India.
17. False. The disease filariasis is caused by worm

Chapter 17. Aids of Health

1. Bordeaux mixture is an antiseptic.
2. Certain meat and fish consists of antibiotics and it work as preservative.
3. Measles is a disease that comes under acquired immunity.
4. The antiserums are prepared from blood serum of animals containing antibodies.
5. The local defence system is dependent on the previous exposure to infection.
6. Immune system deals with the germs after they have entered the body.
7. Antibodies eat up the germs.

8. Human beings can suffer from all those diseases which attack dogs.
9. Anti-venin injection against snake bite is an example of artificially acquired passive immunity.
10. Mother's antibodies may reach the foetus through placenta.
11. A person having once suffered from measles usually gets repeated attacks.
12. Lysol is an antibiotic.
13. Sweat and tears contain germs-killing substances.
14. Salk vaccine is used to prevent tuberculosis.
15. Treatment by the use of chemicals is known as allopathy.
16. Alexander Fleming gave the term "antibiotic" for substances like penicillin.
17. Penicillin was discovered by Alexander Flemming.

Ans. 1. False. It is a disinfectant
2. True
3. True
4. True
5. False. The local defence system is not dependent on the previous exposure to infection.
6. True
7. True
8. False. Human beings are immune to a highly infectious disease of dogs.
9. True
10. True
11. False. A person having once suffered from measles usually does not get repeated attacks.
12. False. Lysol is an antiseptic.
13. True
14. False. Salk vaccine is for poliomyelitis.
15. True
16. False. Selman Waksman coined the term"antibiotic"for substances like penicillin.
17. True

Chapter 19. Waste Generation and Management

1. Some of the electronic wastes may contain valuable metals such as copper.
2. Flyash is the geaseous waste of cement industry.
3. Electricity is one of the primary needs of human society.

Ans. 1. True
2. True
3. False. Electricity is one of the secondary needs of the human society.

Give Biological Reasons

Set 7

Chapter 1. Introducing Biology

1. Biology helps us in leading a healthier life.

Ans. Biology helps us to understand the causes of many diseases. It will not only help to identify various micro organisms that cause diseases, but also helps us to study their mode of living and the different ways in which they spread diseases. The understanding of disease-causing microorganisms has led to the development of new drugs, both natural and synthetic drugs that can kill these organisms.

2. Now-a-days we all can differentiate types of plants and animals based on their characteristics features.

Ans. Taxonomy helps to categorise organisms present on the earth. By using hierarchical classification taxonomy help scientists to understand and organise the diversity of life on our planet.

3. Biology is the science that examines all living organisms.

Ans. The science of biology includes knowledge from many other sciences to describe the behaviour of living things. For example, the science of chemistry describes the various molecular processes that take place inside living cells.

4. Children look a lot alike with their parents.

Ans. Branch of genetics helps in the transmission of body characteristics from parents to offsprings. Therefore, children look a lot alike with their parents because they inherit pairs of genes from their parents. A child gets one set of genes from the father and one set from the mother.

5. Humans have body system which protects them against many diseases.

Ans. Immunology is the study of the immune system which protects us from infection by defending against germs and microorganisms. If the immune system is not functioning as it should, then it can result in disease such as autoimmunity, allergy and cancer.

Chapter 2. Cell: The Unit of Life

1. Cell arises from pre-existing cell.

Ans. According to the cell theory, many of the cells die and form continuously by the division of cell. Thus, all cells arise from pre-existing cell by cell division.

2. The cells of an elephant would be of the same size as the cell of the rats.

Ans. The size of cells does not vary within the organisms. However, the number of cells varies from one organism to another. A larger animal like an elephant will have more number of cells as compared to a smaller animal like a rat. However, the size of the cell will be the same.

3. Lysosomes are called recycling centres.

Ans. Lysosomes break worn-out cells, damaged cells and cell organelles to component molecules for building new cell organelles. So, they are called recycling centres.

4. Mitochondria are the powerhouse of the cell.

Ans. Mitochondria are organelles whose membranes are specialised for aerobic respiration where many of the reactions involved in cellular respiration happen. Mitochondria are the working organelles that keep the cell full of energy. Hence, they are called the power house of cell.

5. Chloroplast is also called the kitchen of the cell.

Ans. Chloroplast being the green pigment of plant cell conducts photosynthesis (a process by which plants make food) by using sunlight, water and carbon dioxide. Thus, chloroplast is called the kitchen of plant cell.

6. Plasma membrane is semi-permeable. **[November, 2019]**

Ans. Plasma membrane is called a selectively permeable membrane because it regulates the movement of substances from body to outside the cell. That means plasma membrane allows the entry of some substances while preventing the movement of some other substance.

Chapter 3. Tissues: Plants and animal tissue

1. Areolar tissues form tendon and ligaments.

Ans. Areolar tissues are strong fibrous tissues which bind different types of tissues. They hold organs in place and attaches epithelial tissue to other underlying tissues. Tendon is a flexible but inelastic cord of strong fibrous collagen tissue attaching a muscle to a bone whereas ligaments is a short band of tough, flexible fibrous connective tissue which connects two bones or cartilages or holds together a joint.

2. Girth of the stem and root increases.

Ans. Girth of the stem and root increases due to presence of lateral meristem. Lateral meristem is responsible for increase in diameter of the organ.

3. Areolar connective tissue is called packing tissue.

Ans. Areolar tissue is a loose connective tissue that can be seen between the skin and muscles in the bone marrow as well as around the blood vessels and nerves. It is also called the packing tissues because fills the spaces between different tissues and organs and connects the skin to the underlying muscles.

4. Xylem and phloem are conducting tissues.

Ans. Xylem and phloem are called conducting tissues as they are involved in the conduction of water and the food.Xylem conducts water absorbed by the roots to different parts of the plant. Phloem conducts food prepared in the leaves to various parts of the plant.

5. Lymph is associated with immune system.

Ans. Lymph is a colourless fluid containing white blood cells, which bathes the tissues and drains through the lymphatic system into the bloodstream. These white blood cells (lymphocyte) help fight infection and disease

Chapter 4. The Flower

1. Sepals and petals are called non-accessory parts of a flower.

Ans. Sepals and petals are called non-accessory parts of a flower because they are sterile and they do not directly participate in the process of sexual reproduction. They are simply the helping parts which either protect the reproductive parts of the flower or make the flower more attractive for the process of pollination.

2. Sweet corn is an incomplete flower.

Ans. Structurally, flowers consist of four main parts: the sepals, petals, stamens and pistils.Sweet corn is an incomplete flower because in this flower one or more set of floral structures are missing.

3. Nectaries are common in brightly coloured flowers.

Ans. Nectaries are nectar-secreting glandular organ in a flower (floral) or on a leaf or stem. It is present in brightly coloured flowers to attract insects and other small organisms so that insects come and sit on them. In this way, pollens fall on the insect. When this insect sits on another flower, where they shed the pollens. This helps in pollination.

4. *Salvia* sepals are also described as petaloid.

Ans. The sepal which is the part of the calyx of *Salvia* flower are described as petaloid because the three sepals are united and are red in colour like petals. These sepals resemble to petals which are modified leaves often bright in colour to attract pollinators.

Chapter 5. Pollination and Fertilisation

1. Seed produced by continuous self-pollination give rise to less vigorous offsprings.

Ans. Self-pollination occurs when the pollen from the anther is deposited on the stigma of the same flower or another flower on the same plant. Continuous self-pollination generation after generation may lead to the production of plants with less genetic diversity. The seeds produced through it are poor in quality and give rise to less vigorous offsprings.

2. Cross-pollination produces healthier offspring and new varieties.

Ans. Cross-pollination produces healthier offspring and new varieties because the seeds are produced in larger number and are more viable. The seeds develop and germinate properly and grow into better plants. It results in new varieties because cross-pollination can be carried out between two different varieties of the same species or even two species.

3. Elephant is the pollinating agent of *Rafflesia* flower.

Ans. Pollination by elephant is called Elephophily.It is found in *Rafflesia* whose flowers are very large and found at ground level. The pollen grains of one flower get attached to the feet of elephants and may be carried to the stigma of another flower.

4. Pollen grains of *Vallisneria* have a mucilaginous covering.

Ans. In *Vallisneria,* the male flowers are released on the water surface. So, to provide protection to these pollen grains, a mucilaginous covering is present.

5. Artificial pollination can produce new varieties.

Ans. Artificial pollination occurs when humans interfere in the process of natural pollination. They carry pollen or plant sperm from one flower to another, allowing the pollen to fertilise the ovaries and create seeds that will develop into fruits and new plants.

6. Sweet pea is an excellent example of insect pollinated flower.

Ans. Sweet pea's flower structure allows only bee to enter because other insects do not have the correct size or mass. The petals of the sweet pea flower are not all alike.

Chapter 6. Seeds: Structure and Germination

1. Seeds sown very deep in the soil fails to germinate.

Ans. Seeds sown very deep in the soil fails to germinate as there is no proper supply of oxygen which is required for respiration and insufficient pushing force in embryonic parts makes them difficult to break through the upper layers of soil.

2. Water is necessary for seeds germination.

Ans. Water is necessary for seed germination as by absorbing water seed swells up and seed coat ruptures so that radicle and plumule can come out and form the root and shoot system respectively. Water activates the enzymes so that they can act on stored food and convert into soluble form so that embryo can utilise it.

3. Maize is a grain instead of a seed or fruit.

Ans. In maize, the seed coat is fused with fruit wall and they together form a protective layer. So instead of fruit or seed, maize is called a grain.

4. Germinated grams are considered highly nutritive.

Ans. Grams are good source of proteins. The germinated grams have active supply of nutrients which is provided to the developing embryo. As there is better availability of nutrients so they are easily utilised by our body hence they are considered highly nutritive.

5. Maize grain is a one seeded fruit.

Ans. Fruit is the ripened ovary and is enclosed by a fruit wall or pericarp. Seed is the ripened ovule surrounded by a seed coat. Inside the fruit, seed is found. But in case of maize the fruit wall and seed coat are fused together forming a protective layer. So maize grain is a one seeded fruit.

6. Seeds kept in very low temperature do not germinate.

Ans. Optimum temperature is required for the for germination of seed which is in the range of 25-35°C so at very low temperature most of the seeds cannot germinate.

7. Ploughing is done in fields before sowing seeds.

Ans. Ploughing aerates the soil which makes oxygen easily available to the seeds and oxygen is one of the important requirements for germination as it is used by germinating seeds for respiration process.

8. Micropyle is needed for a seed.

Ans. Through micropyle water enters the seed and exchange of gases *i.e.,* oxygen and carbon dioxide occurs through this opening, so micropyle is needed for a seed.

9. Seeds should not be planted deep inside the soil. **[November, 2019]**

Ans. Seeds should not be planted deep inside the soil as there would be no proper supply of oxygen and pushing force might not be sufficient in the embryonic parts to break through the upper layers of soil, so they would fail to germinate.

Chapter 7. Respiration in Plants

1. It is difficult to demonstrate respiration in green plants.

Ans. In green plants, photosynthesis occurs where carbon dioxide evolved by respiration process is utilised by leaves to prepare glucose. So it is difficult to demonstrate respiration in green plants.

2. The bell-jar is covered with dark black colour cloth in experiment.

Ans. The experiment done to prove that carbon dioxide is produced in respiration of plants consists of bell-jar covered with black cloth to keep it away from sunlight so that photosynthesis is prevented.

3. Tilling of soil is useful for crops growing in it.

Ans. Tilling of soil creates tiny air spaces around soil particles which provides good source of oxygen for the roots that helps in proper respiration of roots. Hence crops can grow properly.

4. Respiration is a vital process for all living organisms.

Ans. Respiration is a process by which glucose is oxidised to release energy and this energy is required to carry out various metabolic activities in the body of living organisms. So, it is a vital process for all living organisms.

5. Anaerobic respiration is less efficient than aerobic respiration.

Ans. In anaerobic respiration there is incomplete oxidation of glucose, so very less amount of energy is released as compared to aerobic respiration where there is complete break down of glucose releasing huge amount of energy. So, anaerobic respiration is less efficient than aerobic respiration.

6. Respiration occurs at optimum temperature.

Ans. At optimum temperature, respiratory enzymes functions efficiently so respiration occurs properly at optimum temperature. At high or low temperature these enzymes cannot function properly.

7. A little antiseptic is added to flask while analysing respiration in germinating seeds.

Ans. An anticeptic such as carbolic acid is added to prevent the growth of bacteria on dead seeds which would otherwise respire and release carbon dioxide.Thus, antiseptic addition in flask does not show any sign of germination in seed.

8. Respiration is said to be reversal of photosynthesis.

Ans. Respiration is said to be reverse of photosynthesis because:

(i) In respiration oxygen is used up and carbon dioxide is evolved but in photosynthesis, oxygen is evolved and carbon dioxide is used up.

(ii) In respiration glucose is broken down but in photosynthesis glucose is synthesised.

(iii) Photosynthesis results in gain of weight whereas respiration leads to loss of weight.

9. In experiments of respiration, seeds taken as control are boiled.

Ans. Boiled seeds are dead which cannot respire so they are taken as a control in the experiments of respiration.

10. Yeast is used for manufacturing wine.

Ans. Yeast undergo anaerobic respiration and converts glucose into ethanol and carbon dioxide *i.e.*, it helps in fermentation of glucose. So, it is used in wine industry.

Chapter 8. Five Kingdom Classification

1. Blue-green algae and Bacteria are placed in Kingdom Monera.

Ans. Blue-green algae and bacteriaare unicellular and prokaryotic single-celled organisms which lack in well defined nucleus *i.e.*, they do not have nucleolus or nuclear membrane. Chromatin material lie free in cytoplasm. They don't have membrane bound cell organelles. So, they are placed in Kingdom Monera.

2. Protozoans are regarded as primitive animals.

Ans. Protists are the simplest, unicellular eukaryotes. They are microscopic, exhibit cellular level of organisation. They also exhibits a primitive mode of reproduction. So, they are regarded as primitive animals.

3. Bryophytes are called amphibians of plant kingdom.

Ans. Bryophytes can live on land as well as in water, they need water as a source for reproduction so they are called the amphibians of plant kingdom.

4. Scientific names of living beings are considered better than their common names.

Ans. People of different countries with different languages give different local names to plants or animals which creates confusion. So in order to eliminate confusion scientific names of living organisms are considered better than their common names.

5. Chordate are popularly known as vertebrates and other phylum as invertebrates.

Ans. Classification done in kingdom animalia is on the basis of existenceof backbone. Vertebrates have backbone whereas invertebrates donot. Thus, only chordates are classified in vertebrates and others in invertebrates for convenient study.

6. All humans on earth today may differ widely in their facial features, colour, height, etc. yet they belong to a single species *Homo sapiens*.

Ans. Species means an organism of a particular kind whose members can interbreed among themselves to produce fertile young ones. Human beings though differ widely in their facial features, colour, height, etc. yet they belong to a single species *Homo sapiens* as they can interbreed among themselves and produce fertile offsprings.

7. A Bat is a Mammal. **[February, 2020]**

Ans. Bat is a mammal as it gives birth to young ones and suckles the young ones by mamary glands.

8. Earthworms are very useful in agriculture. **[February, 2020]**

Ans. Earthworms loosens and aerates the soil which helps in the respiration of roots of plants. Plants can grow well because they increase the fertility of the soil by interchaning the top layer of soil with inner layers. They are used in vermicomposting. Earthworm's excreta is rich in the nitrogenous matter which helps in the growth of plants thus earthworms are very useful in agriculture.

Chapter 9. Economic Importance of Bacteria and Fungi

1. Tinned or canned food is safe to eat.

Ans. Tinned or canned foods are not always safe to eat because they do not contain good nutrients as compared to freshly cooked food. Sometimes harmful bacteria like *Clostridiumbotulinum* grow in it causing food poisoning like botulism.

2. Antibiotics are given to patients suffering from bacterial diseases.

Ans. Antibiotics are the chemicals produced from microbes which can kill or inhibit the growth of other microbes. They are very effective against bacterial diseases as they inhibit their growth or destroy their cell wall. So, they are given to patients suffering from bacterial diseases.

3. It is generally advised that every living room in the house should get direct sunlight at least for a short time.

Ans. Direct sunlight kills the microbes present in air, thus purifying the air hence, it is generally advised that every living room in the house should get direct sunlight at least for a short time so that we can breathe purified germ free air to stay healthy.

4. We do not categorise bacteria and yeast under plants.

Ans. Bacteria is a prokaryote which lacks membrane bound cell organelles and its genetic material lie free in cytoplasm. So it is placed under separate Kingdom Monera. Similarly yeast has no chlorophyll and it shows heterotrophic mode of nutrition so it is placed under Kingdom Fungi. But plants are eukaryotes, shows autotrophic mode of nutrition and have chlorophyll placed under Kingdom Plantae. Hence, we do not categorise bacteria and yeast under plants.

5. Bioweapons makes an easy means of warfare.

Ans. The highly infectious nature of biological agents their high potency and low cost makes bioweapons an easy means of warfare. For example: Anthrax bacteria is used as a germ bomb which when exploded released this disease causing germs which may become epidemic and cause death through several generations.

6. Denitrification bacteria are a blessing as well as a curse to farmers.

Ans. Denitrifying bacteria converts nitrates present in soil to free nitrogen which is released into atmosphere thus the amount of nitrogen in the soil is reduced that is leading to decreased in fertility of soil. So, it is a curse to farmers.

This bacteria cut down the excess supply of nitrates to plants as excess nitrates may prove harm to plants. Some wild plants can use this excess nitrates and can grow more rapidly as compared to our food crops in crop fields leading to low productivity. So, denitrifying bacteria converts these nitrates to nitrogen and cut down the supply of nitrates to these plants and acts as a boon to farmers.

7. Yeast is used in bakeries and breweries.

Ans. Yeast respire anaerobically and convert glucose into alcohol by alcoholic fermentation process. So, it is used in breweries. Similarly during baking process yeast ferments sugar present in dough and produce carbon dioxide. This gas increases the volume of dough and gives bread light spongy texture. So, it is used in baking industries as well.

8. Spore formation in bacteria is not considered as a form of reproduction.

Ans. Spore formation by bacteria is a process to escape from unfavourable conditions like extreme dryness, high temperature, etc. So, it is not a method of reproduction. Bacteria reproduce asexually by binary fission.

9. Leguminous plants increase soil fertility. **[February, 2020]**

Ans. Leguminous plants have nitrogen-fixing bacteria like Rhizobium in their roots nodules which converts nitrogen form the soil atmosphere into soluble nitrates that increases the fertility of the soil.

Chapter 10. Nutrition

1. Even obese person need some amount of fats in their diet.

Ans. Some vitamins are fat soluble and if no fats are ingested then such vitamins cannot be absorbed.

2. Vitamin B is called Vitamin B complex.

Ans. Vitamin B is group of vitamin B1, B2, B3, B6, B9, B12. Thus, it is known as Vitamin B Complex.

3. Glucose is instant source of energy.

Ans. Glucose is a monosaccharide which is absorbed immediately and sugars burn easily to provide energy.

4. Human canine teeth are not as well developed as cats and tigers.

Ans. We have food habits which do not need eating of raw flesh which needs to be torn by canine.

5. Chapatti when chewed properly becomes sweet in taste.

Ans. Chapatti is prepared from wheat which has carbohydrate. When we chew it properly disaccharides and polysaccharides break to form monosaccharide (which are sweet) .

6. It is advised to consume citrus fruits during illness.

Ans. Citrus fruits are rich in vitamin C. It prevents infection and scurvy. So, it is advised to use citrus fruits.

7. Carrot is useful for proper eyesight.

Ans. Carrot is a rich source of vitamin A. It prevents night blindness. So, carrot must be included in the diet.

8. It is important to take iron during pregnancies.

Ans. Iron is required for the formation of haemoglobin in RBC. It carries oxygen to each and every cell. A pregnant woman has to transport O_2 to her foetus from her body.

9. Water is an important part of our diet. **[November, 2019]**

Ans. Water is an important part of our diet becuase it acts as a solvent for most of the substances present in our body, it helps in transporation of digested food throughout our body, it removes soluble wastes, regulates body termperature.

10. Roughage is an important part of our diet. **[February, 2020]**

Ans. Roughage contains cellulose which is fibrous in nature so it absorbs and retains a lot of water which keeps our faecal matter soft, thus preventing constipation. It makes an easy movement of undigested food through the intestine by stimulating the contraction of muscles of the wall of th intestine. So it is an important part of our diet.

Chapter 11. Digestive System

1. Stomach wall is not digested though meat ingests as food is digested.

Ans. Stomach wall is protected by mucous, which form barrier between the lining of stomach and the contents. Thus, stomach wall is not digested as meat digests.

2. Small intestine is longest part of digestive tract but named small.

Ans. Small intestine is has smaller lumen than large intestine.

3. Ileum is appropriate for absorption.

Ans. Ileum has large surface area due to its length, small lumen and villi which increase absorption area.

4. We should not talk while eating.

Ans. If we talk then epiglottis will not close glottis and food may go into foodpipe.

5. Probiotics are good for health.

Ans. Probiotics (e.g. Yogurt) contain helpful live bacteria which help in keeping out gut healthy.

6. Ileum is well adapted for the absorption of the digestive food. **[November, 2019]**

Ans. Ileum is well adapted for the absorption of the digestive food because it has tiny finger like projections called villi which increase the surface area that facilitates in the absorption of digested food.

Chapter 12. Skeleton-Movement and Locomotion

1. Doctors advice calcium rich diet after fracture of bone.

Ans. Bones have calcium stored in them. It is required for making new bones. Thus, doctors advice calcium rich diet after fracture of bone.

2. Bones become weak in old age.

Ans. Bones become weak in old age because body reabsorb minerals from bones.

3. Our height stops growing at maturity.

Ans. Bones grow till puberty after which they mature and stop growing. Thus, height stops growing at maturity.

4. When we walk more, our legs ache.

Ans. Movement of bones is due to working of muscles. When our muscles get tired due to excessive work, our legs ache.

5. Joints become stiff in old age.

Ans. Lubricating fluid inside joints decreases along with old age and cartilages get thin.

Chapter 13. Skin-"The Jack of all Trades"

1. Hairs turn grey.

Ans. Due to death of pigment cells (melanocytes) air spaces develop in hair which makes they look grey.

2. Skin help in maintaining body temperature during winters.

Ans. Skin contracts to make hair rise which entrap layer of air in them. As air is bad conductor of heat, our body looses less heat and we remain warmer.

3. Skin feels dry during winters.

Ans. Due to low temperature skin contracts and pores get closed. Thus, less sebum comes to surface and skin feels dry.

4. We sweat more during summers.

Ans. Sweat evaporates to cause cooling.

5. Growing children have pimples and acne.

Ans. Due to hormonal changes skin becomes active and produces more of sebum which blocks the pores. Skin infections in clogged pores leads to pimples and acne.

Chapter 14. The Respiratory System

1. We should breathe through nose.

Ans. We should breathe through nose because hair in nose remove dust and warms air before it enters lungs.

2. Mountaineers carry oxygen cylinder with them.

Ans. Mountaineers carry oxygen cylinder because air has low oxygen at higher altitudes.

3. People living at higher altitude have more RBC's in their blood.

Ans. People living at higher altitude have more RBC's in their blood because their body has adapted the change in body by having more RBC's itself to absorb more oxygen even when air is thin at high altitude.

4. Carbon monoxide is highly poisonous.

Ans. Carbon monoxide makes permanent compound with haemoglobin and makes it inactive unlike CO_2 which makes temporary compound carboxy- haemoglobin.

5. Why do we sneeze?

Ans. When any foreign material enters nasal passage, expiratory muscles contract to expel air which causes sneezing.

6. Why does person feel breathlessness at higher altitudes?

Ans. At high altitude, the concentration of oxygen is reduced which is not sufficient for normal breathing movements.

7. The glottis is guarded by epiglottis.

Ans. The glottis is guarded by epiglottis to prevent the entry of food into the trachea during swallowing.

8. The alveoli are covered with blood capillaries.

Ans. Alveoli are covered with blood capillaries for the easy diffusion of resspiratoryy gases, since both alveoli and blood capillaries are thin-walled.

9. The wall of trachea is supported by 'C' shaped cartilaginous rings.

Ans. To prevent the collapsing of trachea, the trachea is supported by cartilaginous rings.

10. Breathing rate increases after a vigorous physical exercise.

Ans. During vigorous physical exercise, the person needs more energy. So, rate of breathing increases during vigorous physical exercise.

11. Breathing through the nose is said to be healthier than through the mouth.

Ans. Nostrils contain hair and Nasal chamber is lined with cillated epithelium and mucus which traps the dust and any other foreign harmful particles from the incoming air making the air clean which is not possible when we breathe through mouth thus breathing through the nose is said to be healthier than through the mouth.

12. Why does gaseous exchange continue in the lungs even during expiration?

Ans. Even after expiration some air is left in the lungs known as residual air. Thus, gaseous exchange continues in the lungs even after expiration.

Chapter 15. Hygiene

1. Hands and skin must be cleaned regularly.

Ans. We are continually handling a variety of things- furniture, books, coins, currency notes, seats, and supporting rods in buses, pets and other domestic animals, tools and machinery in workshops, and so on. All these objects may carry germs which may be picked up by our fingers and transferred over other parts of the body or into the mouth through food.

2. Eye must be cleaned and washed with clear water two or three times every day.

Ans. Trachoma and conjunctivitis are the two common diseases of the eye caused from dust and contamination by hands and towels. Thus, eyes must be cleaned.

3. Physical exercise should be systematic and regular.

Ans. Some physical exercises are necessary for all age groups, especially children, adolescents and young people. Long gentle walks are enough for the old age people. Physical exercise should be systematic and regular because it improves blood circulation.

4. While applying 'kajal' never share the applying sticks.

Ans. The Indian custom of applying "Kajal"may sometimes lead to infections of the eyes through the shared applying sticks, so never share such sticks.

5. One should breathe by nose not by mouth.

Ans. One must always breathe by the nose and never by the mouth. One should blow the nose into a handkerchief and should hold a handkerchief in front of the mouth and nose while sneezing.

6. The living rooms and bed should be well ventilated.

Ans. The living rooms should be well ventilated for fresh air and sunlight. Fresh air brings oxygen and sunlight kills germs.

7. Housefly is designated as public enemy no.1.

Ans. The housefly is a kind of natural scavenger cleaning and consuming the left-overs of the food, etc. But the dangers which the housefly poses to human life are so serious, that it is often designated as Public Enemy No. 1. It spreads numerous diseases including dysentery, cholera, typhoids, etc.

8. Cockroaches and rats are harmful to humans.

Ans. Cockroaches are very common household insects breeding mostly in the connecting manholes and sewers. They eat and spoil food, paper and clothes. They are also suspected to be carriers of certain viral diseases, including possibly a cancer-causing virus. Rats are serious domestic pests. Besides eating grain and other foods they also spread germs of certain diseases. Rat-flea is a parasite which lives on rat's skin. It may carry germs of plague.

9. Industrial waste and agricultural waste can be harmful to human health.

Ans. Industrial waste and agricultural wastes may contain poisonous chemicals and can be harmful to human health such as mercury causes abnormalities in the nervous system.

10. Saline water injection is must for cholera patient.

Ans. The symptoms of cholera are vomiting and no urination due to loss of lot of water and salt from body. Accumulation of urea in the blood become poison for body and may lead to death. So, to fulfill the quantity of water and salt saline water must be injected

11. Eating places must be clean and kept free of flies. **[November, 2019]**

Ans. Eating places must be clean and kept free of flies because many diseases spread by flies like cholera, typhoid, and dysentry. They carry the germs on their body and when they sit on uncovered food, the food gets contaminated and when we eat these foods we became unhealthy. So eating places must be clean and kept free of flies.

12. Living rooms in our houses must receive direct sunlight for a few hours. **[February, 2020]**

Ans. Direct sunlight kills the microbes present in the air, thus purifying the air hence it is generally advised that every living room in the house should get direct sunlight for few hours so that we can breathe purifies gem-free air to stay healthy.

Chapter 16. Disease: Cause and Control

1. The viruses are highly specific.

Ans. The viruses are highly specific because only one type of host and tissues are affected. For example, the virus of myxomatosis is lethal to rabbits but does not affect humans.

2. Anopheles mosquito is the vector of malarial pathogen.

Ans. The malarial pathogen is the Plasmodium which is transmitted through the bite of the female *Anopheles* mosquito along with its saliva.

3. Cancer is a non-infectious disease.

Ans. Cancer is a non-infectious disease because it is not caused by any germ and it does spread from one person to another.

4. Virus is like a gene with no cell its own.

Ans. Virus enable the host cell to produce more virus particles. Eventually, the host cell dies and the particles released from the virus invade new host cells.

5. The disease named as cancer.

Ans. Cancer has been given this name because it adheres to any part of the body which it targets like the well-known crab Cancer in an obstinate manner.

Chapter 17. Aids to Health

1. Tears are mechanical barrier against germs.

Ans. By coating the surface of the eye, tears prevent dryness and also prevent it from external irritants. Due to the presence of certain chemical substances, the foreign body that enters the eye is washed out by tears.

2. Antibody is produced by specialized lymphocytes on exposure to antigen.

Ans. Antibody is produced by B-lymphocytes on exposure to antigen. The antigen when enters the cell the B-lymphocytes activate and produce antibody against the foreign substances.

3. Antibiotics are involved in the killing of bacteria.

Ans. The antibiotics work by affecting things that bacterial cells have but human cells don't. For example, cell wall is present in bacteria but not in human cell. The antibiotics destroy the cell wall and kill the bacteria.

4. Antiseptic kills germs on the body.

Ans. Antiseptics kill the microorganism present on the outer surface of the body, in other words in the skin, and it also slows down the bacterial growth.

5. Human immunodeficiency virus cause AIDS.

Ans. Human immunodeficiency virus affects the immune system by affecting the defence mechanism and slows down the production of antibody even against some mild infection and if it not treated earlier it causes AIDS which is a very dangerous disease.

Chapter 19. Waste Generation and Management

1. Plastic should be banned.

Ans. Plastic should be banned as it is non biodegradable.

2. Electrostatic scrubber in chimneys of industries.

Ans. Electrostatic scrubber removes particulate matter from the chimneys of industries..

3. Broken glass utensils are a kind of non-degradable waste.

Ans. Broken glass utensils are a kind of non-degradable waste as they cannot be decomposed and broken down by living microorganisms. These need to be disposed off in deep dug out pits so that they do not cause any harm. However sometimes broken glass in large quantities can be used in galss industries after melting them.

4. Mine tailing can be used.

Ans. Mine tailing in waste material produced from mining. It can be reused to make masonry tiles, etc.

5. Landfills are coming near to the large cities.

Ans. Landfills are coming up fast near large cities due to huge population growth and lack of using efficient and safe disposal method. As construction activities are very active, new colonies are being raised and construction units have become very active. These produce large quantities of waste stones, pebbles, broken bricks and wood waste. Mostly. These are used as landfills. One can see heaps of landfills in the suburbs.

6. Electric waste should be collected and recycled.

Ans. Electric waste has metal parts which can be recycled.

7. Municipal sewage is first separated into degradable and non-degradable wastes.

Ans. Municipal sewage is first separated into degradable and non-degradable wastes because degradable wastes can be broken down in non-toxic waste in septic tanks, while non-degradable waste needs to be dumped or bried at safe places to avoid any hazardous effect..

8. Every garden should have a compost pit.

Ans. Every garden should have a compost pit because garden waste can be turned into fertiliser by decomposing it.

❑

Match the Following

Set 8

Chapter 1. Introducing Biology

1.

Column I	Column II
(a) Marine biology	(i) Treatment and surgery of animals
(b) Veterinary science	(ii) Production of silk
(c) Sericulture	(iii) Life in the sea
(d) Apiculture	(iv) Honey and beeswax

Ans. (a) – (iii), (b) – (i), (c) – (ii), (d)– (iv)

2.

Column I	Column II
(a) Study of insects	(i) Mycology
(b) Study of fungi	(ii) Ornithology
(c) Study of metabolism of organisms and their parts	(iii) Physiology
(d) Study of birds	(iv) Entomology

Ans. (a) – (iv), (b) – (i), (c) – (iii), (d) – (ii)

3.

Column I	Column II
(a) Branches based on major kinds of organisms	(i) Anatomy, Ecology, Palaeontology
(b) Branches based on special group of organisms	(ii) Space biology, Genomics, Bioengineering
(c) Branches based on the approach on studying it	(iii) Botany, Zoology, Human biology
(d) Branches based on applied biology	(iv) Virology, Mycology, Algology

Ans. (a) – (iii), (b) – (iv), (c) – (i), (d) – (ii)

4.

Column I	Column II
(a) Morphology	(i) Interpretation of structure
(b) Ecology	(ii) Study the effects of radioactivity on living things
(c) Agriculture	(iii) Raising crops and livestock
(d) Nuclear biology	(iv) Relation with environment

Ans. (a) – (i), (b) – (iv), (c) – (iii), (d) – (ii)

5.

Column I	Column II
(a) He rationised the healing and treating the sick and called the father of medicine.	(i) Theophrastus
(b) He mainly studied plant and called the father of botany.	(ii) Aristotle
(c) He was the first person to study animals in sufficient details and called the father of Biology/Zoology	(iii) Hippocrates

Ans. (a) – (iii), (b) – (i), (c) – (ii)

Chapter 2. Cell: The Unit of Life

1.

Column I	Column II
(a) Rudolf Virchow	(i) Every plant is made up of cells
(b) Anton van Leeuwenhoek	(ii) Constructed first microscope
(c) Matthias Scheiden	(iii) All cells arise from pre-existing cells
(d) Theodor Schwann	(iv) All animals are made up of cells

Ans. (a) – (iii), (b) – (ii), (c) – (i), (d) – (iv)

2.

Column I	Column II
(a) Cell	(i) A type of semiautonomous organelle
(b) Plasma membrane	(ii) Fluid Mosaic model
(c) Endoplasmic reticulum	(iii) Robert Hooke
(d) Leucoplast	(iv) Site of protein and lipid synthesis
(e) Lysosome	(v) Green colour plastid
	(vi) Hydrolytic enzymes

Ans. (a) – (iii), (b) – (ii), (c) – (iv), (d) – (i), (e) – (vi)

3.

Column I	Column II
(a) Smallest cells	(i) Orstrich egg
(b) Longest cell	(ii) White blood cells
(c) Largest cell	(iii) Nerve cells
(d) Cell which shows amoeba like movement	(iv) Bacteria and Red blood cells

Ans. (a) – (iv), (b) – (iii), (c) – (i), (d) – (ii)

4.

Column I	Column II
(a) Cytoplasm	(i) Given rigidity and protection to the plant cell.
(b) Cell wall	(ii) Site for most of the enzymatic reactions and metabolic activity of the cell.
(c) Nucleolus	(iii) Carry hereditary information or the genes from parents to off springs.
(d) Chromatin fibres	(iv) Imparts colour to flowers and fruits.
(e) Chromoplasts	(v) Participates in protein synthesis and storing RNA

Ans. (a) – (ii), (b) – (i), (c) – (v), (d) – (iii), (e) – (iv)

5.

Column I (Animals)	Column II (Number of chromosomes)
(a) Humans	(i) 54
(b) Chimpanzee	(ii) 46
(c) Monkey	(iii) 48
(d) Honey bee	(iv) 32

Ans. (a) – (ii), (b) – (iii), (c) – (i), (d) – (iv)

Chapter 3. Tissues: Plants and Animal tissue

1.

Column I	Column II
(a) Cartilage	(i) Fluid connective tissue
(b) Skeletal muscle	(ii) Involuntary striated muscle
(c) Areolar tissue	(iii) Porous tissue with good supply of blood vessels and nerves
(d) Bone	(iv) Surface of joints
(e) Blood	(v) Filling of space inside the organs

Ans. (a) - (iv), (b) - (ii), (c) - (v), (d) - (iii), (e) - (i)

2.

Column I	Column II
(a) Epithelial tissue	(i) Perception and response to stimuli
(b) Connective tissue	(ii) Protection, absorption, secretion, sensory perception, etc.
(c) Muscular tissue	(iii) Movement, contractility, excitability, maintenance of posture and body position
(d) Neural tissue	(iv) Binding, protection, insulation, transportation, storing of reserve fuel

Ans. (a) - (ii), (b) - (iv), (c) - (iii), (d) – (i)

3.

Column I	Column II
(a) Tendon	(i) Form shock absorbing cushions around kidneys and eye balls
(b) Ligament	(ii) Connects bone to bone
(c) Adipose tissue	(iii) Connects skeletal muscle to bone
(d) Meristem	(iv) Stone cells
(e) Sclerenchyma	(v) Actively dividing cells

Ans. (a) - (iii), (b) - (ii), (c) - (i), (d) - (v), (e) - (iv)

4.

Column I	Column II
(a) Conducting tissues	(i) Contract without outside stimulation and do not get tired soon
(b) Permanents tissues of plants	(ii) Protecting, supporting and conducting
(c) Supportive connective tissues of animals	(iii) Xylem and phloem
(d) Cardiac muscles	(iv) Cartilage and bone
(e) Striated muscle	(v) Presence of dark and light band which runs across the fibres

Ans. (a) - (iii), (b) - (ii), (c) - (iv), (d) - (i), (e) - (v)

5.

Column I	Column II
(a) Tracheids	(i) Large thin walled cells and responsible for photosynthesis and storage of food.
(b) Parenchyma	(ii) Responsible for increase in diameter of stem
(c) Collenchyma	(iii) Elongated dead cells which provide mechanical support
(d) Cambial meristem	(iv) Mechanical support

Ans. (a) - (iii), (b) - (i), (c) - (iv), (d) - (ii)

Chapter 4. The flower

1.

Column I	Column II
(a) Gamosepalous	(i) Petals are free
(b) Polysepalous	(ii) Sepals are free
(c) Gamopetalous	(iii) Fused petals
(d) Polypetalous	(iv) Fused sepals

Ans. (a) - (iv), (b) - (ii), (c) - (i), (d) - (iii)

2.

Column I	Column II
(a) Carpels	(i) Outermost, lowermost, non-essential but protective whorl of the flower
(b) Nectar	(ii) Female parts of the flower
(c) Pollen	(iii) Form male gametes
(d) Stigma	(iv) Attracts insects for cross pollination
(e) Calyx	(v) Serve as landing place for pollen during pollination

Ans. (a) - (ii), (b) - (iv), (c) - (iii), (d) - (v), (e) - (i)

3.

Column I	Column II
(a) Androecium	(i) Arranged in single, double and spiral whorl
(b) Gynoecium	(ii) It consists of style, ovary and sitgma
(c) Corolla	(iii) It consists of filament and anther
(d) Calyx	(iv) Usually green but sometimes they are brightly coloured (petaloid) as in Gul Mohar

Ans. (a) - (iii), (b) - (ii), (c) - (i), (d) - (iv)

4.

Column I	Column II
(a) Incomplete or unisexual flower	(i) Papaya, palm
(b) Neuter flower	(ii) Ray florets of sunflower
(c) Polyandrous	(iii) Petunia
(d) Diadelphous	(iv) Pea

Ans. (a) - (i), (b) - (ii), (c) - (iii), (d) - (iv), (e) - (v)

5.

Column I	Column II
(a) Stigma	(i) Tubular slender structure of carpel
(b) Style	(ii) May be tubular, bell-shaped, funnel-shaped or wheel-shaped
(c) Ovary	(iii) Covered with hair or with glandular papillae
(d) Corolla	(iv) Bears one or more ovules attached to a flattened, cushion-like placenta

Ans. (a) - (iii), (b) - (i), (c) - (iv), (d) - (ii)

Chapter 5. Pollination and Fertilisation

1.

Column I	Column II
(a) Intine	(i) Present at the chalazal end

(b) Antipodal cells	(ii) Central part of an ovule, containing the embryo sac
(c) Nucellus	(iii) Inner wall of pollen grain
(e) Polar nuclei	(iv) Form endosperm
(e) Pollen grain	(v) Two sperm nuclei
(f) Generative nuclei	(vi) One egg cell and two synergids
(g) Micropylar end	(vii) Male gamete

Ans. (a) – (iii), (b) – (i), (c) – (ii), (d) – (iv), (e) – (vii), (f) – (v), (g) – (vi)

2.

Column I	Column II
(a) Generative nucleus	(i) Pollen tube
(b) Germ pore	(ii) Fertilisation
(c) Secondary nucleus	(iii) Endosperm nucleus
(d) Egg nucleus	(iv) Male nuclei

Ans. (a) – (iv), (b) – (i), (c) – (iii), (d) – (ii)

3.

Column I	Column II
(a) Unisexuality	(i) In cucumber, the male and female flowers may be borne on the same plant.
(b) Self-sterility	(ii) The pollen grains of flowers do not germinate on the stigma of the same flower but germinate on the stigma of a flower borne by another plant of the same species.
(c) Herkogamy	(iii) A hood covering the stigma acts as a mechanical barrier in Pansy flowers.
(d) Heterostyly	(iv) When stigma and anthers do not grow up to same height.

Ans. (a) – (i), (b) – (ii), (c) – (iii), (d) – (iv)

4.

Column I	Column II
(a) Brightly coloured petals	(i) Even slightest wind can move them
(b) Protruding and easily movable anthers	(ii) Help to trap pollen grains in wind-pollination
(c) Smooth and light pollen	(iii) Attracting insects for cross-pollination
(d) Long and feathery stigma	(iv) Easily carried by wind to enable cross-pollination

Ans. (a) – (iii), (b) – (i), (c) – (iv), (d) – (ii)

Chapter 6. Seed-Structure and Germination

1.

Column A	Column B
(a) Tegmen	(i) Outer seed coat layer
(b) *Vallisneria*	(ii) Inner seed coat layer
(c) Coleoptile	(iii) Protective covering of plumule
(d) Maize grain	(iv) Protective covering of radicle
(e) Epigeal germination	(v) Monocot non-endospermic seed
	(vi) Bean seed
	(vii) Aleurone layer

Ans. (a) – (ii), (b) – (v), (c) – (iii), (d) – (vii), (e) – (vi)

Chapter 7. Respiration in plants

1.

Column A		Column B	
(a)	Lenticels	(i)	Photosynthesis
(b)	Anabolism	(ii)	Ethyl alcohol
(c)	Catabolism	(iii)	Aerial roots
(d)	Anoxybiotic	(iv)	Respiration
(e)	Yeast	(v)	Old woody stems
		(vi)	Kactic acid

Ans. (a) – (v), (b) – (i), (c) – (iv), (d) – (vi), (e) – (ii)

Chapter 8. Five Kingdom Classification

1.

Column A		Column B	
(a)	Thallophyta	(i)	Liverfluke
(b)	Bryophyta	(ii)	Mammals
(c)	Ear pinna	(iii)	Volvox
(d)	*Platyhelminthes*	(iv)	*Planaria*
(e)	Parasite	(v)	Bacteria
		(vi)	Mosses
		(vii)	Aves

Ans. (a) – (iii), (b) – (vi), (c) – (ii), (d) – (iv), (e) – (i)

2.

Column A		Column B	
(a)	Pine	(i)	Fungi
(b)	Earthworm	(ii)	Monera
(c)	Bread mould	(iii)	Prokaryote
(d)	Amoeba	(iv)	Gymnosperm
(e)	Moss	(v)	Plantae
(f)	Bacteria	(vi)	Animalia
		(vii)	Protista
		(viii)	Bryophyta

Ans. (a) – (iv), (b) – (vi), (c) – (i), (d) – (vii), (e) – (viii), (g) – (ii).

3.

Column A		Column B	
(a)	Annelida	(i)	Pigeon
(b)	Porifera	(ii)	Snake
(c)	Mollusca	(iii)	Earthworm
(d)	Reptilia	(iv)	Sponge
(e)	Aves	(v)	Octopus
(f)	Pisces	(vi)	Trout
(g)	Mammal	(vii)	Rabbit

Ans. (a) – (iii), (b) – (iv), (c) – (v), (d) – (ii), (e) – (i), (f) – (vi), (g) – (vii).

Chapter 9. Economic Importance of Bacteria and Fungi

Column A	Column B
(a) Staphylococci	(i) Mushroom cultivation
(b) Streptococci	(ii) Haemophilia A
(c) Spawning	(iii) Bacteria in long chains
(d) Streptomycin	(iv) Haemophilia B
(e) Blood clotting factor VIII	(v) Bacteria in clusters
(f) TAB vaccine	(vi) *Streptomyces griseus*
(g) BCG vaccine	(vii) *Tuberculosis*
(h) Penicillin	(viii) Typhoid
	(ix) Antibiotic

Ans. (a) – (v), (b) – (iii), (c) – (i), (d) – (vi), (e) – (ii), (f) – (viii), (g) – (vii), (h) – (ix).

Chapter 10. Nutrition

1.

Column A	Column B
(a) Protein	(i) Reserve food
(b) Carbohydrate	(ii) Green salad is a source
(c) Fats	(iii) Save from diseases
(d) Vitamins	(iv) Not a nutrient
(e) Water	(v) Required only when we are sick
	(vi) Contains micronutrient
	(vii) Building body
	(viii) Reserve food

Ans. (a) – (vii), (b) – (i) , (c) – (viii), (d) – (iii), (e) – (iv)

2.

Column A	Column B
(a) Kwashiorkor	(i) Formation of haemoglobin
(b) Calcium	(ii) Deficiency of proteins
(c) Iron	(iii) Synthesis of thyroxine
(d) Iodine	(iv) Strong bones and teeth

Ans. (a) – (ii), (b) – (iv), (c) – (i), (d) – (iii)

Chapter 11. Digestive System

1.

Column A	Column B
(a) Erepsin	(i) Sucrose to glucose + galactose
(b) Invertase	(ii) starch to maltose
(c) Lactose	(iii) Maltose to glucose
(d) maltose	(iv) galactose to Lactose
(e) Enterokinase	(v) fats to fatty acids
	(vi) Trypsinogen to trypsin
	(vii) Peptide to Amino acids
	(viii) Lactose to Glucose + galactose

Ans. (a) – (vii), (b) – (i), (c) – (viii), (d) – (iii), (e) – (iv)

2.

Column A	Column B
(a) Bimacromolecules of food	(i) Alimentary canal and associated glands
(b) Human digestive system	(ii) J-shaped bag-like structure
(c) Stomach	(iii) Embedded in jaw bones
(d) Thecodont	(iv) Converted into simple substances

Ans. (a) – (iv), (b) – (i), (c) – (ii), (d) – (iii)

Chapter 12. Movement and Locomotion

1.

Column A	Column B
(a) Bone	(i) Sholder
(b) Pivot joint	(ii) Elbow
(c) Hinge joint	(iii) Neck
(d) Ball and socket joint	(iv) Femur
(e) Gliding joint	(v) Skull
	(vi) Knee cap
	(vii) Wrist

Ans. (a) – (iv), (b) – (v), (c) – (ii), (d) – (i), (e) – (vii)

2.

Column A	Column B
(a) Humerus and shoulder girdle	(i) Partially movable
(b) Two adjacent vertebrae	(ii) Gliding
(c) Skull and upper end of the backbone	(iii) Immovable
(d) Bones of brain box	(iv) Ball and socket
(e) Rib and breast bone	(v) Pivot

Ans. (a) – (iv), (b) – (ii), (c) – (v), (d) – (iii), (e) – (i)

Chapter 13. Skin: "The Jack of all Trades"

1.

Column A	Column B
(a) Stratum malpighian	(i) Provides a touch sensation
(b) Stratum corneum	(ii) Present only in mammalian skin
(c) Epidermis	(iii) Inner most layer of the epidermis
(d) Sebaceous gland	(iv) Outermost layer having dead cells
(e) Hair	(v) Helps to keep the skin moist
	(vi) Prevents the loss of energy
	(vii) Fat is preserved in it
	(viii) Checks the entry of toxic substances

Ans. (a) – (iii), (b) – (iv), (c) – (viii), (d) – (v), (e) – (i)

2.

Column A	Column B
(a) Air spaces	(i) Hair
(b) Mammary glands	(ii) Sebaceous glands
(c) Goose-flesh	(iii) Rudimentary in males
(d) Traces of urea	(iv) Races of hot countries
(e) More numerous sweat pores	(v) Gray hair
(f) Oily secretion	(vi) Sweat

Ans. (a) – (v), (b) – (iii), (c) – (i), (d) – (vi), (e) – (iv), (f) – (ii)

Chapter 14. The Respiratory System

1.

Column A (Process)	Column B (End Product)
(a) Aerobic respiration	(i) CO_2
(b) Photosynthesis	(ii) Glycogen
(c) Respiration	(iii) Lactic acid, ATP, Heat energy
(d) Anaerobic respiration in animals	(iv) Starch
(e) Anaerobic respiration in plants	(v) Eater
	(vi) O_2
	(vii) CO_2, H_2O, ATP, Heat energy
	(viii) CO_2 and ethanol

Ans. (a) – (vii), (b) – (vi), (c) – (i), (d) – (iii), (e) – (viii)

2.

Column A	Column B
(a) Cartilaginous rings	(i) Complemental air
(b) Large surface area	(ii) Bronchi
(c) Voice box	(iii) Epiglottis
(d) Extra inhalation	(iv) Alveoli
(e) Swallowing	(v) Larynx
(f) Breathing movements	(vi) Respiratory pigment
(g) Anaerobic respiration	(vii) ATP molecules
(h) Haemoglobin	(viii) Alveolus
(i) Currency of energy	(ix) Diaphragm
(j) Place where carbon dioxide leaves the blood	(x) Lactic acid

Ans. (a) – (ii), (b) – (iv), (c) – (v), (d) – (i), (e) – (iii), (f) – (ix), (g) – (x), (h) – (vi), (i) – (vii), (j) – (viii)

3.

Column A	Column B
(a) Nasal chamber	(i) Production of voice
(b) Epiglottis	(ii) C-shaped rings
(c) Air-sacs	(iii) Warms air
(d) Lungs	(iv) Drives mucus
(e) Larynx	(v) Closes wind-pipe during swallowing
(f) Trachea	(vi) Network of capillaries
(g) Cilia	(vii) Spongy and elastic

Ans. (a) – (iii), (b) – (v), (c) – (vi), (d) – (vii), (e) – (i), (f) – (ii), (g) – (iv)

4.

Column A	Column B
(a) Cartilaginous	(i) Epiglottis
(b) Large surface area	(ii) Diaphragm
(c) Breathing movements	(iii) Bronchi
(d) Voice	(iv) Alveoli
(e) Complemental air	(v) Larynx
(f) Swallowing	(vi) Extra inhalation

Ans. (a) – (iii), (b) – (iv), (c) – (ii), (d) – (v), (e) – (vi), (f) – (i).

Chapter 15. Hygiene

1.

Column A	Column B
(a) Anopheles	(i) loose motion
(b) Aedes mosquito	(ii) filariasis
(c) Culex	(iii) vomiting
(d) Cholera	(iv) dengue
(e) Dysentery	(v) malaria
(f) Hepatitis	(vi) Shigella
(g) Conjunctivitis	(vii) Eye infection
(h) Bacillary dysentery	(viii) Liver disease
(i) Insecticide	(ix) Amoebic dysentery
(j) Entamoeba histolytica	(x) DDT

Ans. (a) – (v), (b) – (iv), (c) – (ii), (d) – (iii), (e) – (i), (f) – (viii), (g) – (vii), (h) – (vi), (i) – (x), (j) – (ix)

2.

Column A	Column B
(a) Mosquito	(i) DDT
(b) *Plasmodium*	(ii) Dengue
(c) *Entamoeba histolytica*	(iii) Vector
(d) *Aedes* mosquito	(iv) Malaria
(e) Insecticide	(v) Dysentery

Ans. (a) – (iii), (b) – (iv), (c) – (v), (d) – (ii), (e) – (i)

Chapter 16. Disease: Cause and Control

1.

Column A	Column B
(a) *Trypapanosomabruceirhodesiense*	(i) Sleeping sickness
(b) *Wuchereria bancrofti*	(ii) Chicken pox
(c) *Ascarislumbricoides*	(iii) Taeniasis
(d) *Taeniasolium*	(iv) Elephantiasis
(e) Human Immunodeficiency Virus	(v) Cholera
	(vi) Malaria
	(vii) Ascariasis
	(viii) Hepatitis A

Ans. (a) – (i), (b) – (iv), (c) – (vii), (d) – (ii), (e) – (i)

2.

Column A	Column B
(a) Plague	(i) Sporadic
(b) Malaria	(ii) Pandemic
(c) Goitre	(iii) Epidemic
(d) AIDS	(iv) Endemic

Ans. (a) – (iii), (b) – (i), (c) – (iv), (d) – (ii)

Chapter 17. Aids of Health

1.

Column A	Column B
(a) Typhoid	(i) germ killing secretions
(b) Tears	(ii) BCG
(c) Antibody	(iii) TAB
(d) Tuberculosis	(iv) DTP
(e) Immunization	(v) Defence mechanism against foreign body
	(vi) Chemical against snack bite
	(vii) DT
	(viii) Produced after vaccination

Ans. (a) – (iii), (b) – (i), (c) – (v), (d) – (ii), (e) – (viii)

Chapter 18. Health Organisation

1.

Column A	Column B
(a) The organisations concerned with supply of blood for the needy victims of war	(i) Not so crowded
(b) The organisation that looks after the health problems of people of the world	(ii) UNO
(c) Members that are focused on the need for creating international body	(iii) International sanitation regulation
(d) Members state of WHO are bound by	(iv) Red Cross
(e) Small towns are habitations which are	(v) WHO
	(vi) Inhabited by tribal people, mostly dependent on forest products
	(vii) Indian Red Cross Society Indian Red Cross Society

Ans. (a) – (iv), (b) – (v), (c) – (ii), (d) – (iii), (e) – (i)

Chapter 19. Waste Generation and Management

3.

Column A	Column B
(a) Domestic waste	(i) Construction works
(b) Community waste	(ii) Plant stump
(c) e-waste	(iii) Paper scrap
(d) Industrial Waste	(iv) Laboratory samples
(e) Municipal waste	(v) Fertilizers
	(vi) Public toilets
	(vii) Garbage
	(viii) Used mobile phone

Ans. (a) – (vii), (b) – (i), (c) – (viii), (d) – (iii), (e) – (vi).

Odd One Out

Set 9

Q. Identify the odd term in each set and name the category to which the remaining belongs:

Chapter 1. Introducing Biology

1. Bacteriology, Virology, Mycology, Physiology, Phycology

Ans. Odd term : Physiology

Category : All the other branches deals with the study of special group of microorganisms.

2. Sericulture, Apiculture, Pisciculture, Horticulture, Taxonomy

Ans. Odd term : Taxonomy

Category : All the others are branches of applied biology.

3. Pathology, Sericulture, Genetics, Physiology, Embryology

Ans. Odd term : Sericulture

Category : All the other branches deals with approaches for the study of organisms.

4. Increasing population, Industries, factories, Transport, Agriculture

Ans. Odd term : Agriculture

Category : All others are related for creating a problem of pollution.

5. Silk fibre, Gene, DNA, Chromosomes, Chromatin fibre

Ans. Odd term : Silk fibre

Category : All others are related to genomic branch.

6. Butterfly, Ant, Honeybee, Amoeba

Ans. Odd term : Amoeba

Category : All others are related to applied biology.

7. Parrot, Crow, Vulture, Tiger

Ans. Odd term : Tiger

Category : All others are related to ornitology.

8. Earthworm, Snake, Lizard, Alligator

Ans. Odd term : Earthworm

Category : All others are related to herpetology

Chapter 2. Cell: The Unit of Life

1. Cell wall, Chloroplast, Plasmodesmata, Small vacuole, Plastids

Ans. Odd term : Small vacuoles

Category : All the other terms are associated with plant cell.

2. Cristae, Power house, ATP, Matrix, Nucleolus

Ans. Odd term : Nucleolus

Category : All the other terms are associated with mitochondria.

3. Endoplasmic reticulum, Ribosome, Nuclei, Lysosomes, Golgi apparatus

Ans. Odd term : Ribosomes

Category : Others are membrane bound cell organelles.

4. Non:living, Freely permeable, Made of cellulose, Outermost in plant cell, Regulation of cell division

Ans. Odd term : Regulation of cell division

Category : All others are the features of cell wall

5. Nucleolus, Nuclear pore, Chromatin fibre, Nuclear membrane, Xanthophyll

Ans. Odd term : Xanthophyll

Category : Others are the part of nucleus.

Chapter 3. Tissues: Plants and Animal Tissue

1. Red blood cell, Platelets, White blood cells, Plasma, Lymph

Ans. Odd term : Lymph.

Category : The other terms are present in blood.

2. Axon, Dendrite, Myelin sheath, Cyton, Sarcolemma

Ans. Odd term : Sarcolemma

Category : Other terms are associated with neuron or nerve fibre.

3. Parenchyma, Sclerenchyma, Xylem, Collenchyma, Chlorenchyma **[November, 2019]**

Ans. Odd term : Xylem

Category : Other terms are supporting permanent tissue of plant.

4. Squamous tissue, Stratified tissue, Columnar tissue, Glandular tissue, Areolar tissue

Ans. Odd term : Areolar tissue

Category : Others are types of epithelial tissues.

5. Wall of intestine, Muscle of iris of eye, Skin, Lining of the blood vessels, Walls of heart

Ans. Odd term : Walls of heart

Category : Other are the location of unstriated muscles.

Chapter 4. The Flower

1. Calyx, Corolla, Stamens, Carpels, Inflorescence

Ans. Odd term : Inflorescence

Category : All the other terms are floral parts.

2. Style, Stigma, Ovary, Anther, Ovule

Ans. Odd term : Anther

Category : All other terms are parts of carpels (gynoecium).

3. Outer essential whorl of flower, Male reproductive organ, Stamen, Staminode, inner essential whorl of flower

Ans. Odd term : Inner essential whorl of flower

Category : Other features are associated with androecium.

4. Sepals, Petals, Perianth, Sepaloid, Stamens

Ans. Odd term : Stamen

Category : Other terms are non:essential parts of flower.

5. Maize, Cucumber, Pumpkin, Papaya, Pine

Ans. Odd term : Papayas

Category : Other are the examples of monoecious plants.

6. Monoadelphous, monoecious, diadelphous, polyadelphous

Ans. Odd term : Monoecious

Category : All others are related to stamens variation in flower.

Chapter 5. Pollination and Fertilisation

1. Large and brightly coloured flower, Large and loosely attached anther, Sticky pollen grain, Sticky stigma, Presence of nectar

Ans. Odd term: Large and loosely attached anther
Category : All the other features are associated with insect pollinated flowers.

2. Autogamy, Cleistogamy, Herkogamy, Geitonogamy, Homogamy

Ans. Odd term : Herkogamy
Category : All others are associated with self:pollination.

3. Seed, Tube nucleus, Fruit, Micropyle, Embryo sac,

Ans. Odd term : Tube nucleus
Category : Others terms are associated with ovule.

4. Maize, Grass, Rice, China rose, Wheat,

Ans. Odd term : China rose
Category : All others are the examples of wind pollinated flowers

5. *Bignonia, Canna, Rafflesia, Vallisneria*, Maize

Ans. Odd term : Maize
Category : Others are the examples of water pollinated flowers

Chapter 6. Seeds: Structure and Germination

1. Pea, gram, bean, grass

Ans. Odd term : Grass
Category : All others are dicot seeds

2. Testa, tegmen, nucellus, cotyledon

Ans. Odd term : Nucellus
Category : All others are the parts of a dicot seed

3. Aleurone layer, scutellum, endosperm, synergids

Ans. Odd term : Synergids
Category : All others are the parts found in a maize grain

4. Poppy, millets, cereals, palm

Ans. Odd term : Poppy
Category : All others are the parts of a monocot albuminous seeds

5. Pea, wheat, maize, castor

Ans. Odd term : Pea
Category : All others are the parts of albuminous seeds

6. Gram seed, Bean seed, Pea seed, Rice **[November, 2019]**

Ans. Odd term : Rice
Category : All others are dicotyledonous seeds.

Chapter 7. Respiration in Plants

1. Stomata, lenticels, root hairs, alveoli

Ans. Odd term : Alveoli
Category : These are the structures in plants through which gaseous exchange occurs.

2. Glycolysis, TCA cycle, urea cycle, electron transport chain

Ans. Odd term : Urea cycle
Category : Steps of respiration

3. Pyurvate, carbon dioxide, ethanol, ammonia

Ans. Odd term : Ammonia

Category : Products of respiration

4. Ethanol, carbon dioxide, yeast cells, lactic acid

Ans. Odd term : Yeast cells

Category : Products of anaerobic respiration

5. Carbon dioxide, water, 2 ATP, 38 ATP

Ans. Odd term : 2 ATP

Category : Products of aerobic respiration

Chapter 8. Five Kingdom Classification

1. Phylum, order, genus, family, spirogyra

Ans. Odd term : Spirogyra

Category : All other terms are related to taxon

2. Mucor, Agaricus, Albugo, Rhizopus, moss

Ans. Odd term : Moss

Category : All other terms are related to Kingdom Fungi

3. Pila, hydra,catla, brittle star, sycon

Ans. Odd term : Catla

Category : All other terms are related to Invertebrates

4. Frog, toad, turtle, lizard, crocodile

Ans. Odd term : Crocodile

Category : All others have three chambered heart

5. Volvox, pinus, cycas, cereals, pulses

Ans. Odd term : Volvox

Category : All others are Phanerogams

6. Amoeba, nucleus, tentacles,food vacuole

Ans. Odd term : Tentacles

Category : All other terms match with amoeba.

7. Invertebrates, hydra, Cnidaria, Crustacea

Ans. Odd term : Crustacea

Category : All other terms have match with hydra

8. Gills, ear drum, fish, paired fins

Ans. Odd term : Ear drum

Category : Others come under class pisces

9. Earthworm, annelid, insect, invertebrates

Ans. Odd term : Insect

Category : All other terms match with earthworm

10. Lateral line, external ear, sweat glands, monkey

Ans. Odd term : Lateral line

Category : All others are related to class Mammalia

11. Bat, Duck, Crow, Parrot **[November, 2019]**

Ans. Odd term : Bat

Category : Others are classes of Aves.

12. Thallophyta, Annelida, Mollusca, Arthopoda **[November, 2019]**

Ans. Odd term : Thallophyta

Category : All others are invertebrates of kingdom Animilia.

13. Whale, Seal, Shark, Dolphin [February, 2020]

Ans. Odd term : Shark

Category : All others are mammals.

Chapter 9. Economic Importance of Bacteria and Fungi

1. Azotobacter, Clostridium, Rhodospirillum, Rhizobium

Ans. Odd term : Rhizobium

Category : Others match with free living bacteria.

2. Xanthomonas, Psudomonas, Albugo, Shigella

Ans. Odd term : Albugo

Category : All others are bacterias.

3. Anthrax, Automycosis, Tetanus, Plague

Ans. Odd term : Automycosis

Category : Others are bacterial diseases in human beings.

4. Agaricus, Volvox, Volvariella, Saccharomyces

Ans. Odd term : Volvox

Category : All others are mushroom's variety.

5. Late blight of potato, citrus ranker, leaf blight of rice, wilt disease of potato.

Ans. Odd term : Late blight of potato

Category : All others are the diseases in plants caused by bacteria.

Chapter 10. Nutrition

1. Rice, Sugar, Wheat, Milk

Ans. Odd term: Milk

Category : All others are carbohydrate sources.

2. Oranges, lemon, Sweet lemon, Apple

Ans. Odd term: Apple

Category : All others are sources of Vitamin C.

3. Protein, Carbohydrate, Fats, Roughage

Ans. Odd term: Roughage

Category : All others are nutrient.

4. protein, carbohydrate, fats, minerals

Ans. Odd term: Minerals

Category : All others are macronutrient.

5. Vitamin A, Vitamin C, Vitamin D, Vitamin E

Ans. Odd term : Vitamin C

Category : All others are fat soluble vitamins.

6. Scurvy, Rickets, Goiter, weak eyesight.

Ans. Odd term : Goiter

Category : Other diseases are due to deficiency of vitamins.

7. Fats, Carbohydrate, proteins, Vitamin.

Ans. Odd term : Vitamin

Category : Other nutrients are required in small amount.

8. Sāmbhar with rice, *Khichdi,* chapatti with vegetable, Cheese roll.

Ans. Odd term : Cheese roll

Category : All other terms are related to balanced diet.

9. Raw fruits, leafy vegetables, salad, milk

Ans. Odd term: Milk

Category : All others are plant products.

10. Night blindness, common cold, retarded growth, Marasmus

Ans. Odd term: Common cold

Category : All others are deficiency diseases.

11. Xerophthalmia, Marasums, Pellagra, Scurvy

Ans. Odd term: Marasums

Category : Vitamin deficiency diseases.

12. Riboflavin, Thiamine, Folic acid, Iodine

Ans. Odd term: Iodine

Category : Vitamin B group

13. Vitamin C, Vitamin K, Vitamin A, Vitamin E **[February, 2020]**

Ans. Odd term: Vitamin C

Category : All others are fat soluble vitamins.

Chapter 11. Digestive System

1. Amylase, pepsin, ATPase, Invertase.

Ans. Odd term : ATPase

Category : All others are digestive enzymes.

2. Liver, Pancreas, Gastric glands, thyroid

Ans. Odd term : Thyroid

Category : All others are digestive glands.

3. Mouth, Stomach, Small intestine, Rectum

Ans. Odd term : Rectum

Category : All others help in digestion of food.

4. Ileum, jejunum, duodenum, small intestine.

Ans. Odd term : Small intestine

Category : All others are parts of small intestine.

5. Starch, Sugar, Polypeptide, Polysaccharide,

Ans. Odd term: Polypeptide

Category : All others are carbohydrates.

6. Potato, sweet potato, rice, onion.

Ans. Odd term : Onion

Category : All others are rich in starch.

7. Caecum, Colon, pancreas, Rectum.

Ans. Odd term: Pancreas

Category : All other terms are the parts of large intestine.

8. Pepsin, Gastric lipase, Ptyalin, Renin.

Ans. Odd term: Ptyalin

Category : All others are gastric enzymes .

9. Lactose, Glucose, Glycerol, Galactose.

Ans. Odd term : Glycerol

Category : All others are intestinal enzymes.

10. Alkaline, emulsify fats, cleaves nucleic acids, neutralise acid

Ans. Odd term : Cleaves nucleic acid

Category : All others are characteristics of Bile.

Chapter 12. Skeleton: Movement and Locomotion

1. Thorasic, Lumbar, Sternum Sacral,

Ans. Odd term : Sternum
Category : All others are vertebrae.

2. Bone, Cartilage, Ligament, Blood

Ans. Odd term : Blood
Category : All others are skeletal support.

3. Fibula, Humerus , Radius, Ulna

Ans. Odd term : Fibula
Category : All other terms are bones of arm.

4. Centrum, Cartilage, Vertebral arch, Notches

Ans. Odd term : Cartilage
Category : All others are parts of vertebra.

5. Ilium, Ischium , Scapula, Pubis

Ans. Odd term : Scapula
Category : All others are bones of ear.

6. Cranium, Facial bones, Carpal, Teeth

Ans. Odd term : Carpal
Category : Bones of skull.

7. Locomotion, Protection, Transport of materials, Storage of minerals.

Ans. Odd term : Transport of materials
Category : Functions of skeleton system.

8. Multinucleated, unbranched, voluntary, central nucleus.

Ans. Odd term : Central nucleus
Category : Characteristics of striated muscles.

9. Blood vessels, lungs, skin, heart.

Ans. Odd term : Heart
Category : All others have smooth muscles.

10. Hinge, Gliding, Fixed, Pivot

Ans. Odd term : Fixed
Category : All other joints show movement.

11. Ilium, Sacrum, Ischium, Rebis **[February, 2020]**

Ans. Odd term : Sacrum
Category : All others are bones of hip.

Chapter 13. Skin: "The Jack of all Trades"

1. Leucoderma, Albinism, Dermatitis, Diphtheria

Ans. Odd term : Diphtheria
Category : All others are skin diseases.

2. Sebaceous, Mammary, Sweat, Thyroid

Ans. Odd term : Thyroid
Category : All others are skin glands

3. Heat regulation, protection, water proofing, Hibernation .

Ans. Odd term : Hibernation
Category : All others are functions of skin.

4. Cornified layer, granular layer, Malpighian layer , hypodermis

Ans. Odd term: Hypodermis

Category: All others are parts of epidermis.

5. Sebaceous gland, sweat gland, hair follicle, muscles

Ans. Odd term: Muscles

Category: All others are parts of dermis.

6. Sebaceous, Sweat, Mucous, Adipose

Ans. Odd term: Adipose

Category: All others are glands.

7. Leucoderma, Leukemia, Leucocytes, Leucopenia

Ans. Odd term : Leucocytes

Category : All others are diseases.

8. Protect, Secrete, Cooling, Transport material.

Ans. Odd term : Transport material

Category : All others are functions of skin.

9. Skin, Hair, Nail , Teeth

Ans. Odd term : Teeth

Category : All others are integumentary system.

10. Sweat, Wax, Milk, Hair papilla

Ans. Odd term : Hair papilla

Category : All others are dead parts.

11. Saliva, Water, Sweat, Tear **[February, 2020]**

Ans. Odd term : Water

Category : Other terms are secretion from glands.

Chapter 14. The Respiratory System

1. Trachea, Bronchiole , Diaphragm , Heart

Ans. Odd term : Heart

Category : All others are respiratory organs.

2. Carbon dioxide, urea, uric acid, starch

Ans. Odd term : Starch

Category : All others are excretory products.

3. Nostril, mucous, hair, blood

Ans. Odd term : Blood

Category : All others help in breathing.

4. Oxygen, carbon dioxide, moisture, nitrogen

Ans. Odd term : Nitrogen

Category : All others are gases exchanged during respiration.

5. Epiglottis, bronchi, bronchioles, trachea

Ans. Odd term : Epiglottis

Category : All others are directly involved in respiration.

6. Glycolysis, Pyruvate oxidation, Citric acid cycle, oxidative phosphorylation

Ans. Odd term : Glycolysis

Category : All others are steps of aerobic respiration .

7. Lenticels, leaf, stomata, root hair

Ans. Odd term : Leaf

Category : All others are respiratory organs.

8. Air sacs, wind pipe, nasal chamber, buccal cavity

Ans. Odd term : Buccal cavity

Category : All others are respiratory organs.

9. Dust free, warm, moist , pure

Ans. Odd term : Pure

Category : All others are qualities of inhaled air.

10. Nasal cavity ,tracheae, diaphragm , lungs

Ans. Odd term : Lungs

Category : All others are breathing organs.

11. Trachea, bronchus, alvolus, diaphragm

Ans. Odd term : Diaphragm

Category : Other terms are respiratory organs.

12. Ethyl alcohol, carbon dioxide, starch, oxygen absence

Ans. Odd term : Starch

Category : Others are products of respiration in absence of oxygen in plants.

13. Diffusion, respiratory gases, alveoli, capillary network

Ans. Odd term : Diffusion

Category : All other terms are related to respiration

14. Oxyhaemoglobin, Carbaminohaemoglobin, hypoxia, carbonoxyhaemoglobin

Ans. Odd term : Hypoxia

Category : Others are related to various gases like oxygen, carbon dioxide and carbon monoxide when combined with haemoglobin.

15. Hairy, moist nostril, vocal cord

Ans. Odd term : Vocal cord

Category : Other terms are related to nasal cavity.

Chapter 15. Hygiene

1. Cholera, typhoid, tuberculosis, asthma

Ans. Odd term : Asthma

Category : All others are bacterial diseases.

2. Natural scavenger, dysentery, trachoma, hepatitis

Ans. Odd term : Hepatitis

Category : Other terms are related to housefly.

3. Malaria, elephantiasis, dengue, trachoma

Ans. Odd term : Trachoma

Category : All other diseases are caused by mosquitoes.

4. Malaria, amoebic dysentery, cataract, sleeping sickness

Ans. Odd term : Cataract

Category : All other diseases are caused by protozoa.

5. Arthritis, Ascariasis, taeniasis, filariasis

Ans. Odd term : Arthritis

Category : All other diseases are caused by worms.

6. Severe diarrhea, urea in blood, mild fever, no urination.

Ans. Odd term : Mild fever

Category : Other symptoms related to cholera.

7. Body ache, enlarged liver, vomiting, deep yellow urine.

Ans. Odd term : Vomiting

Category : All other symptoms are related to hepatitis

Chapter 16. Disease: Cause and Control

1. Cholera, TB, ascariasis, typhoid.

Ans. Odd term : Ascariasis (parasitic worm)

Category : bacterial diseases

2. *Plasmodium falciparum, Taeniasolium, Entamoeba histolytica, Trypanosoma bruceirhodesiense.*

Ans. Odd term : *Taeniasolium*(parasitic worm)

Category : Protozoa

3. AID, Chicken pox, Cancer, TB.

Ans. Odd term : Cancer (non:infectious disease)

Category : Infectious disease

4. Cholera, Ascariasis, typhoid, Malaria.

Ans. Odd term : Malaria (affect blood cells)

Category : intestinal diseases

5. Endemic, pandemic, infectious, sporadic

Ans. Odd term : infectious

Category: extent of occurrence

6. AIDS, Measles, Rabies, Typhoid

Ans. Odd term : Typhoid

Category : All are viral diseases

7. Malaria, Amoebic dysentery, Sleeping sickness, Filaria

Ans. Odd term : Filaria

Category : All are diseases caused by protozoa

8. Thalassemia, Syphilis, Cholera, Pneumonia

Ans. Odd term : Thalassemia

Category : All are diseases caused by bacteria

9. Ascaris, Tapeworm, Pinworm, Earthworm

Ans. Odd term : Earthworm

Category : All are related to ascariasis.

10. AIDS, Sexual contact, Sharing clothes, Shared injection needles

Ans. Odd term : Shaking hands

Category : Extent of occurrence of AIDS

Chapter 17. Aids of Health

1. Carbolic acid, Boric acid, DDT, Benzoic acid

Ans. Odd term: DDT

Category : All others are ntiseptic.

2. TAB, Salk's vaccine, BCG, Imovax

Ans. Odd term: BCG

Category : All other comes under killed germ.

3. Skin, tear, saliva, toxoids

Ans. Odd term : toxoids

Category : Protective mechanical barriers.

4. Lymphocytes, antibody, antigen, mitochondria

Ans. Odd term : Mitochondria

Category : Immune cells.

5. Measles, chicken pox, malaria, cow pox

Ans. Odd term: Malaria

Category : Caused by virus and vaccine is available.

6. Lysol, benzoic acid, DDT, mercurochrome

Ans. Odd term: DDT

Category : All other ome under antiseptics.

7. Formalin, iodine, Lysol, phenol

Ans. Odd term : Formalin

Category : All other comes under disinfectants.

8. BCG, DTP, ATP

Ans. Odd term : ATP

Category : All other comes under vaccine.

9. Tears, skin, nasal secretion, HCL.

Ans. Odd term : skin

Category : All other comes under germ-killing secretions.

Chapter 18. Health Organisation

1. Diarrhea, Typhoid, Malaria, Gastroenterities

Ans. Odd term : Malaria

Category : All others are food borne disease.

2. UNO, WHO, NICD, Red cross

Ans. Odd term : UNO

Category : All others are related to Health Organization.

3. Small population, poultry farming, multi-storeyed buildings, dairies

Ans. Odd term : Multi-storeyed buildings

Category : All other terms are related to village.

4. Injection needles, Polythene covers, Dried leaves, mobile batteries. **[November, 2019]**

Ans. Odd term : Dried leaves.

Category : All others are non-biodegradable wastes.

Chapter 19. Waste Generation and Management

1. Broken glass, food wrapping foil, dead leaves, used hand towel.

Ans. Odd term : Dead leaves

Category : All others are domestic waste.

2. Rubble, concrete, sludge, broken bricks, plaster pieces

Ans. Odd term : Sludge

Category : All others are construction waste.

3. Expired medicines, used bandages, broken needles, plastic bottles

Ans. Odd term : Plastic bottle

Category : All others are biomedical waste.

4. Oily sludge, tree cutting, waste lubricant, batteries

Ans. Odd term : Tree cutting

Category : All others are hazardous waste.

5. Garbage, Compost, Carbon dioxide, Recyclables

Ans. Odd term : Carbon dioxide

Category : All others are solid waste.

6. Polythene bags, Paper kite, plastic bucket, lamination sheet.

Ans. Odd term : Paper kite

Category : All others are non biodegradable

7. Coal, Ash, Acid, Pesticide.

Ans. Odd term : Pesticide

Category : All others are related to industrial waste

8. Garbage, Office waste, Plastic, Smoke.

Ans. Odd term : Smoke

Category : All others are related to solid waste

9. Plant, Animal, Paper, Metal.

Ans. Odd term : Metal

Category : All others are biodegradable

10. Computer part, CDs, TV, Paint

Ans. Odd term : Paint

Category : All others are e-waste

11. Pesticides, Old computers, Tissue paper, Electronic toys. **[February, 2020]**

Ans. Odd term : Tissue paper

Category : All others are non-biodegradable wastes.

❑

Give Examples

Set 10

Chapter 1. Introducing Biology

1. The study dealing with the inherent characteristics in a child from his parents.
2. Study of snake and lizards, etc.
3. Biometrics
4. An advancing branch of biotechnology
5. Use of biotechnology in applied biology

Ans. 1. Genetics 2. Herpetology 3. Pattern of iris in the eye
4. Cloning 5. Biogas production

Chapter 2. Cell: The Unit of Life

1. Double membrane organelle
2. Give turgidly to the cell
3. Kitchen of the cell
4. Largest cell
5. Protein synthesising organelle

Ans. 1. Nucleus, Mitochondria, Chloroplast 2. Vacuoles 3. Chloroplast
4. Ostrich egg 5. Ribosome

Chapter 3. Tissues: Plants and Animal Tissue

1. Tissue found in the skin and the cornea.
2. Smooth, unstriped, involuntary animal tissue.
3. Actively dividing plant tissue.
4. Living parenchyma cells closely associated with sieve tube cells.
5. Tissue concerned with perception and response.

Ans. 1. Squamous epithelium 2. Unstriated muscular tissue
3. Meristematic tissue 4. Companion cells
5. Nerve tissue

Chapter 4. The Flower

1. Androecium in which stamens are free.
2. Tissue that attaches the ovules to the wall of the ovary.
3. Pistillate flower
4. Polyadelphous androecium
5. Nectaries
6. Monadelphous androecium
7. Diadelphous androecium

8. Monoecious plants
9. Dioecious plants
10. Gymnosperm [November, 2019]

Ans. 1. *Petunia* 2. Placenta 3. Begonias 4. *Bombax*
5. Nasturtium 6. China rose 7. Pea 8. Cucumber
9. Papaya 10. Cycas

Chapter 5. Pollination and Fertilisation

1. Flower pollinated by birds.
2. Pollination affected by elephants
3. Water pollinated flower
4. Plant whose anther of flower matures earlier than stigma
5. Plant that can propagate by using non-opening, self-pollinating flowers (a condition called cleistogamy)
6. A wind pollinated flower.
7. A seed dispersed by explosive mechanism.
8. A flower which shows homogamy.
9. A water pollinated flower.
10. A seed dispersed by wind.

Ans. 1. *Bignonnia* 2. *Rafflesia* 3. *Vallisneria* 4. Sweet pea
5. Pansy 6. Maize 7. Poppy 8. *Mirabilis*
9. Lotus 10. Maize

Chapter 6. Seeds: Structure and Germination

1. Monocot seeds
2. Dicot albuminous seeds
3. Monocot ex-albuminous seeds
4. Seeds in which viviparous germination occurs.
5. A chemical which absorbs oxygen.

Ans. 1. Cereals, millets 2. Poppy, custard apple
3. Vallisneria, orchids 4. Rhizophora, Sonneratia
5. Pyrogallic acid

Chapter 8. Five kingdom Classification

1. Phylum having large number of species in animal kingdom.
2. An animal with jointed legs
3. Animals with hairy skin
4. Animals with feathers
5. Animals with three pairs of legs
6. Coelenterates.
7. Nematodes
8. Reptiles [November, 2019]
9. Pteridophytes
10. Vertebrates

11. Amphibian [November, 2019]
12. Bryophyte [November, 2019]
13. Poriferan [November, 2019]

Ans. 1. Crayfish, crabs, spider, scorpion (arthropoda)
2. Cockroach
3. Monkey, apes, man
4. Duck, pigeon, crow, sparrow
5. Housefly, mosquito, grasshopper
6. Hydra, jellyfish, corals, sea anemone
7. Hookworms, Ascaris, pinworm
8. Lizard, turtle, alligator
9. Selaginella, Marsilea, Dryopteris
10. Rohu, toad, salamander, pigeon, monkey
11. Frog
12. Mosses
13. Sycon

Chapter 9. Economic Importance of Bacteria and Fungi

1. Fungi used as food.
2. Fungi that cause diseases in plants.
3. Antibiotics derived from bacteria.
4. Bacteria that cause diseases in humans.
5. Antibiotics derived from fungi.
6. Two common moulds popularly called blue and green mould
7. Curdling of milk by bacteria.

Ans. 1. Agaricus bisporous, Volvariella
2. *Albugo candida, Phytophthora*
3. Streptomycin, Erythromycin
4. *Vibrio cholerae, Mycobacterium tuberculosis*
5. Penicillin, Calvacin
6. *Penicillium* and *Aspergillus*
7. *Lactobacillus*

Chapter 10. Nutrition

1. Disease caused by lack of vitamin B12.
2. Vitamin which we get from fish liver oil.
3. Source of Vitamin A.
4. Disease cause due to lack of Thiamine.
5. A source of Potassium.
6. Mineral present in blood protein.
7. Food recommended for patient of bone fracture.
8. A wholesome food.

9. End product of starch.
10. A carbohydrate made of many units of sugar.
11. Fat soluable vitamin. **[November, 2019]**
12. Water soluable vitamin. **[November, 2019]**
13. Protein deficiency disease. **[November, 2019]**

Ans. **1.** Perinicious anemia **2.** Calciferol (Vitamin D) **3.** Carrot / yellow fruits **4.** Beri- Beri **5.** Banana or potato **6.** Iron **7.** Milk product **8.** Milk **9.** Monosaccharide **10.** Cellulose or Glycogen **11.** Vitamin A **12.** Vitamin B **13.** Kwashiorkor

Chapter 11. Degestive System

1. Nutrient which needs no digestive enzyme.
2. Monosaccharide
3. Water soluble vitamin
4. Nutrient which help in absorption of other nutrient
5. Starch rich food
6. Ring like sphincter which allow food into stomach
7. Protective covering on inner side of stomach
8. Reagent used for protein test
9. Singular of villi
10. Largest gland of human body
11. A protein found in milk that forms curd, it ends to be difficult for infants to digest.
12. A lymphatic vessel in an intertinal villus, it aids in the absorption of facts.
13. Sphere formed by bile salts that surround lipids and make them temporarily water soluble.
14. Fine microscopic, finger-like projections lining the small intestine that serve to increase the absorptive surface area.

Ans. **1.** Vitamin **2.** Glucose **3.** C **4.** water **5.** Potato **6.** Cardiac Sphincter **7.** Mucus **8.** Biuret solution **9.** villus **10.** Liver **11.** Casein **12.** Lacteal **13.** Micelle **14.** Villi

Chapter 12. Skeleton: Movement and Locomotion

1. Immovable joint.
2. Only moveable bone of axial skeleton.
3. Type of freely moveable joint.
4. Movement in hinge joint.
5. Organ in which ball and socket joint is present.
6. Bones in which gliding joint is present.
7. Organ in which pivot joint is present.
8. Saddle joint.
9. Antagonistic muscle.
10. Location of striated muscle.
11. Ball and socket joint. **[November, 2019]**

Ans. 1. Skull
2. Hyoid
3. Synovial joint
4. One side
5. Shoulder
6. Metacarpal and carpal
7. Neck
8. Carpometacarpal joint
9. Biceps and triceps
10. Muscle attached to bones
11. Shoulder joint

Chapter 13. Skin-"The Jack of all Trades"

1. Organ where skin is thick.
2. Modifies sweat gland.
3. Colour pigment of skin.
4. Vitamin synthesized by skin.
5. Protein present in hair.
6. Derivative structure of skin.
7. Heat regulating centre of brain.
8. Normal human body temperature.
9. Animal which show summer sleep.
10. Excretory product present in sweat.

Ans. 1. Sole
2. Milk gland
3. Melanin
4. Vitamin D
5. Keratin
6. Hair or nails
7. Hypothalamus
8. 98.6 °F or 37 °C
9. Frog (example may vary)
10. Urea or salts

Chapter 14. The Respiratory System

1. End product of anaerobic respiration in plants.
2. Respiratory organ of plants.
3. Chemical which absorbs oxygen in air.
4. Similarity in respiration and burning.
5. Currency of energy.
6. Other name for anaerobic respiration in plants.
7. Chemical which cause muscle cramp.
8. Common reaction of aerobic and anaerobic cellular respiration.
9. Series of chemical reactions for oxidation of Acetyl CoA.
10. Other name for Krebs cycle.

Ans. 1. Ethanol
2. Lentils/ stomata
3. Pyrogallate of potash
4. both use oxygen, produce CO_2 and energy
5. ATP
6. fermentation
7. Lactic acid
8. Glycolysis
9. Krebs cycle
10. Citric acid cycle

Chapter 15. Hygiene

1. Healthy habits
2. Water borne disease
3. Mosquito disease
4. Bacterial disease
5. Viral disease
6. Protozoan disease

7. Eye disease
8. Vector borne disease
9. Rat vector disease
10. Insecticides

Ans. 1. Taking food at regular hours 2. Typhoid 3. Malaria
4. Cholera 5. Hepatitis 6. Amoebic dysentery 7. Conjunctivitis
8. Dysentery 9. Plague 10. DDT

Chapter 16. Disease: Cause and control

1. Bacteria that attack intestinal tract.
2. Disease caused by bites of mosquitoes.
3. Diseases that spread due to blood transfusion.
4. Illness caused by parasitic worms.
5. Disease that does not spread from one person to another.
6. Eating infected beef causes disease.
7. The worm that present in the host's intestine.
8. Infectious disease that have short incubation period.
9. Disease that cause mostly in children.
10. Disease that is pandemic.
11. Viral disease. **[November, 2019]**

Ans. 1. Vibrio cholarae 2. Filariasis 3. AIDS 4. Filariasis
5. Cancer 6. Taeniasis 7. Ascaris lumbricoides
8. Cholera 9. Hepatitis A 10. AIDS 11. Chicken pox

Chapter 17. Aids of Health

1. Innate immunity
2. Vaccine
3. Mechanical barrier
4. Antibiotics
5. Toxoids
6. Sexually transmitted disease
7. Living fully poisonous germs
8. Disinfectants
9. Antivenin
10. Immune cell
11. Germ killing body secretion
12. Germ trapping body secretion
13. Mechanical barrier that prevents the entry of germs into the body

Ans. 1. Body's immune system 2. DTP, BCG, Measles vaccine
3. Saliva, HCL, sweat. 4. Streptomycin, penicillin.
5. Diphtheria and tetanus (DT). 6. AIDS
7. BCG 8. DDT, 40% formalin

9. Crotalidae
10. B-lymphocytes
11. Tears
12. Mucus
13. Skin

Chapter 18. Health Organisation

1. The habitation where too many vehicles and lots of industries are found.
2. The organisation which concerned about the health and well beings of people.
3. The health problems that are found in each kind of habitation.
4. The habitation where tribal people are found thriving on forests and forest products.
5. The disease which is cause due to the consumption of harmful minerals present in water.
6. The agency which is a national as well as International bodies.
7. The agency which is a specialized agency of UNO.
8. The organisation that help the victims of flood.
9. The habitation where less crowded of population and vehicles are seen.
10. The organisation that organise campaigns for the control of epidemic and endemic diseases.

Ans. 1. Big cities 2. Red Cross 3. Food borne disease 4. Remote areas
5. Diarrhoea 6. Red Cross 7. WHO 8. Red Cross
9. Small town 10. WHO

Chapter 19. Waste Generation and Management

1. An organic waste.
2. An industrial waste.
3. A domestic waste.
3. A commercial waste.
5. A biomedical waste.
6. An environment friendly practice which reduce waste production.
7. A radioactive waste.
8. A method to minimize use of plastic.
9. A plant nutrient made from garden waste.
10. An item which can be recycled

Ans. 1. Paper 2. Coal ash 3. Vegetable peels
4. Paper packing 5. Used cotton / syringes 6. Recycling
7. Nuclear power plant 8. Use cloth bags 9. Humus
10. Paper bags/ News paper

Diagram Based Questions | Set 11 |

Chapter 1. Introducing Biology

1. A student observes the given organism under the microscope. Study the given diagram carefully and answer the following questions:

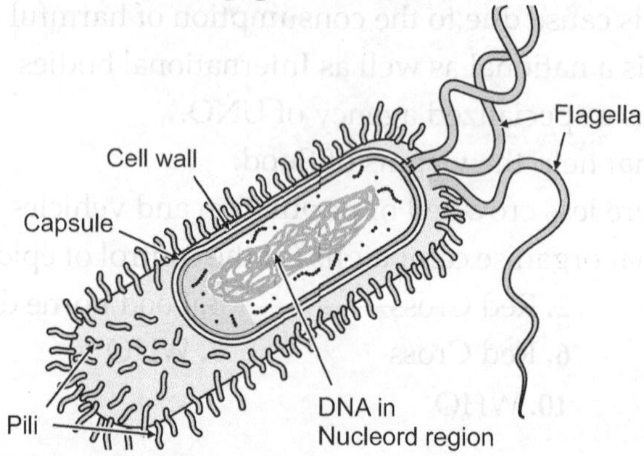

(i) Name the branch which studies the above organism.

(ii) Give the importance of the branch which helps to study the above organism.

(iii) Bacteriology is a branch of __________.

Ans. (i) Bacteriology.

(ii) Bacteriology focus on disease-causing bacteria in human beings. It requires the detection and identification of various bacterial pathogens and the mechanism through which they cause various diseases.

(iii) Bacteriology is a branch of microbiology that is concerned with the study of special microorganisms like bacteria and related aspects such as their characteristics.

2. Study the given figure carefully which represent a branch of applied biology and answer the following questions:

(i) If Mohan want to study the life animals shown in the above figure then he has to opt for which branch of applied biology?

(ii) Define the above branch taken by Mohan.

(iii) What is the key difference between aquaculture and pisciculture?

(iv) Give some examples of branches related to applied biology.
(v) Define any two branches given in (iv).

Ans. (i) Marine biology
(ii) Marine biology is the scientific study of marine organisms, their behaviours and interactions with the environment.
(iii) Aquaculture is the process of breeding, rearing and harvesting aquatic flora and fauna with commercial value in salt water or fresh water while pisciculture is the culturing of fish (fish farming) to obtain fish and fish products as food.
(iv) Exobiology, bioinformatics, horticulture, veterinary science, etc.
(v) Exobiology: It is the branch of science that deals with the possibility and likely nature of life on other planets or in space.
Bioinformatics: It is the application of statistical analysis to biological data.

3. Given below are four diagrams. Study them thoroughly and answer the following questions:

(a)

(b)

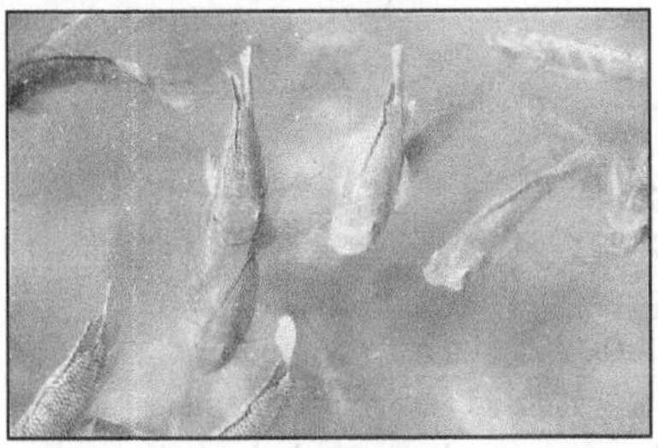

(c)

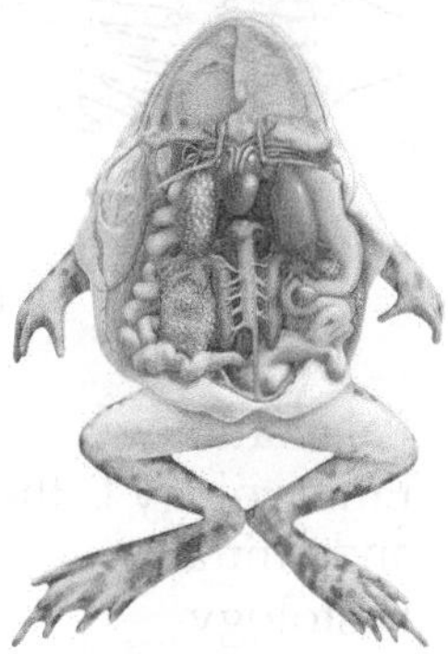

(d)

(i) Mention the type of biology branch shown by the given diagram.
(ii) Which figure deals with the study of the organisms for the manufacturing of honey and beeswax? Define the type of study.
(iii) Which two figures deals with the study for the kind of organisms?
(iv) Define the branch of biology shown by figure (c) and (d).

Ans. (i) Figure (a), (b), (c) and (d) shows the branches of ornithology, apiculture, ichthyology and anatomy respectively.
(ii) Figure (b) deals with the study of organisms for the manufacturing of honey and beeswax. Study of honey bee for the manufacturing of honey and beeswax is called apiculture.
(iii) Figure1(birds) and (b) (honey bee) deals with the study for the kind of organisms
(iv) Ichthyology (figure (c)) is the study of fishes. Anatomy (figure (d)) is concerned with the study of the structure of organisms and their parts.

4. (i) Draw a diagram showing the principal sub-divisions of biology.
(ii) Based on the diagram drawn in (i) answer the following questions:
(1) Study of fossils is called____.
(2) Define embryology.
(3) Which branch of biology deals with the microscopic structure of tissues?
(4) Which branch provides useful evidence on the interdependence between people and the natural world, as well as the consequences of human activity on the environment?

Ans. (i) Diagram for principal sub-division of biology

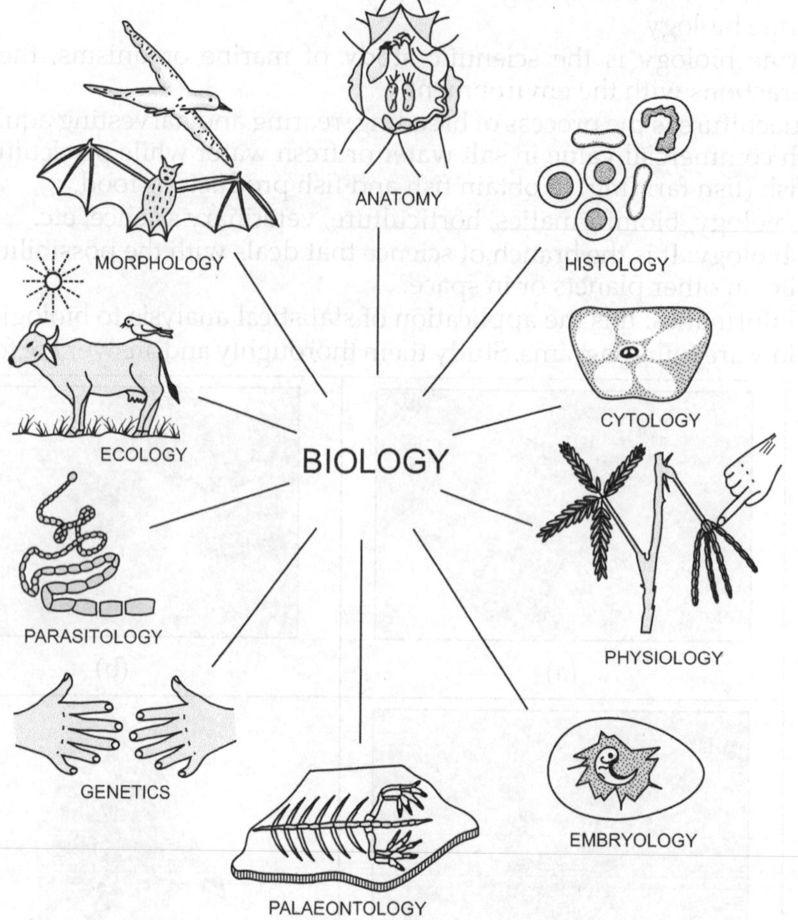

(ii) (1) Palaeontology

(2) Embryology is the study of the formation and development of embryo of plants and animals.

(3) Histology

(4) Ecology is the branch of biology that deals with the relations of organisms to one another and to their physical surroundings

Chapter 2. Cell: The Unit of Life

1. Refer the given figure (cell 1 and cell 2) and answer the following questions.

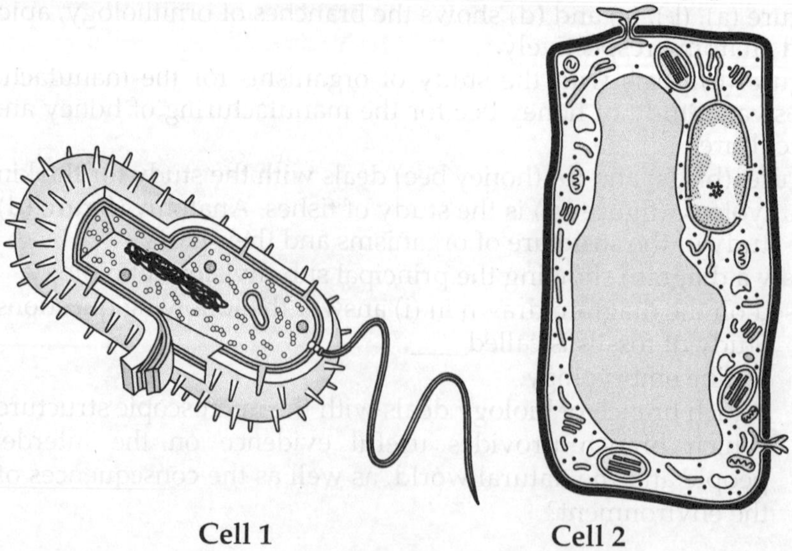

Cell 1 Cell 2

(i) Identify the structures which are common to both cells.
(ii) Which of these cells belongs to prokaryotes and eukaryotes?
(iii) In cell 2, name the organelle responsible for making food.
(iv) Nucleoid is a region of cytoplasm where the chromosomal DNA is located. It is present in which cell?
(v) Mention one difference between cell 1 and cell 2.

Ans. (i) Ribosomes and Cell wall
(ii) Cell 1: Prokaryotes and Cell 2: Eukaryotes
(iii) Chloroplast
(iv) Cell 1
(v) Eukaryotic cells contain membrane-bound organelles such as the nucleus, while prokaryotic cells do not.

2. Refer the given figure and answer the following question.

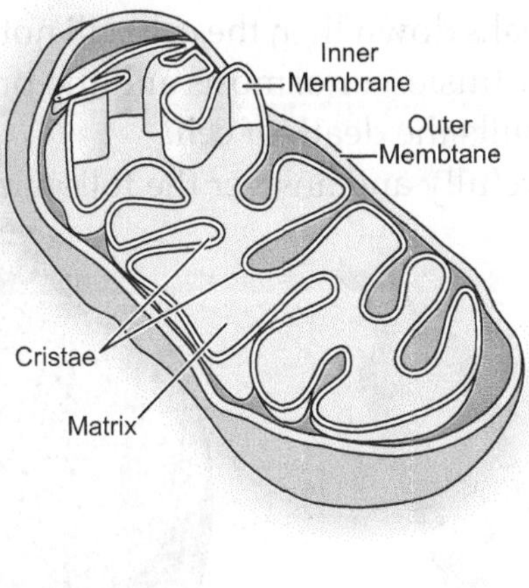

(i) Identify the given figure.
(ii) Why is the given figure called power house of the cell?
(iii) Which among the two membranes (outer and inner) is fairly smooth?
(iv) Define matrix and cristae.
(v) What molecule do mitochondria use to make energy?

Ans. (i) Mitochondria
(ii) Mitochondria synthesise and store the energy rich molecules ATP (Adenosine triphosphate)during aerobic respiration. So, they are called "Power houses of the cell".
(iii) The inner membrane of mitochondria is highly convoluted into folds (called cristae) while its outer membrane is fairly smooth.
(iv) A crista is a fold in the inner membrane of a mitochondrion. It increases the surface area of the inner membrane. Matrix is the space surrounding the cristae, which also includes molecules that participate in cellular respiration as well as ribosomes and mitochondrial DNA.
(v) ATP

3. Study the given figure carefully and answer the following questions.

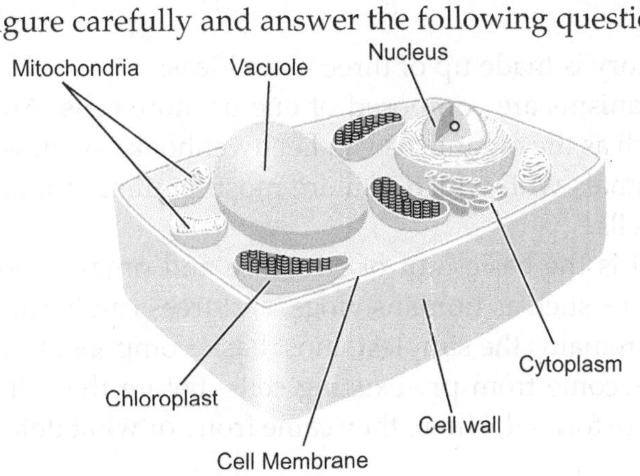

(i) Which organelle from the above figure would help you determine if the image shown comes from a plant or an animal?
(ii) Identify the figure.
(iii) Which structure will determines the shape of the cells and provides a strong structural support to prevent the bacterium from bursting or collapsing?
(iv) Identify the organelle which are absent in animal cell.
(v) What would happen if the cell membrane ruptures or breaks down?

Ans. (i) All the organelles can be used to determine the given figure which comes from a plant or animal.
(ii) It is a plant cell.
(iii) Cell wall determines the shape of the cells and provides a strong structural support to prevent the bacterium from bursting or collapsing.
(iv) Cell wall, Chloroplast and Vacuole.
(v) If cell membrane breaks down then the cell will not be able to exchange materials from its surrounding by diffusion or osmosis and the protoplasmic material of the cell will disappear which results the death of cell.

4. Study the given figure carefully and answer the following questions.

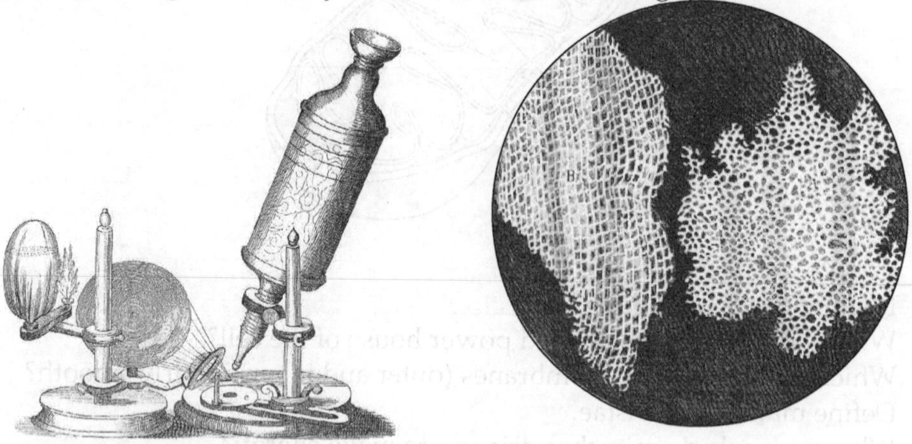

(a) (b)

(i) What was discovered on the basis of the given figure?
(ii) Who discovered the cell and how?
(iii) Explain any three main ideas about cell theory.
(iv) Name the scientists who discovered the cell theory.
(v) Who noted that all cells come from other cells?

Ans. (i) Cell
(ii) Cell was discovered by Robert Hooke. Robert Hooke used a compound light microscope to study cork and the dead cells of oak bark. In cork, Hooke observed small geometric shapes, like those shown in figure. Hooke gave these box shaped structures the name cells.
(iii) The cell theory is made up of three main ideas:
(1) All organisms are composed of one or more cells. An organism may be a single cell, such as the organisms van Leeuwenhoek saw in water. Others, like the plants and animals with which you are most familiar, are multicellular, or made up of many cells.
(2) The cell is the basic unit of structure and organisation of organisms. Although organisms such as humans, dogs, and trees can become very large and complex, the cell remains the simplest, most basic component of any organism.
(3) All cells come from pre-existing cells. Before the cell theory, no one knew how cells were formed, where they came from, or what determined the type of cell they

became. The cell theory states that a cell divides to form two identical cells.

(iv) Theodor Schwann and Matthias Jacob Schleiden

(v) Rudolf Virchow (1821–1902) noted that all cells come from other cells.

5. (i) Draw a well labelled diagram of nucleus.

(ii) Nucleus controls most of the cellular activities. Explain.

(iii) How nucleus is different from nucleoid?

(iv) Give two examples in which nucleus are not well defined.

(v) In which part of nucleus chromatin fibres are found?

Ans. (i) Structure of nucleus

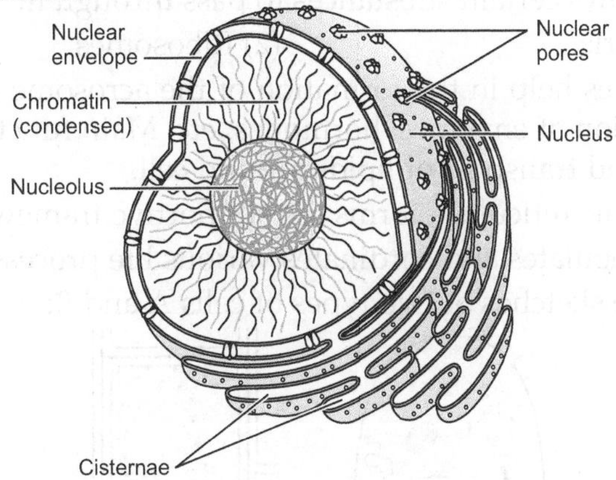

(ii) Nucleus controls most of the activities of the cell, therefore it is called control centre of the cell. It contains genetic information for structure, reproduction, development, metabolism and behaviour. The genetic information is then transferred to daughter cells. It also controls cellular activities through directing synthesis of particular proteins and enzymes.

(iii) Nucleus is a membrane bound organelle (also called true nucleus). It contains the cell's hereditary information and controls the cell's growth and reproduction. Whereas nucleoid is an unidentified nuclearregion present in prokaryotic cells. In prokaryotes, nucleoid is present in the form of a DNA strand liesin the cytoplasm.

(iv) Bacteria, blue green algae.

(v) Chromatin fibres are found in the nucleoplasm.

6. The diagram given below show a few parts of a cell. **[February, 2020]**

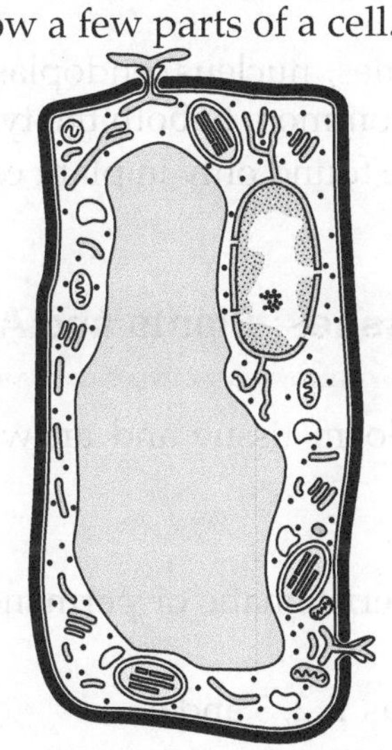

(i) Is it a plant or an animal cell? Give reason to support your answer.

(ii) Name a pigment which can be present in the part lablled X.

(iii) State the difference between the parts labelled A and B on the basis of permeability.

(iv) Name the cell organelles concerned with the following function.

(1) Production of energy (2) Protein synthesis

(v) State the functions of the following cell organelles:

(1) Golgi bodies (2) Endoplasmic reticulum

(3) Nucleus

Ans. (i) It is a plant cell as large, centrally placed vacuole and cell wall are present.

(ii) Anthocyanins

(iii) A is cell wall which is freely permeable allowing all substances to enter and leave the cell without any hindrance. B is cell membrane which is semi-permeable *i.e.,* it selectively allows certain substances to pass through it.

(iv) (1) Mitochondria (2) Ribosomes

(v) (1) Golgi bodies help in the formation of the acrosome of sperm, helps in synthesis and secretion of enzymes, hormones, etc. Main function is packing and sorting of proteins and transport of lipids around cell.

(2) Endoplasmic reticulum forms the supportive framework for the cell.

(3) Nucleus regulates and cordinates various life processes of the cell.

7. Given below are the sketches of two types of cells A and B:

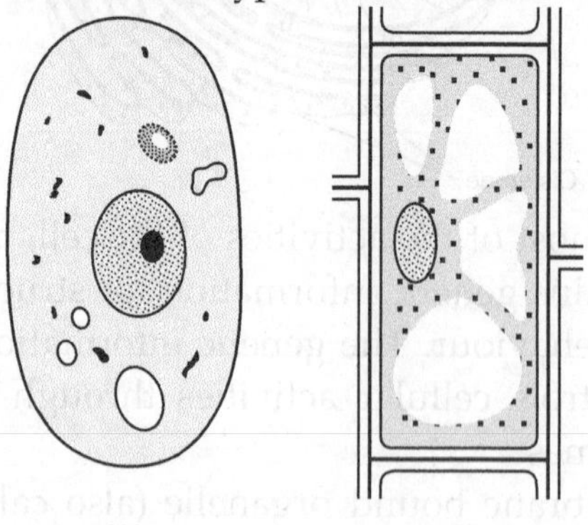

(A) **(B)**

(i) Which one of these is a plant cell? Give reason in support of your answer.

(ii) List the cell structures which are common to both the types.

(iii) Name the structures found only in plant cells and those found only in animal cells.

Ans. (i) Fig. B is a plant cell. It has a cell wall and a large vacuole which pushes the nucleus towards the periphery.

(ii) Cell membrane, ribosomes, nucleus, endoplasmic reticulum, lysosomes, golgi body and mitrochondria are common to both the types.

(iii) Plastids and cell wall are found only in plant cell. Centrosome is found only in animal cell.

Chapter 3. Tissues: Plants and Animals Tissue

1. Refer the given figure of phloem tissue and answer the following questions.

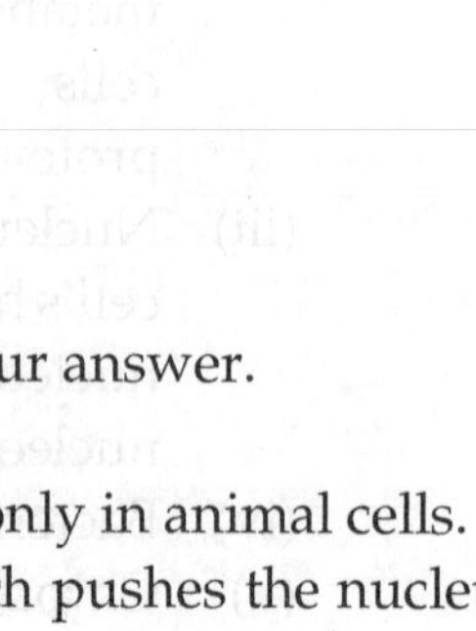

(i) Define phloem tissue.

(ii) Which type of tissue (meristematic or permanent) is shown by the figure?

(iii) Label the parts marked as 1, 2, 3 and 4.

(iv) Mention the function of 1 and 3.

(v) Where isthis tissue (phloem) likely to be found in the plant?

Ans. (i) Phloem tissue is a food-conducting vascular tissue in plants which conducts sugars and other metabolic products downwards from the leaves to the various parts of plant.

(ii) The given figure is a type of permanent plant tissue.

(iii) Label 1, 2, 3 and 4 are respectively sieve plate, sieve tube, phloem parenchyma and companion cells.

(iv) Function of sieve plate: Perforation in the sieve plates allows water and dissolved organic solutes to flow along the sieve tube.

Function of phloem parenchyma: They are responsible for storing food like starch, fats and other organic food materials.

(v) Phloem is likely to be found in the leaves and stem of plants to carry the food manufactured in the leaves to various parts of the plant.

2. Study the given figure of blood and their composition and answer the following questions.

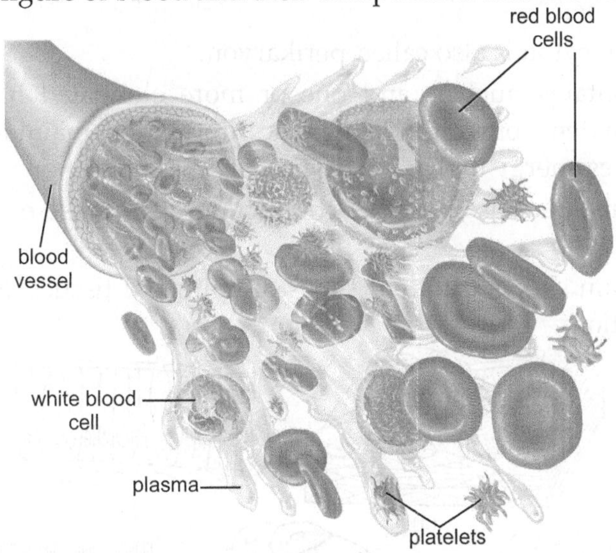

(i) Define blood and write their function.

(ii) Which part of the blood helps in blood clotting?

(iii) Which part of the blood are the cells of the immune system?

(iv) Which part of the blood is also called erythrocyte and is responsible for the transport of respiratory gases?

(v) Identify the type of tissue to which the given figure belongs.

Ans. (i) Blood is a specialised body fluid. It has four main components: plasma, red blood cells, white blood cells, and platelets. Blood transports oxygen and nutrients around the body and removes cellular waste.

(ii) Platelet plays an important role in the process of blood clotting.

(iii) White blood cells are the cells of immune system and these cells are involved in protecting the body against many infectious diseases and foreign invaders.

(iv) Red blood cells are also called erythrocyte and are responsible for the transport of respiratory gases.

(v) Blood is a fluid connective tissue.

3. Answer the following questions:

(i) Draw a well labelled diagram of neuron or nerve cell.

(ii) Where is this cell likely to be found in the human body?

(iii) Which part of neuron is also called as perikaryon?

(iv) Briefly explain the structure and function of perikaryon.

(v) What is the importance of the given figure in the human body?

Ans. (i) Diagram of neuron or nerve cell

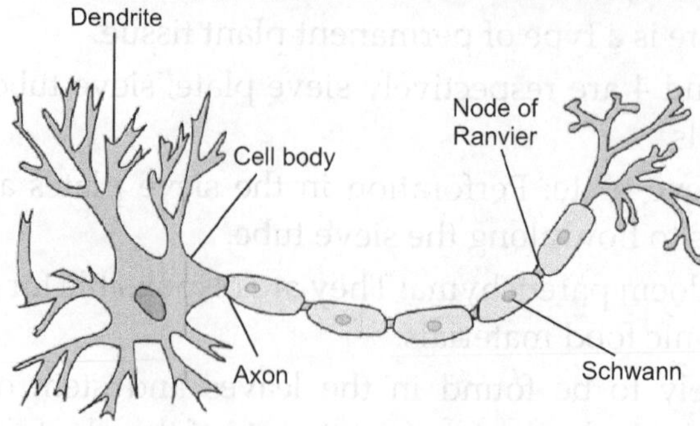

(ii) This cell is found in the nervous system. It is the structural and functional unit of nervous tissue.

(iii) Cell body or cyton is also called perikaryon.

(iv) Neuron contains nucleus and one or more elongated hair like extensions called dendrites or dendrons. It carries genetic information, maintains the neuron's structure, and provides energy to drive activities.

(v) Neurons are specialised for the conduction of impulse over great distance at great speed.

4. The given diagram shows different types of epithelial tissues. Study them carefully and answer the questions.

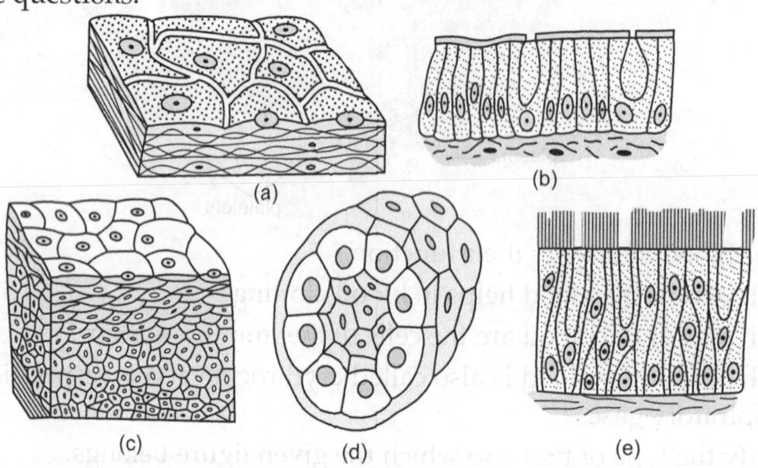

(i) Identify the tissues labelled as (a), (b), (c), (d) and (e).

(ii) Which epithelial tissues have threadlike protoplasmic projections at their free ends?

(iii) Epithelial tissue (c) are found in which parts of the human body?

(iv) Give a short note on epithelial tissue (a).

(v) Identify the organs (given below) in which epithelial tissue (d) is found.
"Salivary gland, Pancreatic duct, Inner lining of stomach, Intestines, Sweat gland, Blood vessels, Cornea"

Ans. (i) The tissues labelled as (a), (b), (c), (d) and (e) is respectively squamous epithelium, stratified epithelium, cuboidal epithelium, columnar epithelium and ciliated columnar epithelium.

(ii) Epithelial tissues (e) have thread like protoplasmic projections at their free ends.

(iii) Epithelial tissue (c) is cuboidal epithelium. This tissue is found in some parts of kidney tubules, salivary glands, and pancreatic duct.

(iv) Epithelial tissue (a) is squamous epithelium. Squamous epithelium is formed of single layer of thin and flat cells, found in the lining the mouth and nasal cavities, blood vessels, etc.

(v) Epithelial tissue (d) is columnar epithelium. This tissue is found in inner lining of the stomach and intestine only.

5. Refer the given figure which shows several typical plant tissues. Study the figure carefully and answer the following questions.

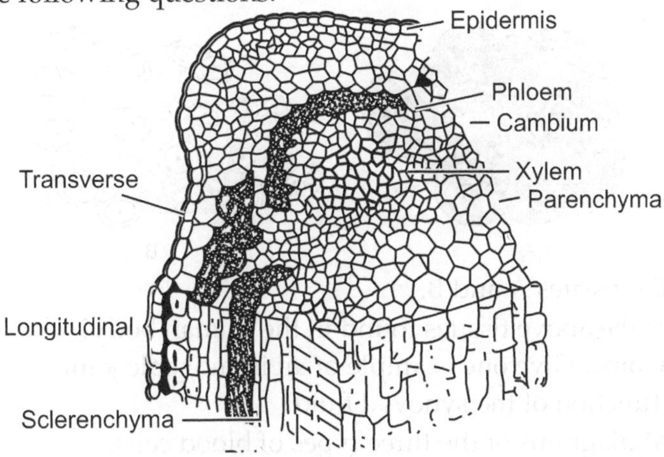

(i) Describe the structure of sclerenchyma with the help of diagram.

(ii) Identify the type of tissue from the above figure which secretes water proof waxy material.

(iii) Name the components of xylem elements.

(iv) Define tracheary elements and identify the components of xylem which are collectively called tracheary elements?

(v) Give brief account of tracheids.

Ans. (i) Sclerenchyma is the chief mechanical tissue of plants. It is a permanent tissue. The cells are made up of sclerenchymatous tissue. The cells are usually long, narrow, pointed at both ends, become dead and uniformly thickened by the deposition of lignin without any space in between the cells.

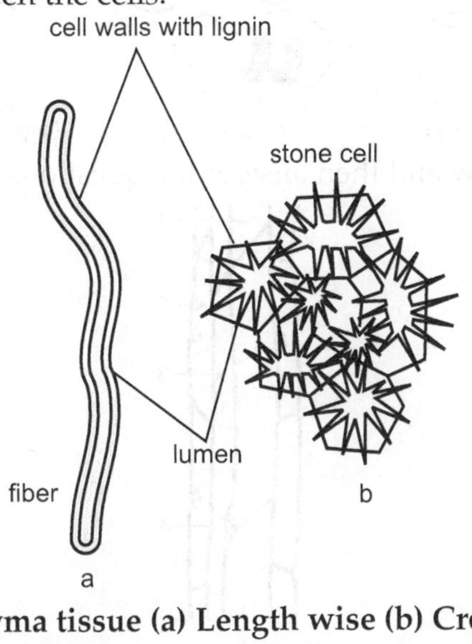

Sclerenchyma tissue (a) Length wise (b) Cross section

(ii) Epidermis secretes a waxy substance(called cutin) that coats the outside of the leaf.

(iii) The xylem consists of tracheids, vessels, and xylem parenchyma and xylem fibres.

(iv) Tracheary elements are cells in the xylem that are highly specialised for transporting water and solutes up the plant. Tracheids and vessels are collectively called tracheary elements.

(v) Tracheids are primitive conducting elements of xylem. They are unicellular and having narrow lumen. Tracheids are dead and lignified cells. The deposition of lignin on cell wall is responsible to form a different type of thickenings. Usually bordered pits are present at end wall of tracheids.

6. The diagram given below shows the supportive connective tissue. Answer the questions that follow: **[February, 2020]**

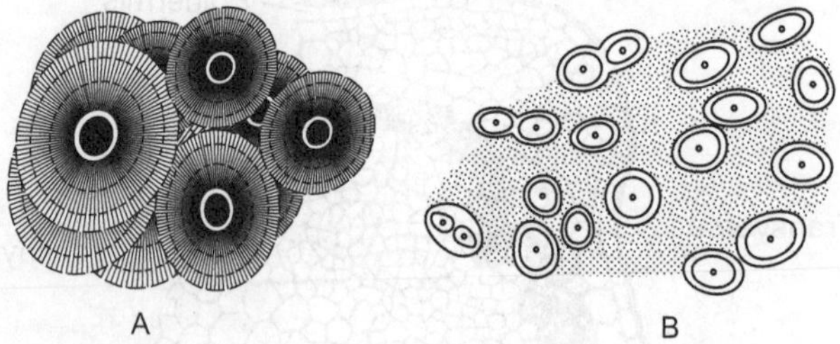

(i) Identify the tissues A and B.
(ii) Where are the above tissues found in the human body?
(iii) What is a joint? Give one example of an immovable joint.
(iv) State the function of the synovial fluid.
(v) Draw neat diagrams of the three types of blood cells.

Ans. (i) A is bone; B is cartilage
(ii) Cartilage is found in the tip of the nose, ear pinna, trachea, between vertebrae, end of long bones.
Bones forms the hard framework of the skeletal system.
(iii) Joint is the point at which two seperate bones meet. sutures between the bones of skull box are example of an immovable joint.
(iv) Synovial fluid is a lubricating fluid which reduces the friction between the two bones.
(v) The three types of blood cells are RBCs, WBCs and platelets.
Components of blood

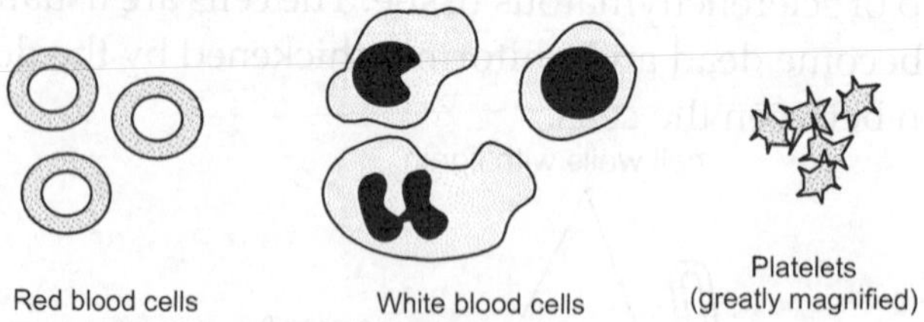

7. Study the diagram below and then answer the questions that follow:

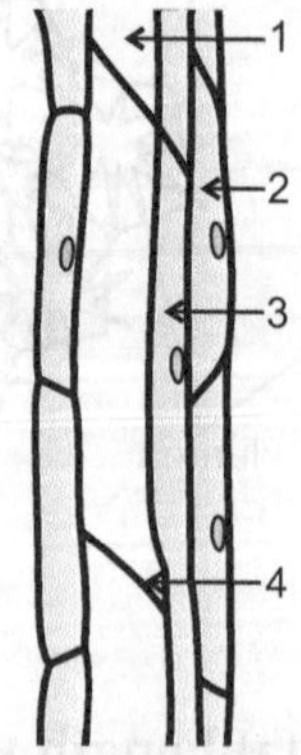

(i) Indentify the tissue and give a reason to support your answer.
(ii) Name the parts labelled 1, 2, 3 and 4.
(iii) Where is this tissue likely to be found in the plant?
(iv) State the function of the parts labelled 1, 2, 3 and 4.

Ans. (i) The given diagram is of the phloem tissue because that the cells show celluelar contents unlike the xylem tissue which contains hollw cells without any cellular contents.
(ii) 1 — Sieve cell
2 — Phloem parenchyma cell
3 — Companion cell
4 — Sieve plate

(iii) The phloem is a food conducting tissue and is likely to be found in the leaves and stem of plants to carry the food manufactured in the leaves to various parts of the plants.

(iv) 1 — Sieve cells: Help in the transport of food from leaves to storage organs and other parts of the plant.

2 — Phloem parenchyma cells: Storage of starch, fat and other organic food material.

3 — Companion cells: Help in the functioning of sieve tuber cells.

4 — Sieve plate: perforation in the sieve plates allow water and dissolved.

8. Study the diagram given below and then answer the questions that follow:

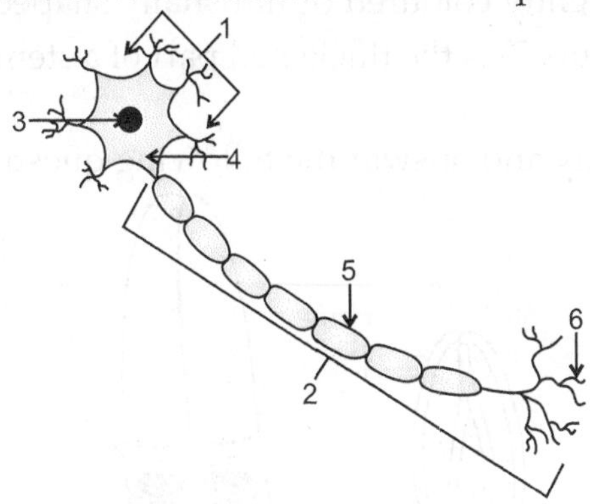

(i) Identify the cell.

(ii) Name the parts labelled 1, 2, 3, 4, 5 and 6.

(iii) Where is this cell likely to be found in human body and what is its function?

Ans. (i) The given diagram shows a nerve cell or neuron.

(ii) 1 — Dendrites/Dendrons

2 — Axon

3 — Nucleus

4 — Cyton/Perikaryon

5 — Nearolemma

6 — Axon endings

(iii) The nerve cell is likely to be found in the nervous system of the human body. The function of the nerve cell is to transmitting message from one part of the body to another. It is associated with perception and responses of animals.

Chapter 4. The Flower

1. Refer the given figure of bisexual flower and answer the following questions.

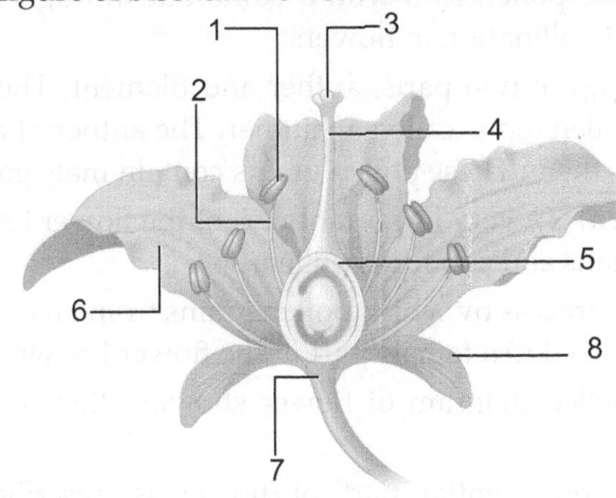

(i) Label the structure marked as 1 to 8.

(ii) Which labelled structures are the components of gynoecium?

(iii) Which labelled structure containsthe male reproductive cell of plants?
(iv) Give brief description of structure marked as 6.
(v) Define structure 7.

Ans. (i) Structure marked as 1, 2, 3,4, 5, 6, 7 and 8 are anther, filament, stigma, style, ovary, petal, receptacle and sepal respectively.
(ii) Structure 3, 4 and 5 named as stigma, style and ovaryare the components of gynoecium.
(iii) Structure 1 (anther) contains the male reproductive cells of plants (called pollen grain).
(iv) Petals (structure 6) are modified leaves that surround the reproductive parts of flowers. They are often brightly coloured or unusually shaped to attract pollinators.
(v) Receptacle (structure 7) is the thickened part of a stem (pedicel) from which the flower organs grow.

2. Study the figure carefully and answer the following questions.

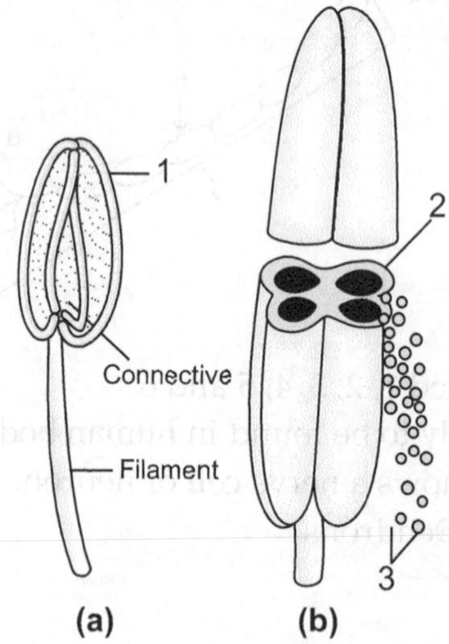

(i) Identify the figure (a) and the structure labelled as 1, 2, and 3.
(ii) How does the content of the pollen sacs would come out?
(iii) Briefly describe the structure shown in the figure (a).
(iv) Figure(a) is an important part of the flower. Why?
(v) Define pollination.

Ans.

(i) The given figure (a) represents the stamen and the structure labelled as 1, 2 and 3 are anther, pollen sac and pollen grains respectively.
(ii) The content of the pollen sacs would come out through the agents like air, wind, insects leading to pollination in flowers.
(iii) Stamen is made up of two parts, anther and filament. The stalk of stamen is called filament and swollen top is called an anther. The anther of a stamen produced pollen grains and stores them. These pollen grains contain male gametes of the plant.
(iv) Figure (a) (Stamen) is a very important part of the flower because it contains the male reproductive organs and produces pollen.
(v) Pollination is the process by which pollen grains from an anther, the male portion of a flower are transferred to a female part in the flower known as the stigma.

3. (i) Draw a well labelled diagram of flower showing their essential and non-essential parts.
(ii) Which part of non-essential part of flower is described as polysepalous and gamosepalous?

(iii) Define polysepalous and gamosepalous with examples.

(iv) Which part of the flower is also known as pistil?

(v) Define bisexual flowers with examples. Which two parts are present in bisexual flowers?

Ans. (i)

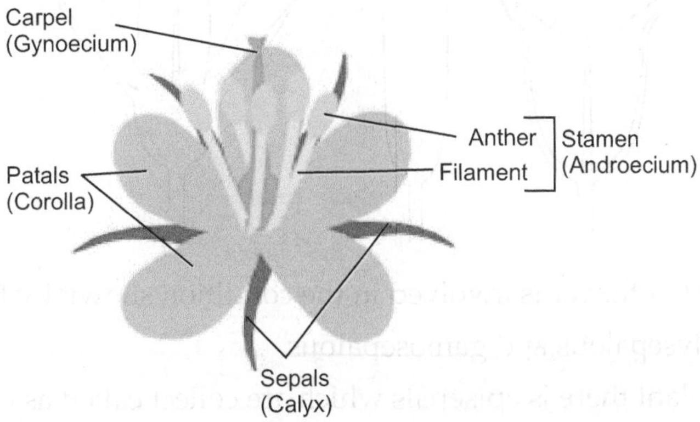

(ii) The calyx is described as polysepalous and gamosepalous.

(iii) When the sepals of a flower are fused, they are called gamosepalous, e.g. *Hibiscus* and when the sepals of a flower are free, they are called polysepalous,e.g., rose.

(iv) Gynoecium is also known as pistil.

(v) The flowers which have both male (carpel) and female sex organs (stamen) are called bisexual flower. Flowers of *hibiscus* and mustard plant are called bisexual flowers.

4. Refer the figure (a), (b), and (c) given below and answer the questions.

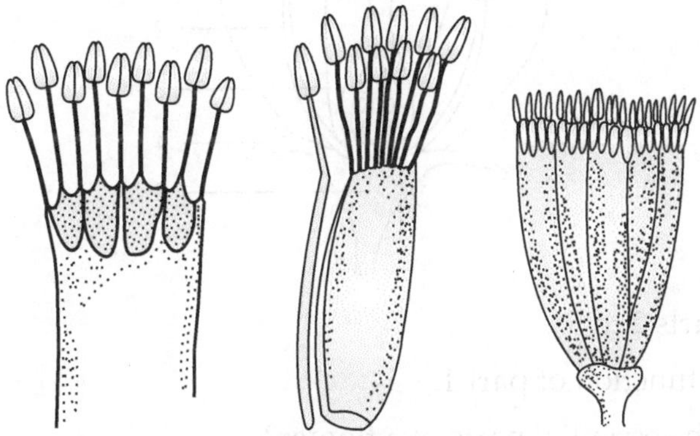

(a) (b) (c)

(i) What is represented by the given figure?

(ii) Identify the figure (a), (b) and (c).

(iii) In which figure stamens are joined in several groups?

(iv) Define figure (a) and (b).

Ans. (i) The given figures represent the cohesion (joining) of stamens with the help of filaments.

(ii) Figure (a), (b) and (c) are monadelphous, diadelphous and polyadelphous respectively.

(iii) In figure (c), stamens are joined in several groups.

(iv) Monadelphous: In this androecium, filaments are fused into single bundle.

Diadelphous: In this andoecium filaments are used into two bundles.

5. Refer the diagram of a condition shown by calyx. Study them carefully and answer the following questions.

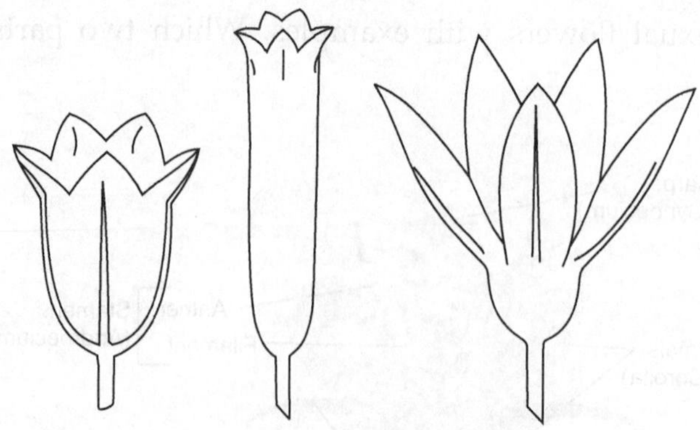

(i) Which part of flower is involved in the condition shown by the figures?

(ii) Define polysepalous and gamosepalous.

(iii) In which plant there is episepals which are collect called as epicalyx?

Ans. (i) Sepal is involved in the condition shown by the figures.

(ii) Polysepalous: It is a flower that possesses a calyx consisting of separate sepals.
Gamosepalous: When the sepals of a flower are fused, they are called gamosepalous.

(iii) *Hibiscus*

6. The diagram below shows a vertical section of a flower. **[November, 2019]**
Answer the questions that follow:

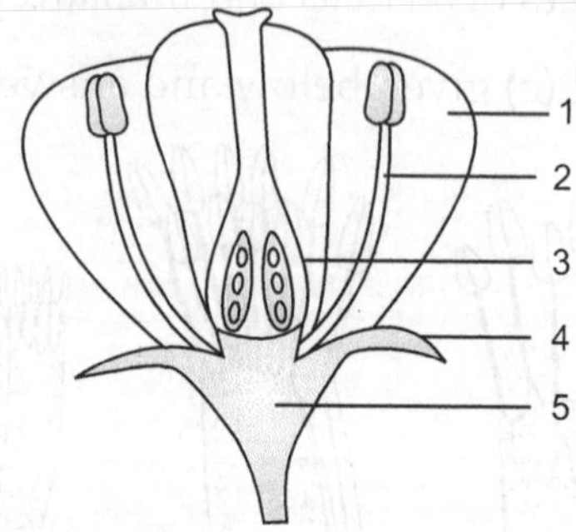

(i) Label the parts 2 to 5.

(ii) What is the function of part 1.

(iii) Which are the essential parts of a flower?

(iv) Define the term 'Inflorescence'.

(v) Draw a neat, labelled diagram of a pollen grain?

Ans. (i) 2 — Filament [stamen]
3 — Ovary
4 — Sepal
5 — Receptacle

(ii) Part 1 is petal which is beautiful coloured that attracts insects for pollination.

(iii) Essential part of a flower are directly involved in reproduction process. They include stamen and carpel which are male and female parts of a flower respectively.

(iv) Inflorescence is the mode of arrangement of flowers on the axis of the plant.

(v) Labelled diagram of pollen grain.

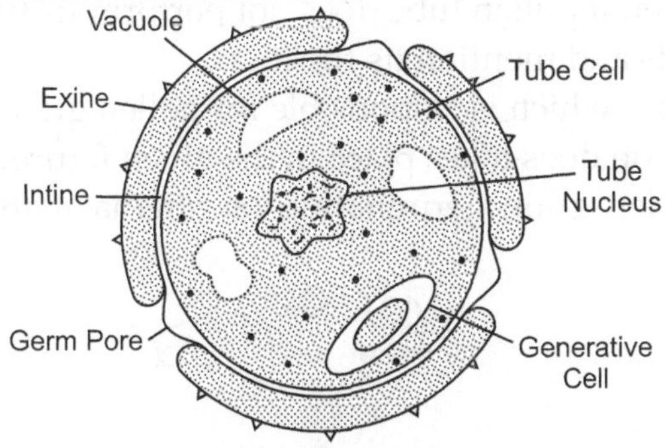

Chapter 5. Pollination and Fertilisation

1. The given diagram shows different types of pollination marked as X, Y, and Z between two plants (a) and (b) of the same species.

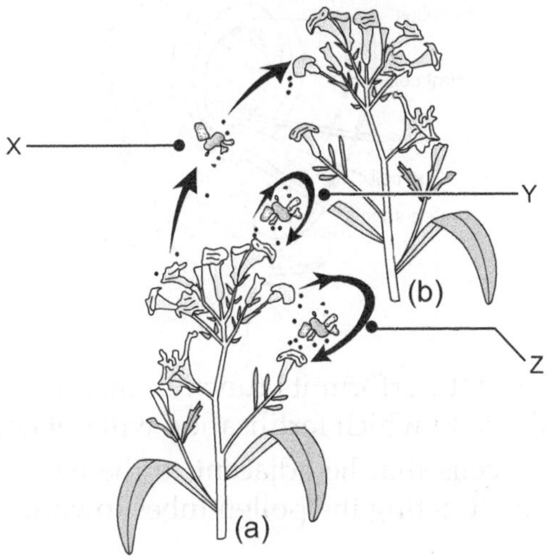

(i) Identify the types of pollination shown by the X, Y and Z.
(ii) Define all the three types of pollination.
(iii) Give two advantages of cross pollination.
(iv) Cross pollination requires production of large number of pollen grains. Explain.
(v) Name the conditions which favour self-pollination.

Ans. (i) The types of pollination marked as X, Y and Z are allogamy, autogamy and Geitonogamy respectively.

(ii) Allogamy: It is the process of transfer of pollen grains from anther of one flower to the stigma of a genetically different flower.
Autogamy: It is the type of pollination in which a flower is pollinated by its own pollen grains.
Geitonogamy: It is the type of pollination in which pollen grains of one flower are transferred to the stigma of another flower on the same plant.

(iii) Advantages of cross pollination:
(1) The seeds produced are abundant and viable.
(2) The offsprings are healthier.

(iv) Most of pollen grains are lost in their journey from one flower to another flower. Thus, larger in numbers are required for cross-pollination.

(v) Bisexuality, homogamy and cleistogamy.

2. (i) Draw a well labelled diagram of pollen grain germinating on stigma with one pollen tube has entered ovule.

(ii) What will happen if pollen tube does not perform its function?
(iii) Define synergids and mention its function.
(iv) Identify the factor which is responsible for pollen grain to fall on the stigma.
(v) Where fertilisation does takes place in the given figure?

Ans. (i) Diagram of pollen grain germinating on stigma with one pollen tube has entered ovule

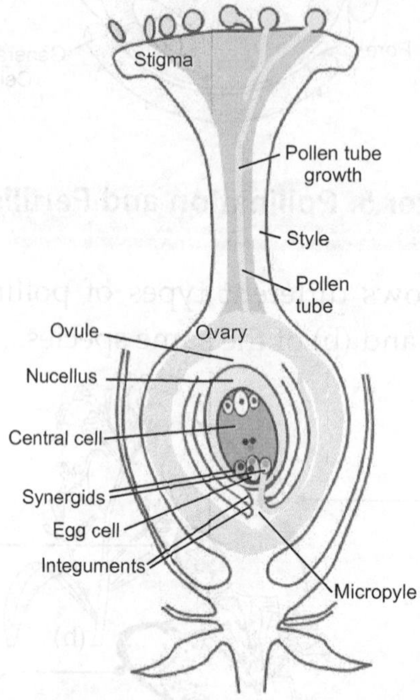

(ii) If pollen tube does not perform its function then the sperm from the pollen will not reach the egg cell due to which fertilisation will not occur to form the seed.
(iii) Synergids are two cells that lie adjacent to the egg cell in the female gametophyte. These cells help in attracting the pollen tubes towards the embryo sac for the process of fertilisation.
(iv) The secretion of sugar by the stigma is responsible for pollen grain to fall on the stigma.
(v) Fertilisation takes place in the ovary.

3. Study the given figure carefully and answer the following questions.

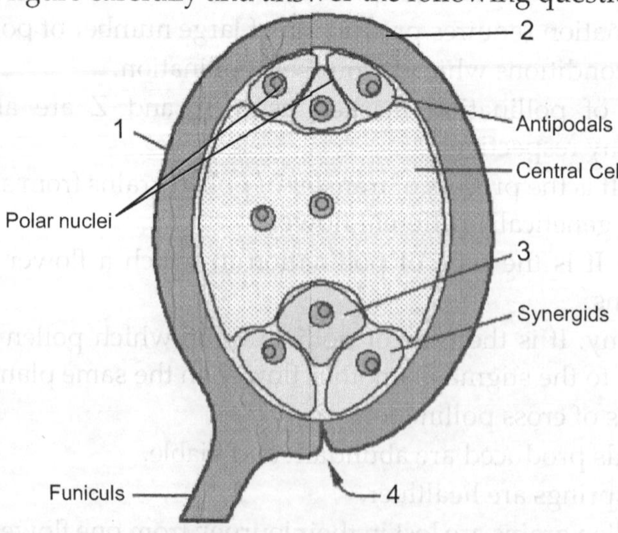

(i) Identify the labelling marked as 1, 2, 3 and 4.
(ii) In which part of the embryo pollen tube arrives and enter the embryo sac?
(iii) Define antipodal cells.
(iv) What will happen if labelling marked as 4 is blocked?

(v) Define double fertilisation.

Ans. (i) The figure marked as 1, 2, 3 and 4 are respectively integuments, nucellus, egg and micropyle.

(ii) The pollen tube arrives in the micropyle of the integument of the ovule, and enters the embryo sac.

(iii) Antipodal cells are three haploid cells in the mature embryo sac of flowering plants that are situated at the opposite end to the micropyle.

(iv) Figure 4 is micropyle. Micropyle is an opening in the integument of an ovule and if it is blocked then the male gamete from the pollen grain does not enter the ovule.

(v) Double fertilisation is a process of two fusions (*i.e.*, syngamy and triple fusion) takes place. In this one sperm fertilises the egg cell and form a diploid zygote, while the other sperm fuses with the two polar nuclei, forming a triploid cell that develops into the endosperm.

4. The figure represent a step-wise process [marked as (a), (b) and (c)] which occurs in flowering plant. Study the figure carefully and answer the following questions.

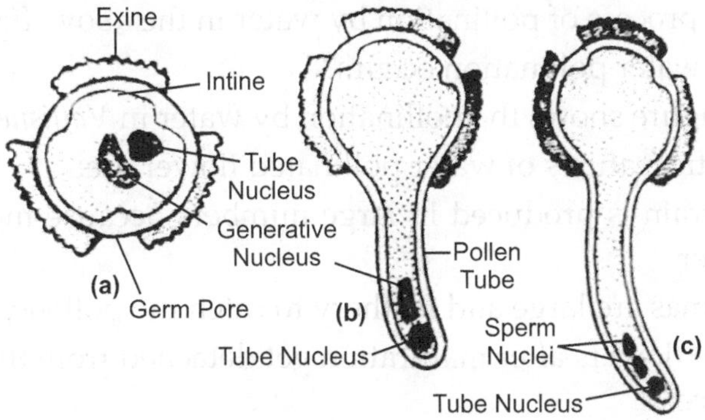

(i) Identify the process represented by the given figure.

(ii) Mention the name of the steps marked as (a), (b) and (c).

(iii) Explain the process.

(iv) What does exine and intine represent? Explain.

(v) Define germ pore.

Ans. (i) The given figure represents the germination of pollen grain.

(ii) (a) - A mature pollen grain

(b) - Pollen grain begins to germinate

(c) - Germinated pollen grain.

(iii) Pollen germination starts with the pollen grain landing on stigma. The cell of the pollen grain divides into two cells- one of these is smaller in size and called generative cell and bigger one is called tube cell. The intine grows out through the germ pore and form pollen tube. It grows into the tissue of the style towards the ovary. By this time the tube nucleus degenerates and the generative nucleus divides to form two male gametes. The pollen tube bursts open and the male gametes are released, one of which then fuse with the egg cell and other nuclei with two polar nuclei.

(iv) Exine and intine represent the layers of pollen grain. Exine constitutes the hard outer layer, made up of sporo pollenin (the most resistant material) and intine is the thin inner layer of the pollen grain made up of cellulose and pectin.

(v) Germ pore is a thin-walled area through which the germ tube or pollen tube emerges on germination.

5. Refer the given figure and answer the following questions.

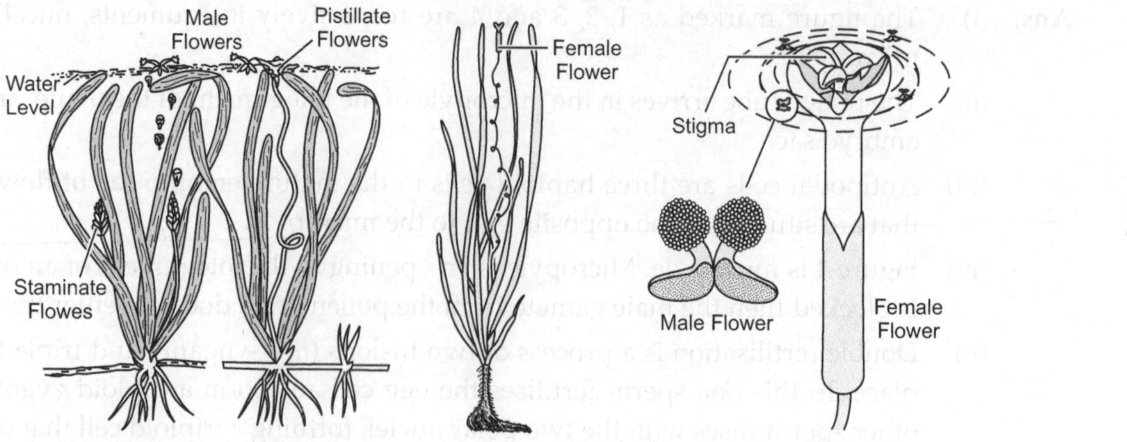

(i) Identify the given figure.
(ii) Write down the characteristic features of water pollinated flower.
(iii) Give two examples of water pollinated plants.
(iv) Explain the process of pollination by water in the above figure.
(v) When does water pollination occurs?

Ans. (i) The given figure shows the pollination by water in *Vallisneria*.

(ii) Characteristic features of water pollinated flower are:

(1) Pollen grain is produced in large numbers because most of them get lost by the flow of water.

(2) The stigmas are large and feathery to catch the pollens.

(3) The male flowers after maturation get detached from the plant and float above the water surface.

(4) No animal transporters are needed for pollination.

(5) The stalks of the pollinated flowers are usually very long to project their stigmas above the water surface.

(iii) Examples of water pollinated plants are *Ceratophyllum, Vallisneria*. .

(iv) In *Vallisneria*, flowers are unisexual. Male flowers get detached from the plant and float on water surface. Female flowers are produced on long coiled stalks. These stalks uncoil and bring mature female flowers to water surface. They come in contact with floating male flowers through water currents and receive pollen.

(v) Water pollination occurs when pollen drifts on water in rivers, lakes, ponds, etc. and comes in contact with flowers directly, thereby allowing certain plants to reproduce.

6. The diagram given below represents a process occuring in a flower. **[February, 2020]**

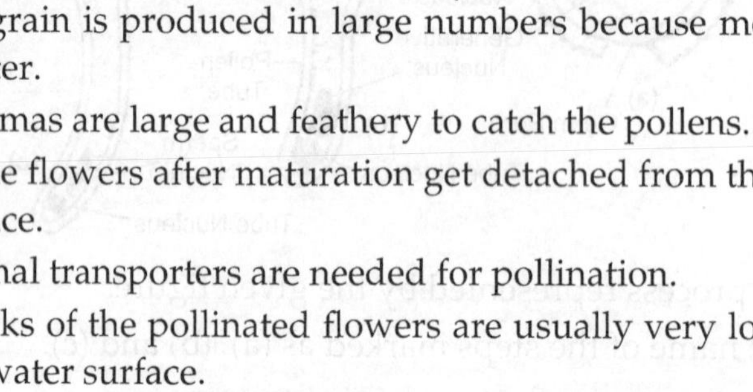

(i) Name the process and explain it.
(ii) Mention the agent involved in the above process.
(iii) What is 'Double Fertilisation'?
(iv) Define the term 'Inflorescence'.

(v) Draw a neat, labelled diagram of a carpel.

Ans. (i) Pollination

(ii) It is a process of transfer of pollen grain from the anther to the stigma of the same flower or different flower of the same plant or in different plants.

(iii) Double fertilization means two fertilization have occurred simultaneously. First is one sperm nucleus fueses with the egg cell nucleus to form zygote and second is another sperm nucleus fuses with two polar nuclei together to form endosperm.

(iv) The mode of arrangement of flower on the axis of the plant is called an inflorescence.

(v)

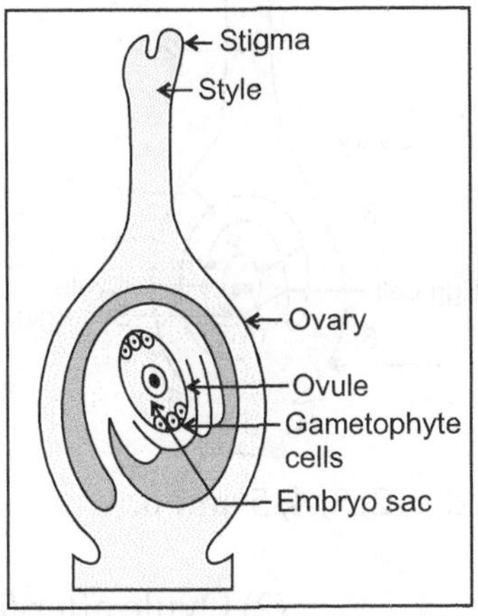

7. Given ahead is a diagrammatic sketch of the sectional view of germinating pollen grain. Study the same and then answer the questions that follow:

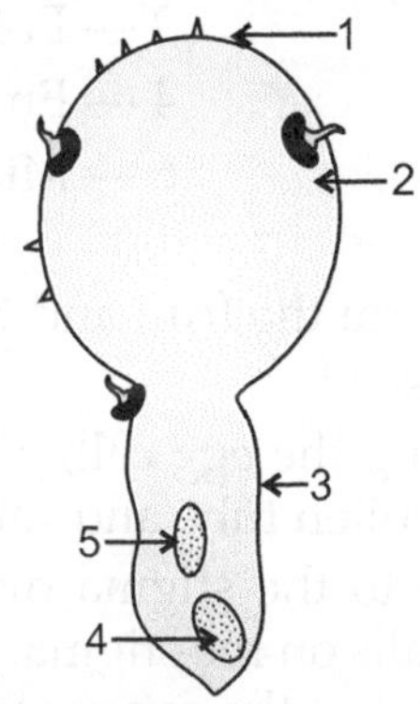

(i) Name the parts labelled 1, 2, 3, 4 and 5.

(ii) Where does the germination of the pollen grain take place and how?

(iii) What is the function of part labelled 4.

(iv) What happens to the part labelled 5 during the process.

Ans. (i) 1 — Exine

2 — Intine

3 — Pollen tube

4 — Tube nucelus

5 — Generative nucleus

(ii) Germination of the pollen grain takes place only after it falls on the stigma of the same plant species. The pollen grain is stimulated to germinate due to the secretion of sugers by the stigma.

(iii) The function of part 4 (tube nucleus) is to direct the growth of the pollen tube towards the ovary.

(iv) During germination of the pollen grain, part 5 (generative neucleus) present at the tip of the pollen tube tube divides into two sperm nuclei. The pollen enters one of the synergids and releases its two sperm nuclei. Of these, one sperm nucleus enters the egg cell and fuses with its nucleus, while the other sperm nucleus moves towards the two polar nuclei in the central cell and fuses with them.

8. Given ahead is a diagrammatic representation of the process of fertilisation. Study the same and then answer the questions that follow:

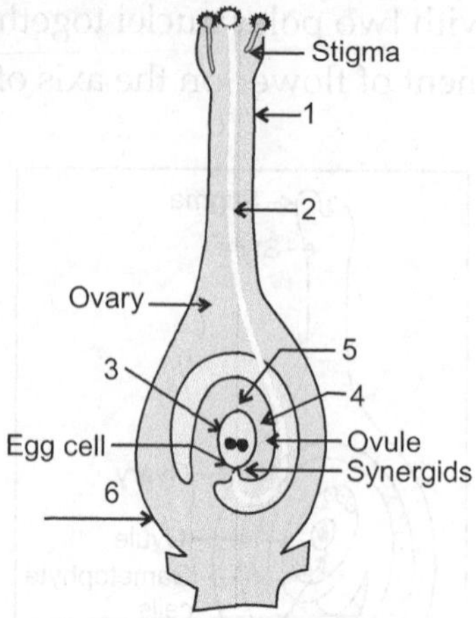

(i) Name the parts labelled 1, 2, 3, 4, 5 and 6.

(ii) What happens to

(1) Ovary (2) Ovule after fertilisation

(iii) What is the function of the synergids?

(iv) What part does the stigma play in the process of fertilisation?

Ans. (i) 1 — Style 2 — Pollen tube

3 — Poller nuclei 4 — Embryo sac

5 — Antipodal cells 6 — Micropyle

(ii) After fertilisation

(1) The overy enlarges to form the fruit and the ovarian wall forms the fruit wall.

(2) The ovule becomes the seed.

(iii) Synergids help in nourishing the egg cell, guidding the pollen tube towards the egg, proper functioning of the pollen tube and release of sperm nuclei.

(iv) Pollen grain is transferred to the stigma during pollination. Germination of pollen grain takes place only if it falls on the stigma. After germination, the pollen tube grows through the stigma and reaches the ovary for the fertilisation of the egg cell.

Chapter 6. Seeds-Structure and Germination

1. Study the diagram and answer the following questions:

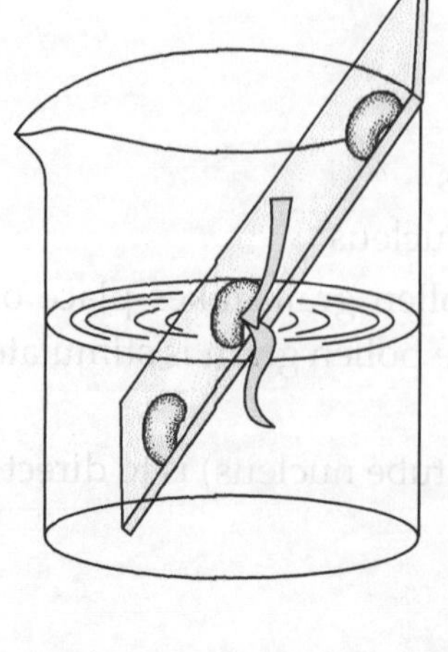

(i) What are the conditions necessary for germination of seed?

(ii) Name a solution which absorbs oxygen.

(iii) Why in the figure shown middle seed germinates but not top and bottom seed?

(iv) Define germination.

(v) Give a brief for the types of germination.

Ans. (i) Suitable temperature, oxygen and water.

(ii) Pyrogallic acid

(iii) The middle seed gets all the conditions that are favourable for germination *i.e.,* water, oxygen and suitable temperature so it germinates. But the top seed does not get water and bottom seed does not get adequate oxygen hence they cannot germinate.

(iv) Germination is the process of formation of seedling developed from the embryo.

(v) Germination are of three types:

(1) Epigeal germination: The germination in which cotyledons are pushed above the ground due to elongation of hypocotyl. Example, bean, castor, etc.

(2) Hypogeal germination: In this germination, cotyledons remain underground and elongation of epicotyl occurs like in maize, pea, etc.

(3) Viviparous germination: It is a special germination seen in mangrove plants. In viviparous seed germinates inside the fruit when it is still attached to the plant.

2. Study the diagram and answer the following questions:

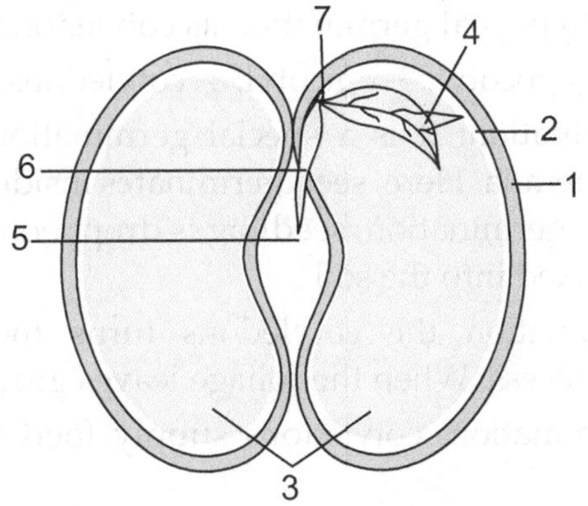

(i) Identify the seed. Is it a monocotyledonous or dicotyledonous seed?

(ii) Label the parts.

(iii) Give some examples of this category of seed.

(iv) What is the function of the part labelled as 3, 4,5.

(v) What do you mean by embryonal axis and what are its parts?

Ans. (i) It is a dicot seed with two cotyledons. It is a dicotyledonous seed.

(ii) 1- testa; 2- tegmen; 3- cotyledons; 4- plumule; 5- radicle; 6- hypocotyl; 7- epicotyl

(iii) Pea, gram

(iv) 3- Cotyledons stores food for embryo and protects it.

4- Plumule forms the shoot system.

5- Radicle forms the root system.

(v) Embryonal axis is a young miniature plant present between the two cotyledons. It consists of radicle, plumule, epicotyl and hypocotyl.

3. Study the diagrams and answer the following:

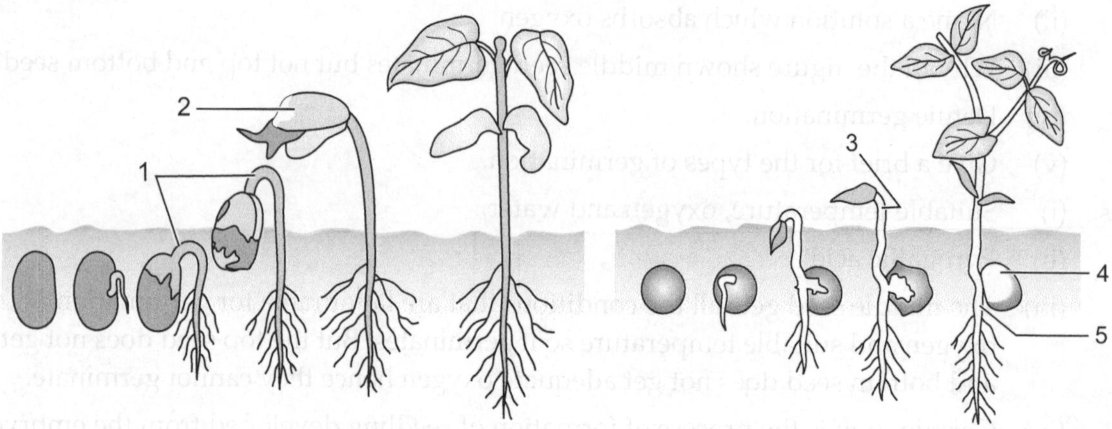

(i) Identify the type of germination shown in figure (a) and figure (b) stating suitable reason for each.

(ii) Label the parts 1, 2.

(iii) Name the special type of germination which is seen in Sonneratia and describe this germination process.

(iv) What are the functions of cotyledons in the type of germination shown in figure (a) and figure (b)?

(v) What do you meant by seedling?

Ans. (i) Figure A depicts epigeal germination as cotyledons are pushed above the ground.
Figure B depicts hypogeal germination as cotyledons remain underground.

(ii) 1- cotyledons; 2-hypocotyl; 3- epicotyl; 4- cotyledons; 5- radicle

(iii) Viviparous germination: It is a special germination seen in mangrove plants like Rhizophora, Sonneratia. Here, seed germinates inside the fruit when it is still attached to the plant. After germination, seedling is dropped into the soil which develops root system and gets fixed into the soil.

(iv) In epigeal germination the cotyledons turns the first green leaves and start photosynthesis process. When the foliage leaves grow then these cotyledons fall off.
In hypogeal germination, cotyledons supply food till the seedling is able to exist independently.

(v) Seedling is a stage in the growth of a plant from a seed before it has become wholly independent of the food stored in it.

4. Study the diagram and answer the following questions:

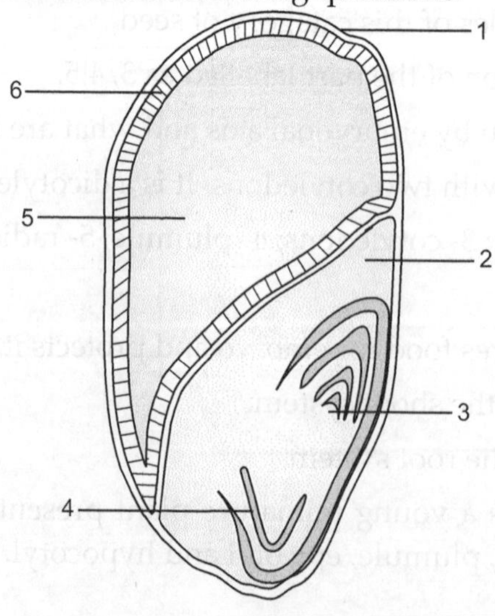

(i) Name the structure shown and label its parts 1-6.
(ii) Is it an albuminous or non-albuminous and justify your answer.
(iii) Name the protective coverings of parts labelled 3 and 4.
(iv) Name the layer which separates endosperm and embryonic part in this structure.
(v) How is it different from a bean seed?

Ans. (i) It is a maize grain.

(1) Fused pericarp and testa	(2) Scutellum
(3) Plumule	(4) Radicle
(5) Endosperm	(6) Aleurone layer

(ii) It is an albuminous seed as it contains an endosperm, a nutritive tissue.
(iii) Plumule is covered by a protective sheath coleoptile and radicle by coleorhiza.
(iv) Epithelium
(v) Maize grain has endosperm but a bean seed does not.
It has one cotyledon but a bean seed has two cotyledons.
In maize grain seed coat is fused with fruit wall but a bean seed has a distinct seed coat.

5. Given below is an experimental set up to study seed germination, study the same and answer the following questions:

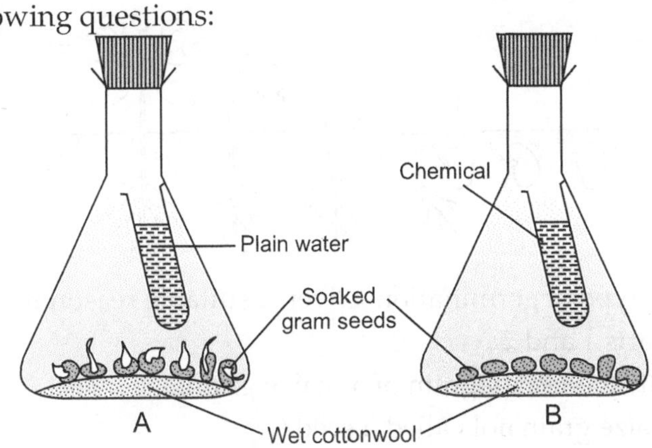

(i) Name the chemical present in test tube of set-up (b) and what is its role?
(ii) Why seeds germinate in set-up (a) but not in set-up (b)?
(iii) What is the role of water in seed germination?
(iv) Besides water state the essential factors required for seed germination?
(v) What is the aim of the above experimental set up?

Ans. (i) Pyrogallic acid. It absorbs oxygen.
(ii) In set-up (a) oxygen is present so seed germinates but in set-up (b) due to presence of pyrogallic acid, oxygen is absent hence, seed cannot germinates.
(iv) Water activates enzymes which converts stored food into diffusible form which is utilised by embryo. Seed absorbs water and swells up thus seed coat ruptures so that radicle and plumule can emerge out to form root and shoot system respectively.
(iv) Oxygen and optimum temperature.
(v) This experiment is set up to analyse that oxygen is required for seed germination.

6. (i) Draw a neat and labelled diagram of a maize grain showing its longitudinal section.
(ii) Based on the diagram drawn in (i) give a suitable technical term for each of the following descriptions:
(1) Seeds with thin and membranous cotyledons.
(2) A modified cotyledon in the embryo.

(3) Outermost layer that is rich in proteins.
(4) The protective sheath that protects the root of germinating seed.
(5) The protective sheath that protects a young shoot tip.

Ans. (i) Diagram of maize grain in longitudinal section:

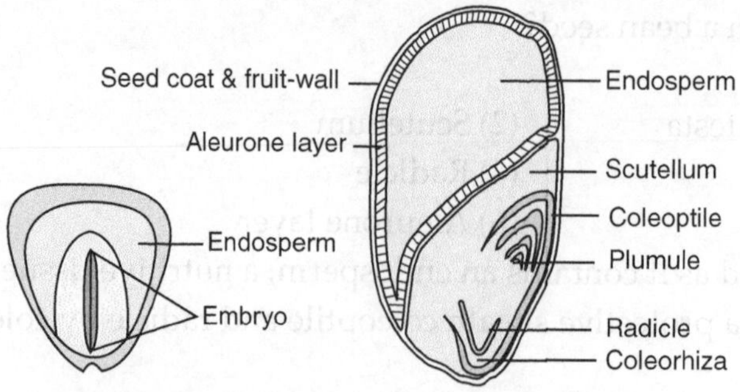

(ii) (1) Albuminous seeds (2) Scutellum
(3) Aleurone layer (4) Coleorhiza
(5) Coleoptile

7. The diagram below represents the germination of a bean seed. Study the diagram and answer the following questions: **[November, 2019]**

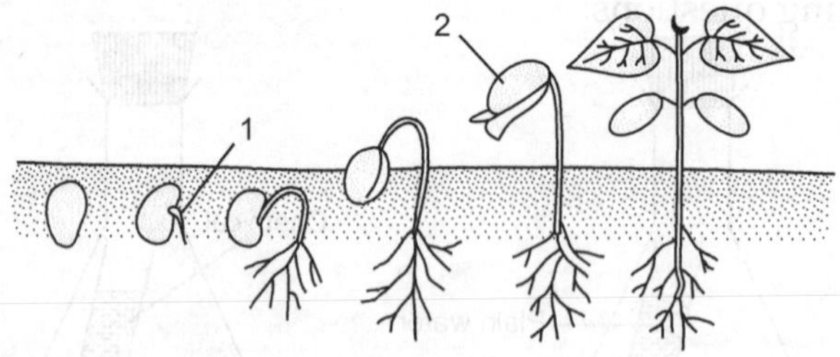

(i) Identify the type of germination. Given a suitable reason for your answer.
(ii) Label the parts 1 and 2.
(iii) Draw a neat labelled diagram of a maize grain.
(iv) Why is a maize grain not called a seed?
(v) Write any two condiations needed for germination.

Ans. (i) Epigeal germination. In this case cotyledons are pushed above the ground, hypocotyl elongates so it is an example of epigeal germination.
(ii) 1 – Radicle; 2 – Cotyledons.
(iii)

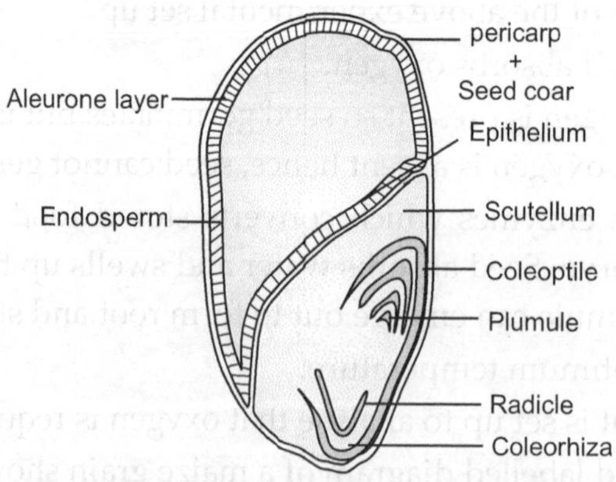

(iv) Maize grain is not called a seed as the fruit wall and seed coat are fused together to form a protective layer.
(v) Two conditions needed for germination are:
(a) Suitable temperature (b) Water and oxygen.

Diagram Based Questions

8. The diagram given below shows three bean seeds placed at different levels. Answer the following questions: **[February, 2020]**

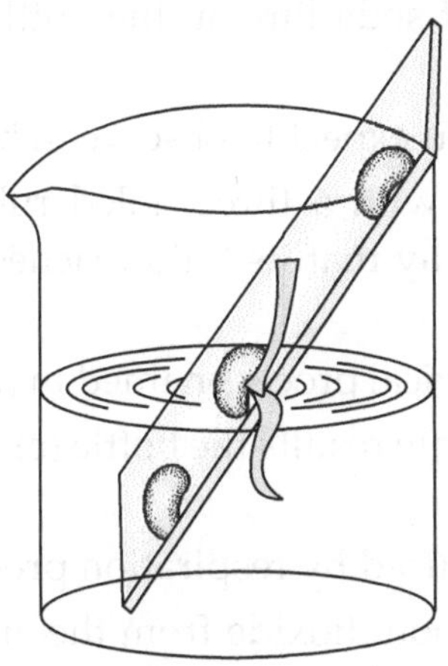

(i) What changes will you observe in the seeds A, B and C after a few days? Explain with suitable reasons.
(ii) Name the part of the seed that provides nutrition for the growing seedling.
(iii) Draw a neat, labelled diagram showing the internal structure of a bean seed.
(iv) What type of germination does a bean seed show?
(v) Mention the parts of a flower which form the fruits and seeds.

Ans. (i) Only seed B shows germination after few days but not seeds A and C. The middle seed B gets all the condition that are favourable for germination—water, oxygen and suitable temperature so it germinates. But the top seed C does not get water and bottom seed A does not get adequate oxygen hence they cannot germinate.
(ii) Cotyledons
(iii)

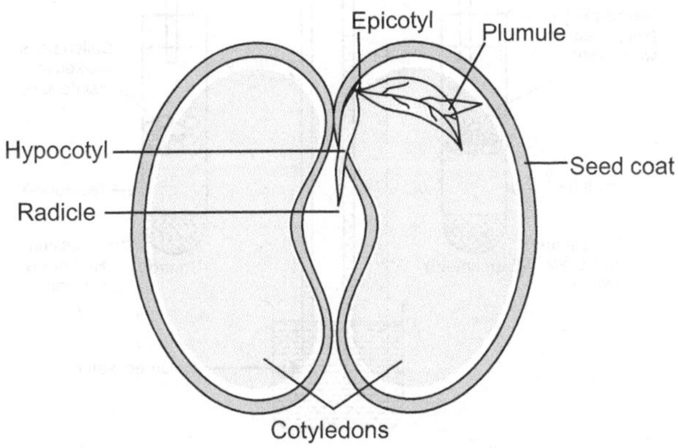

(iv) Epigeal germination.
(v) Ovary forms fruit and ovules form seeds.

Chapter 7. Respiration in Plants

1. The following set up was arranged to prove a particular physiological process in plants.

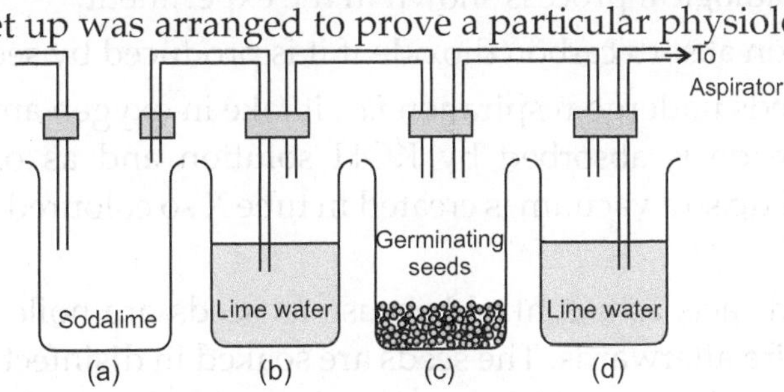

(i) Name the physiological process being studied.

(ii) What is the function of soda lime in the bottle (a) and why is lime water placed in bottle (b)?

(iii) What change would you expect to observe in bottle (b)?

(iv) If bottle (c) was fitted with a three holed rubber stopper and a thermometer was introduced in such a way that its bulb reaches close to the germinating seeds, what would you observe?

(v) Represent the physiological process named in (i) above in form of a chemical equation.

(iv) In order to obtain accurate results the bottle (c) should be covered with a piece of black cloth. Why?

Ans. (i) Carbon dioxide is produced by respiration process from germinating seeds.

(ii) Soda lime removes carbon dioxide from the air which enters the tube. Lime water is placed in bottle (b) to check whether the incoming air is free from carbon dioxide or not as it turns milky in presence of carbon dioxide gas.

(iii) Lime water turns milky as carbon dioxide is given out by germinating seeds by respiration process.

(iv) As heat is also evolved during respiration so if a thermometer is inserted in bottle (c) there would be rise in mercury level.

(v) $C_6H_{12}O_6 + 6O_2 \rightarrow 6CO_2 + 6H_2O + \text{Energy}$

(vi) To prevent photosynthesis process bottle (c) should be covered with a black cloth as this process may interfere with the results of respiration process.

2. The following diagram refers to an apparatus which is used to demonstrate a physiological process.

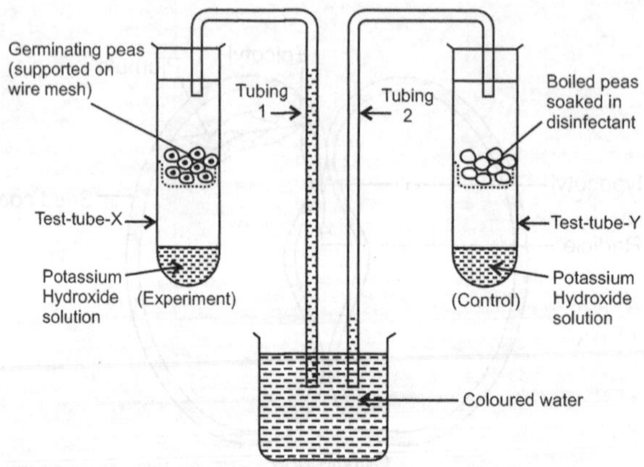

(i) What is the purpose of keeping KOH solution in the test tubes X and Y?

(ii) Why has coloured water risen in tubing 1?

(iii) What is the purpose of keeping boiled peas soaked in disinfectant in test tube Y?

(iv) Name the biological process which causes the above rise.

(v) Define the biological process shown in the experiment.

Ans. (i) KOH solution absorb carbon dioxide if it is produced by seeds.

(ii) In tube 1 seeds undergo respiration *i.e.*, it take in oxygen and releases carbon dioxide. Carbon dioxide is absorbed by KOH solution and as oxygen present in tube is slowly used up so a vacuum is created in tube X so coloured water from beaker rises in tubing 1.

(iii) Boileds seed acts as a control because as seeds are boiled they are dead and they cannot respire afterwards. The seeds are soaked in disinfectant so that there would be no growth of microbes which may interfere with the results of the experiment.

(iv) Respiration

(v) Respiration is a catabolic process by which glucose is oxidised to release energy in presence of oxygen and this energy produced is used to carry out various life processes.

3. Study the figure and answer the following questions:

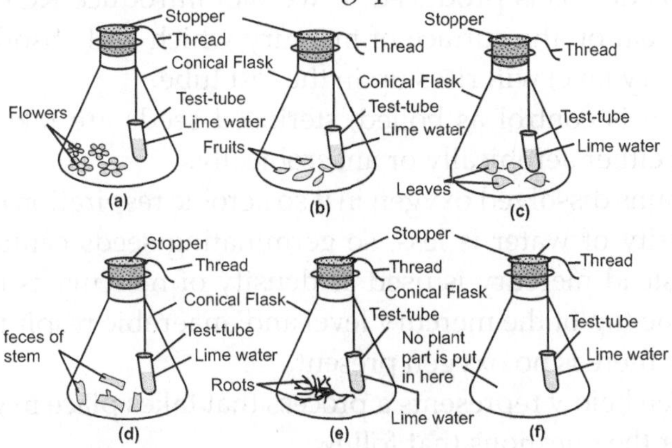

(i) In how many flasks the different plant parts are kept in observation?

(ii) What is the purpose of keeping a test-tube containing lime water in each flask?

(iii) In which test-tubes lime water will turn milky?

(iv) What is the purpose of set-up (f)?

(v) What conclusion do you draw from this experiment?

Ans. (i) Five flasks (a)-(e).

(ii) To see if carbon dioxide is produced in each flask as lime water turns milky in presence of carbon dioxide gas.

(iii) All five flasks (a)-(e) contain plant parts which on respiration produce carbon dioxide gas that will turn lime water milky.

(iv) Set up F acts as a control where all conditions are similar as in all other flasks except plant parts are missing in flask (f).

(v) We can conclude that carbon dioxide is produced by respiration from plants.

4. Study the figure below and answer the following questions:

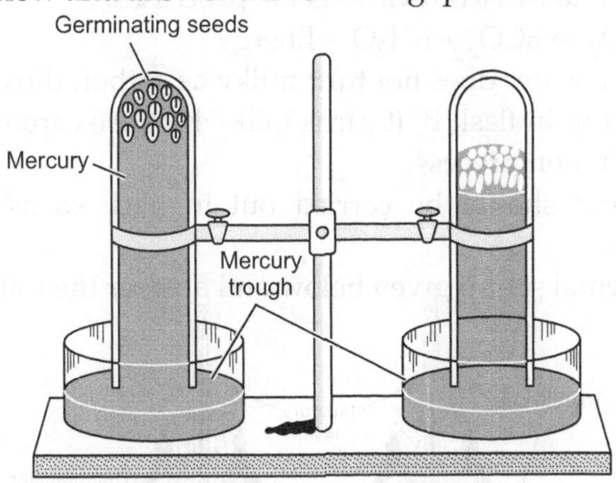

(i) What does the experiment demonstrate?

(ii) Why level of mercury in test tube B falls after few days?

(iii) Which gas is liberated by this process and how can you prove its presence?

(iv) Which is the control in this experiment and why?

(v) Why can water not be used instead of mercury in the test tube and beaker in this experiment?

Ans. (i) This experiment demonstrates anaerobic respiration in germinating seeds.

(ii) After few days mercury level falls off due to anaerobic respiration where carbon dioxide is produced which will occupy the space in the top of test tube and mercury level falls down.

(iii) Carbon dioxide gas is produced. If we will introduce KOH solution in test tube (b) KOH will float on the surface of mercury which will absorb carbon dioxide gas and again mercury level will rises up in the test tube.

(iv) Test tube (a) is control as boiled, sterilised seeds are taken which cannot undergo respiration either aerobically or anaerobically.

(v) Water contains dissolved oxygen in it so aerobic respiration can occurs by germinating seeds. Density of water is less, so germinating seeds cannot reach the top of water surface. Instead mercury is used as density of mercury is more than water so seeds can reach the top of the mercury level and anaerobic respiration can be demonstrated properly as there is no oxygen present.

5. The diagram given below represents a process that takes place in green plants when kept in the dark. Answer the questions that follow: **[November, 2019]**

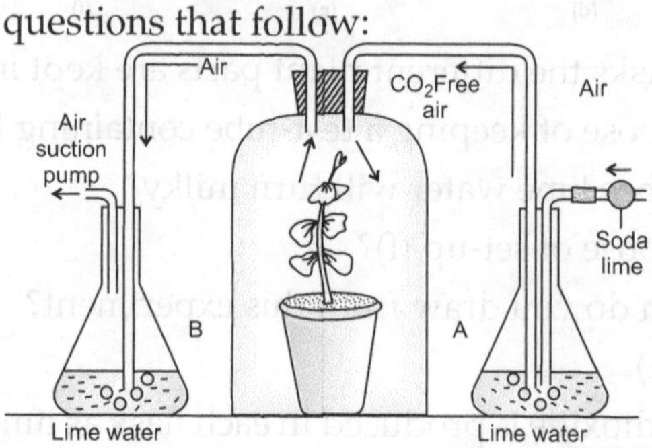

(i) Name the phyisological process tha occurs in the experiment.

(ii) Define the above mentioned process.

(iii) Represent the process in the form of a balanced chemical equation.

(iv) What happen ot the line water in flasks A and B.

(v) Why should the experiment be carried out in the dark?

Ans. (i) Respiration.

(ii) Respiration is a catabolic process where oxidation of glucose occurs to release energy in form of ATP to carry out various life processes.

(iii) $C_6H_{12}O_6 + 6O_2 \rightarrow 6CO_2 + 6H_2O + \text{Energy}$

(iv) In flask A lime water does not turn milky as carbon dioxide present in air is trapped by soda lime but in flask B, it turns milky because carbon dioxide is given by plants through respiration process.

(v) The experiment should be carried out in dark so as to prevent the process of photosythesis.

6. Study the experimental set-up given below and answer the following questions:

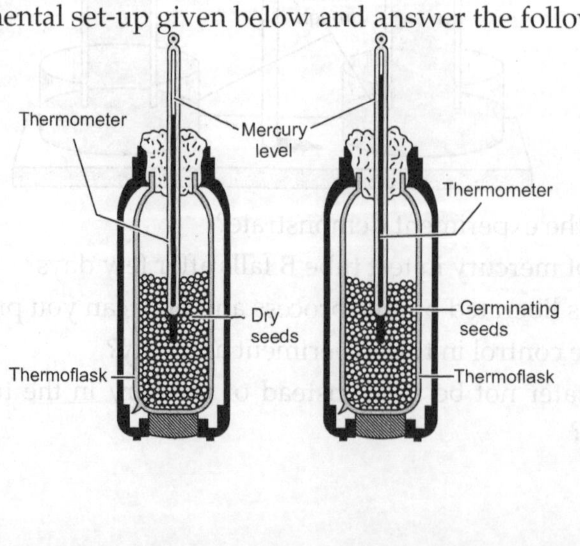

(i) What is the aim of the experiment shown above?
(ii) What is your observation after a few hours for flask (a) and (b)?
(iii) Name the chemical used to prevent bacterial growth. Explain how the bacteria would interfere with the experiment.
(iv) Why do we use thermos flasks specifically for the experiment?
(v) Which is the control set-up and why?

Ans. (i) Heat is liberated by respiration process from germinating seeds.

(ii) In flask (a) as seeds are alive so they undergo respiration and heat is evolved which is marked by rise in mercury level of thermometer. But in flask (a) as seeds are dead, so no respiration and no heat is produced. So, mercury level in thermometer will not rise.

(iii) Carbolic acid.

Bacteria will undergo respiration and will produce heat that will cause rise in level of mercury of thermometer. Thus, the bacterial growth may interfere with results of the experiment.

(iv) Thermos flask acts as an insulator. It does not allow outside environmental temperature to interfere with the results of experiment.

(v) Flask (b) is control set up as seeds are dead, so there is no respiration hence no production of heat.

7. Give a diagrammatic representation to show how the energy liberated in respiration process is used up?

Ans. During respiration process, energy is liberated. A small amount of energy is lost to the environment in form of heat. Rest amount of energy is converted into ATP (chemical energy). This ATP is required to carry out various metabolic activities.

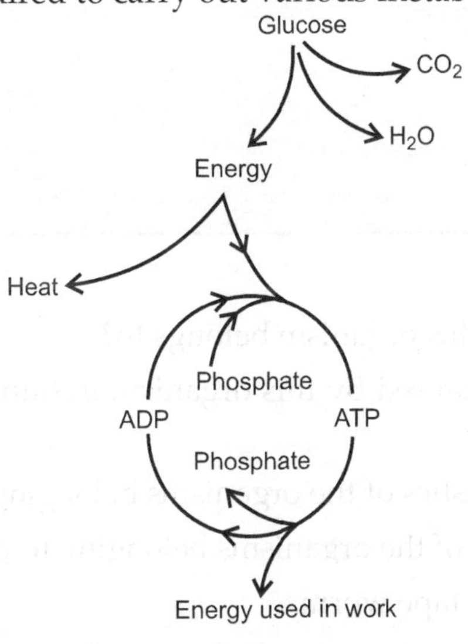

8. Given below is an experimental set-up to demonstrate a particular process in animal. Anser the questions that follow: **[February, 2020]**

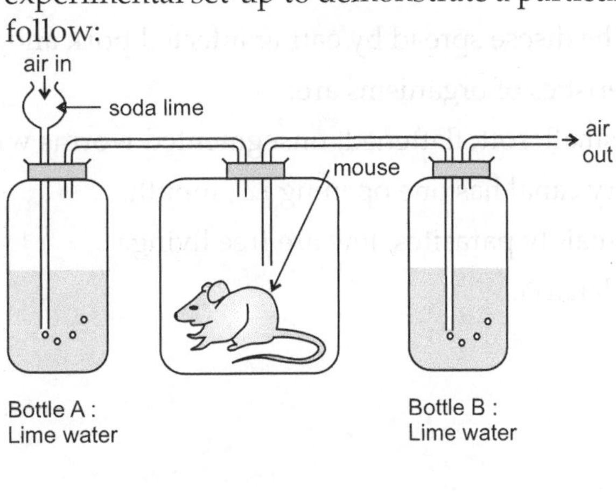

(i) Name the phyisological process being studied.

(ii) Define the process mentioned in (i).

(iii) What is the function of soda lime.

(iv) After a few hours what will happen to the lime water in Bottle A and in Bottle B? Give suitable reasons for your answer.

(v) Write a balanced chemical equation for the process carried out by the mouse.

Ans. (i) Respiration.

(ii) Respiration is a catabolic process where energy is released by breaking down of glucose to carry out various life processes.

(iii) Soda-lime absorbs carbon dioxide gas form the incoming air.

(iv) Lime water in bottle A will not turns milky as soda-lime absorbs carbon dioxide gas form incoming air. But lime water in bottle B will turns milky as carbon dioxide gas is given out by respiration process in the mouse.

(v) $C_6H_{12}O_6 + 6O_2 \rightarrow 6CO_2 + 6H_2O$ + Energy

Chapter 8. Five kingdom classification

1. Study the figure below and answer the following:

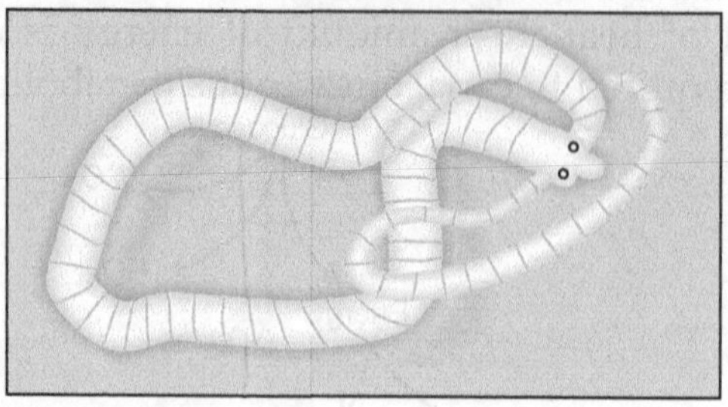

(i) Identify the figure.

(ii) To which phylum this organism belongs to?

(iii) Name the disease caused by this organism in human beings. How does this disease mainly spread?

(iv) Give few characteristics of the organisms belonging to phylum mentioned in part (ii).

(v) Give few examples of the organisms belonging to phylum mentioned in part (ii).

Ans. (i) It is *Taenia Solium* a tape worm.

(ii) Platyhelminthes

(iii) Taeniasis is the disese spread by eating infected pork and beef.

(iv) Few characteristics of organisms are:

(1) They are small, soft, flattened, unsegmented worms without a body cavity.

(2) Alimentary canal has one opening *i.e.*, mouth.

(3) They are mainly parasites, few are free living.

(v) Liverfluke, Planaria.

2. Study the diagram and answer the following questions:

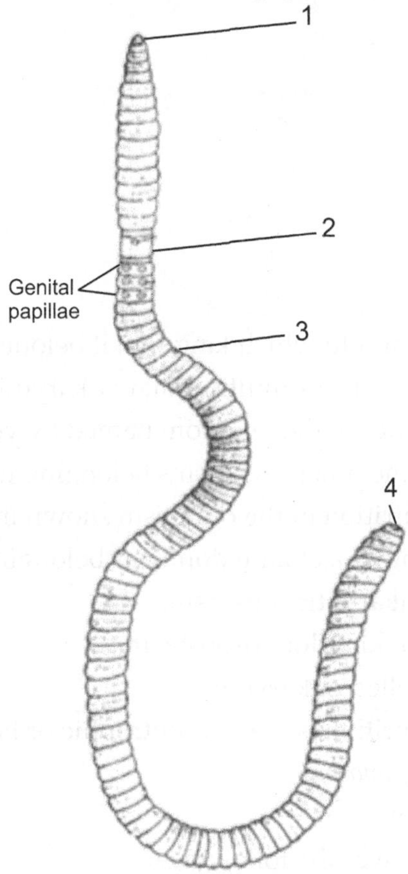

(i) Identify the organism. Give its scientific name.

(ii) Name the phylum to which it belongs. Give some characteristics features of organisms belonging to the phylum named by you.

(iii) Label parts 1-4.

(iv) What is the function of part 2?

(v) Give the economic importance of this organism.

Ans. (i) It is an Indian earthworm. Scientific name is *Pheretimaposthuma*.

(ii) Figure belongs to phylum Annelida. Some characteristics features of the organisms belonging to this phylum are:

(1) Their body is cylindrical and divided into ring-like segments.

(2) They have well developed digestive system with alimentary canal open at mouth and anus.

(3) They have a true body cavity.

(iii) 1- Mouth; 2- clitellum; 3- body segment; 4- anus

(iv) Clitellum plays an important role in reproduction process.

(v) Some economic importance of this organism are:

(1) They help in loosening and aeration of soil.

(2) It increases the fertility of the soil by interchanging top layer of soil with inner layers.

(3) It is used in vermiculture.

(4) Its excreta is rich in nitrogenous matter which is essential component for plant's growth.

3. Study the below figure and answer the following questions:

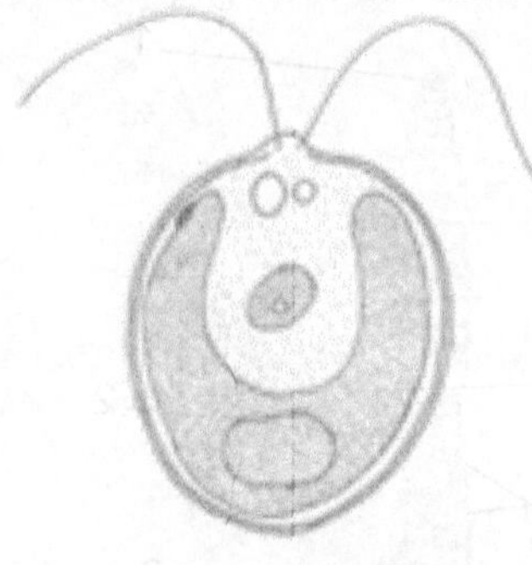

(i) Name the organism and to which kingdom it belongs to.
(ii) Is it unicellular eukaryotic or multicellular eukaryotic organism?
(iii) Give few characteristics of the kingdom named by you in part (i) above.
(iv) Give examples of some other organisms belonging to this category.
(v) Name the type of nutrition in the organism shown in diagram.

Ans. (i) The diagram showing is a chlamydomonas belonging to kingdom Protista.
(ii) This is unicellular eukaryotic organism.
(iii) Few characteristics of kingdom protista are:
(1) They are single celled eukaryotes.
(2) Their mode of nutrition is either autotrophic or heterotrophic.
(iv) *Euglena, paramecium, amoeba*
(v) Autotrophic nutrition.

4. Study the diagram and answer the following-

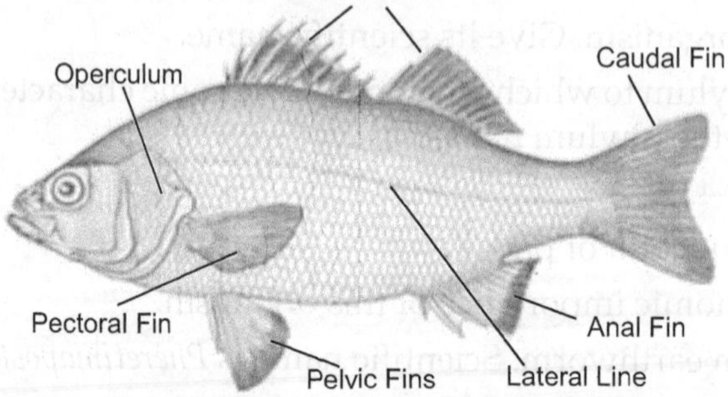

(i) Identify the figure. State its class, phylum, kingdom.
(ii) What type of adaptation does this creature have?
(iii) What are its types? Differentiate between these types.
(iv) Is it cold blooded or warm blooded?

Ans. (i) The diagram showing is a bony fish which belongs to class pisces, phylum chordata and kingdom Animalia.
(ii) They breathe by means of gills, have streamlined body, body covered with scales, have fins. They are cold blooded *i.e.,* temperature of body directly fluctuates with temperature of environment.
(iii) They are mainly of two types:
(1) Cartilaginous fishes whose skeleton is made up of cartilage. Gill slits are exposed without gill cover. Examples; sharks, dogfish, etc.
(2) Bony fishes whose skeleton is made up of bones. Gill slits are covered with operculum. Examples; carps, roaches, etc.
(iv) No, it is cold blooded organism.

5. Study the diagrams and answer the following questions:

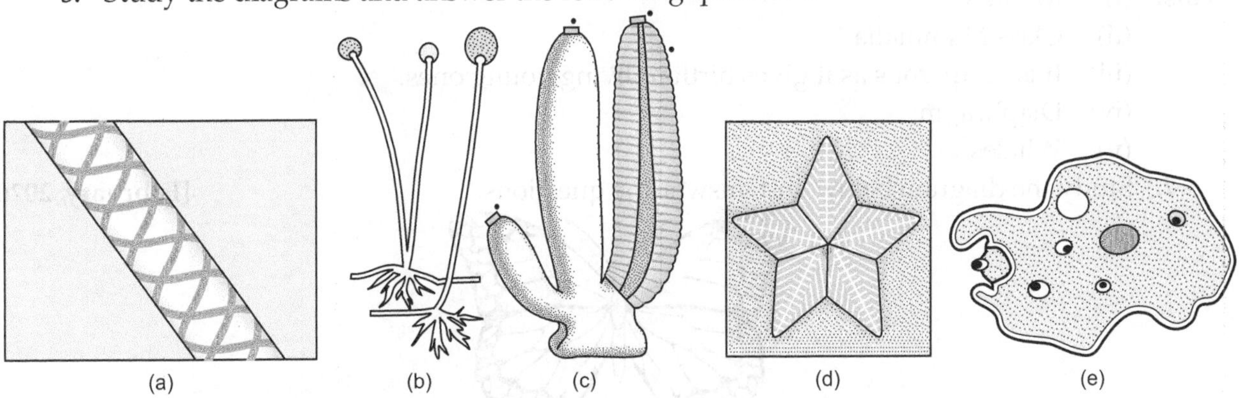

(a) (b) (c) (d) (e)

(i) Identify the figures
(ii) State the phylum/kingdom/class they belong to.
(iii) Give some unique features of these organisms shown in figure.

Ans. (i) (a) Spirogyra
(b) Rhizopus
(c) Sycon
(d) Starfish
(e) Amoeba

(ii) (a) Spirogyra belongs to Kingdom Plantae, division Thallophyta
(b) Rhizopus belongs to kingdom Fungi.
(c) Sycon belongs to Kingdom Animalia and Phylum Porifera
(d) Starfish belongs to Kingdom Animalia and phylum Echinodermata
Amoeba belongs to Kingdom Protista

(iii) (a) Spirogyra has a thallus like body and its chloroplast is spiral ribbon like shape.
(b) Rhizopus has thread like structures called hyphae and is heterotrophic in nature.
(c) Sycon has many pores present on its body surface and its skeleton is made up of spicules.
(d) Starfish has water vascular system and moves by tube feet and is exclusively marine organism.
(e) Amoeba is unicellular eukaryote which moves by forming pseudopodia.

6. Study the diagram below and answer the questions: **[November, 2019]**

(i) Identify the animal
(ii) Name the class to which it belongs.
(iii) Is it oviparous or viviparous? Explain your answer.
(iv) Mention the structure that separates the thorax form the abdomen.
(v) Give one example of an animal which belongs to the same class as mentioned in (ii) but leads an aquatic life.

Ans. (i) Monkey
(ii) Class Mammalia
(iii) It is viviparous as it gives birth to living young ones.
(iv) Diaphragm
(v) Whales.

7. Study the diagram below and answer the questions: **[February, 2020]**

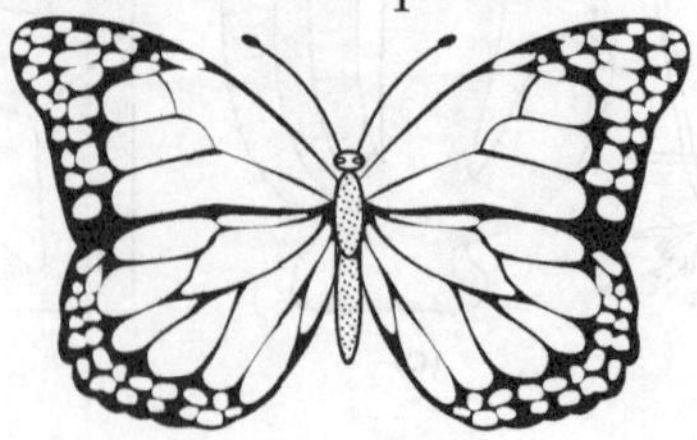

(i) Identify the organism.
(ii) Name to phylum it belongs to.
(iii) What is its exoskeleton made up of.
(iv) Explain the term 'Moulting'.
(v) What are Vertebrates?

Ans. (i) Butterfly
(ii) Arthropoda
(iii) Chitin
(iv) Moulting is the process of casting off and re-growing of the exoskeleton.
(v) Chordates which posses a back bone are called vertebrates.

Chapter 9. Economic importance of Bacteria and Fungi

1. Study the diagram given below and answer the following questions:

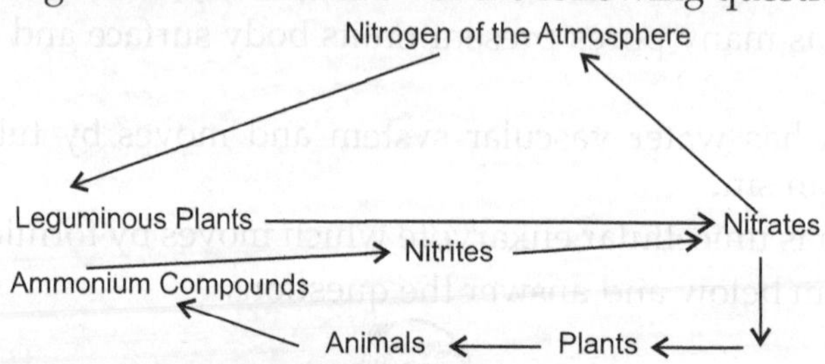

(i) What does the diagram represent?
(ii) How leguminous plants help in nitrogen fixation?
(iii) Name the bacteria that converts:
(1) Ammonium compounds to nitrites
(2) Nitrites to nitrates
(iv) How nitrates in the soil converted to nitrogen?
(v) How plants and animals play an important role in formation of ammonium compounds?

Ans. (i) Nitrogen cycle
(ii) Leguminous plants contain *Rhizobium* in their root nodules which pick up free nitrogen from atmosphere and convert into soluble nitrates which can be used by plants. This process is called nitrogen fixation.
(iii) (1) Nitrosomonas
(2) Nitrobacter
(iv) Denitrifying bacteria are present in soil which converts nitrates present in soil into nitrogen gas which is released into atmosphere.

(v) Nitrogenous wastes are present in dead remains of plants and animals which are converted into ammonia by bacteria. Then ammonia is converted into ammonium compounds. In this way plants and animals play an important role in formation of ammonium compounds.

2. Study the figure and answer the following questions:

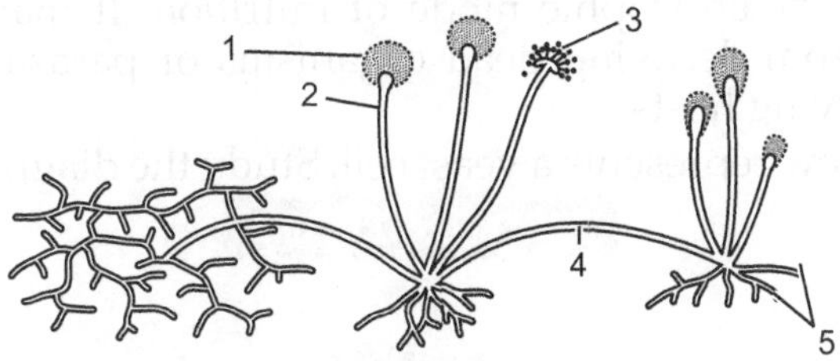

(i) Identify the figure.

(ii) To which kingdom does it belongs to?

(iii) Label the parts 1-5.

(iv) What is the function of part labelled 3 and 5.

(v) How does it reproduces?

Ans. (i) It is a bread mould or *Rhizopus.*

(ii) *Rhizopus* belongs to kingdom fungi.

(iii) 1- Sporangium; 2- Sporangiophores; 3- Spores; 4- Stolon; 5- Hyphae

(iv) Sporangium bursts to release the spores which gets scattered by wind and under suitable conditions like temperature, substratum, moisture the spores germinate to give rise to new mycelium.

Hyphae penetrate into the substratum like bread and secrete some enzymes which digest the food in the bread and then it is absorbed by rhizopus.

(v) *Rhizopus* reproduces by asexual reproduction.

3. The figure given below shows a cell. Study the figure and answer the following questions:

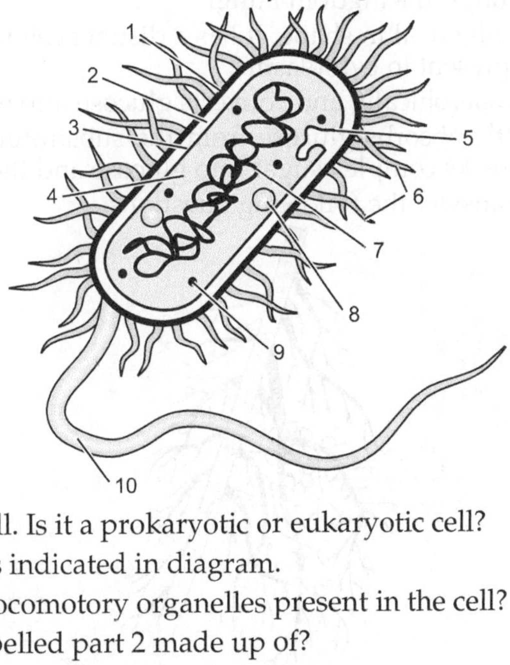

(i) Identify the cell. Is it a prokaryotic or eukaryotic cell?

(ii) Label the parts indicated in diagram.

(iii) What are the locomotory organelles present in the cell?

(iv) What is the labelled part 2 made up of?

(v) How does this cell obtain its nourishment?

Ans. (i) The diagram is of a bacterial cell. It is a prokaryotic cell.
(ii) 1- Capsule; 2- Cell wall; 3- Cell membrane; 4- Cytoplasm; 5- Mesosome; 6- Pilli; 7- Bacterial chromosome; 8- Plasmid; 9- Glycogen granules; 10- Flagellum
(iii) Flagellum is the locomotory organelle. Lashing movement of flagellum provide locomotion of bacteria in a liquid environment.
(iv) Cell wall of bacteria is made up of peptidoglycan.
(v) Bacteria shows heterotrophic mode of nutrition. It may be saprotrophic *i.e.,* draws nourishment from decaying dead organisms or parasitic *i.e.,* it draws nourishment from body of living hosts.

4. The figure given below represents a yeast cell. Study the diagram and answer the following-

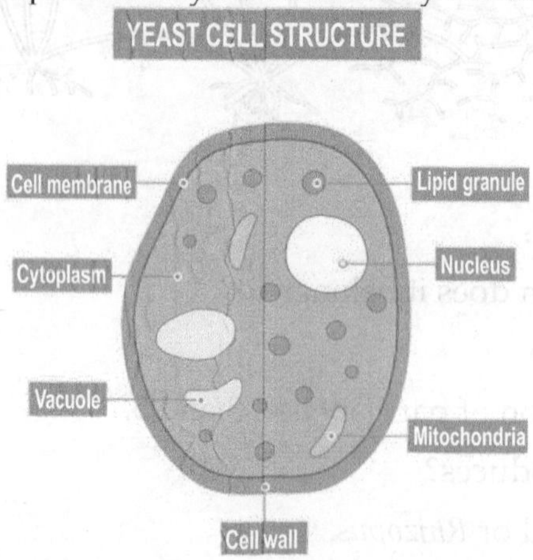

(i) What is the importance of this organism to mankind?
(ii) Where are they found and to which kingdom does it belongs to?
(iii) Describe the structure of yeast.
(iv) How does it respire?
(v) How does it obtain its nutrition?

Ans. (i) Yeast has great importance to mankind due to its property of fermentation. It respires anaerobically and converts glucose into ethanol. So, it is widely used in wine industries. It is also used in baking industry as carbon dioxide evolved by yeast during anaerobic respiration makes the dough to increase in volume and gives the bread a light spongy texture.
(ii) Yeast are found freely in atmosphere and they can grow easily in all kinds of sugary solutions. It belongs to Kingdom Fungi.
(iii) Yeast cell is usually oval in shape. It has a distinct cell wall, nucleus. There are one or more vacuoles present in cytoplasm.
(iv) Yeast respires anaerobically and converts glucose into ethanol and carbon dioxide.
(v) Yeast cell directly absorbs glucose from the substratum where it grows. It releases enzymes that breaks complex sugar into glucose and then it absorbs it.

5. Study the figure and answer the following questions:

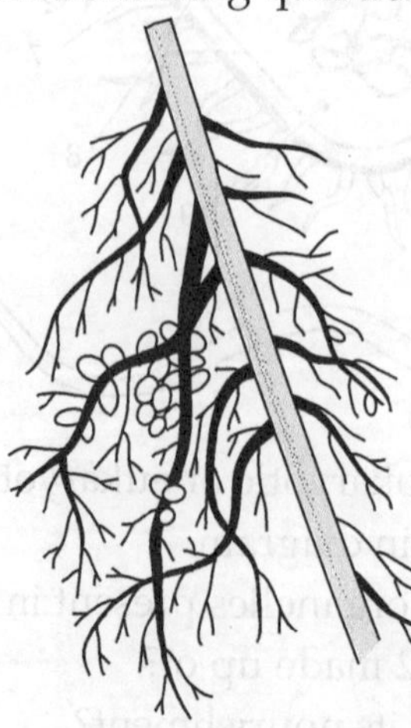

(i) What does the figure represent?

(ii) Name the bacteria present in it and write its importance.

(iii) Give some examples of plants where this bacteria is present and name the relationship that exists between this bacteria and plant.

Ans. (i) It represents root nodules of leguminous plants.

(ii) Nitrogen fixing bacteria like *Rhizobium* is present in the root nodules. They convert free nitrogen of the soil atmosphere into soluble nitrates so that it can be utilised by plants easily. This process is called biological nitrogen fixation.

(iii) This bacteria is found in leguminous plants like beans, peas, etc.

This type of relationship is known as symbiotic relationship where both bacteria and plants are mutually benifitted from each other.

6. (i) Identify the figure given below and label the parts as indicated in diagram.

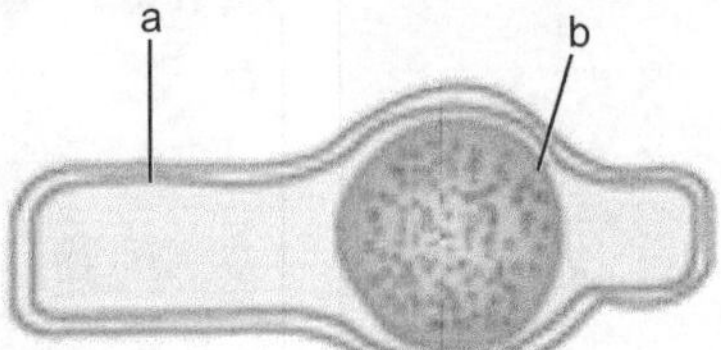

(ii) What is the significance of this structure?

Ans. (i) The figure indicates a spore of bacteria.

(1) Bacterial wall; (2) Spore wall

(ii) During unfavourable conditions like dryness, high temperature etc the content of the bacterial cell is drawn up into a spherical mass which gets surrounded by a thick protective wall and this is called a spore so that the bacterial cell can withstand the unfavourable conditions.

7. The figure shows growth of mould on a piece of moist bread which was covered under a small bell jar and kept at a warm place for few days. Answer the following with reference to this observation.

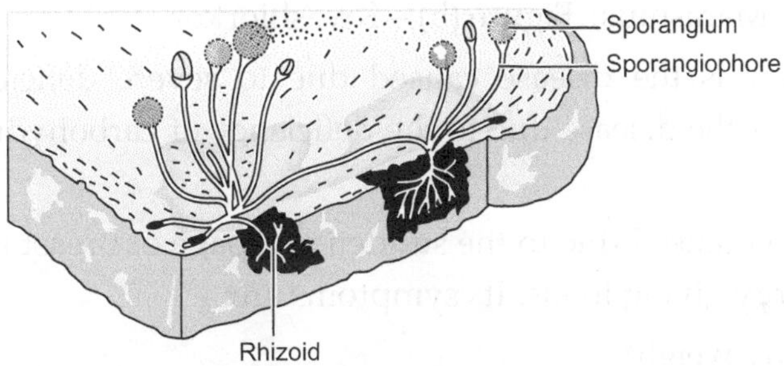

(i) How did the mould get inside the bell jar?

(ii) What would happen if the bread was not covered by the bell jar?

(iii) What would happen if moist bread was placed in a refrigerator?

(iv) What appears first on the bread- the mycelia or the spores?

(v) How does bread mould obtain nourishment? State the type of nourishment of the bread mould?

Ans. (i) Moulds are present everywhere in air, water, etc. So they were already present on the surface of moist bread.

(ii) If moist bread was not covered by a bell jar mould would still have grown but it would have taken more longer time for their growth. The covered bell jar provides suitable warmth and humid conditions so mould grew on the bread in a shorter period of time.

(iii) There would be no growth of mould on moist bread as mould could not grow in lower temperature.

(iv) Mycelia first appears on the bread.

(v) Bread mould obtain its nourishment by extracellular digestion from the substratum on which it grows. This nourishment is called saprophytic mode of nutrition.

Chapter 10. Nutrition

1. Study the diagrams and answer the following questions:

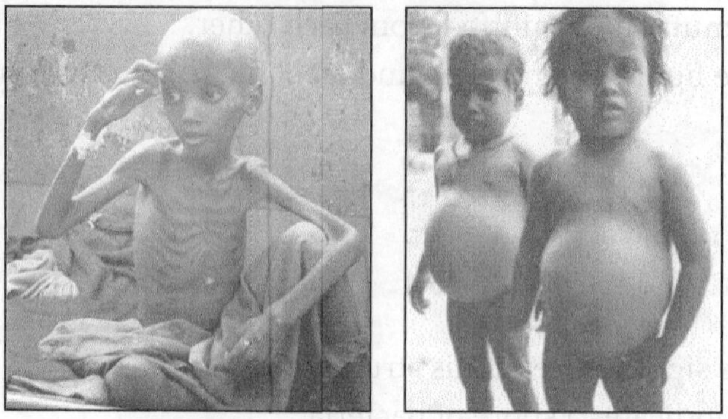

Figure (a) Figure (b)

(i) Identify the kind of disease both the children are suffering from?

(ii) These diseases are due to deficiency of which nutrients?

(iii) What are the causes and symptoms of disease shown in figure (a)?

(iv) What are the causes and symptoms of disease shown in figure (b)?

(v) What are the control measures for both the diseases?

Ans. (i) Figure (a) - Marasmus; Figure (b) – Kwashiorkor

(ii) Kwashiorkor is the disease caused due to severe deficiency of proteins in diet. Marasmus is the disease due to the deficiency of carbohydrates, fats and proteins in diet.

(iii) Marasmus is caused due to the sudden stoppage of breast feeding, followed by poor diet in energy giving foods. Its symptoms are:

(1) Less body weight

(2) Thin face, thinning of limbs, ribs are readily noticeable.

(3) Retarded mental and physical growth

(iv) Kwashiorkor is caused when mothers stop beast feeding their babies at an early age and children get poor diet in proteins. Its symptoms are:

(1) Underweight

(2) Belly protruding out

(3) Repeated diarrhoea

(4) Enlarged liver and anaemia

(v) The children suffering form kwashiorkor must intake proteins in their diet like pulses, eggs, fish, meat, etc. The children suffering from Marasmus should intake proteins and carbohydrates in their diet.

Diagram Based Questions

2. Study the flow chart related to nutrients and answer the following questions:

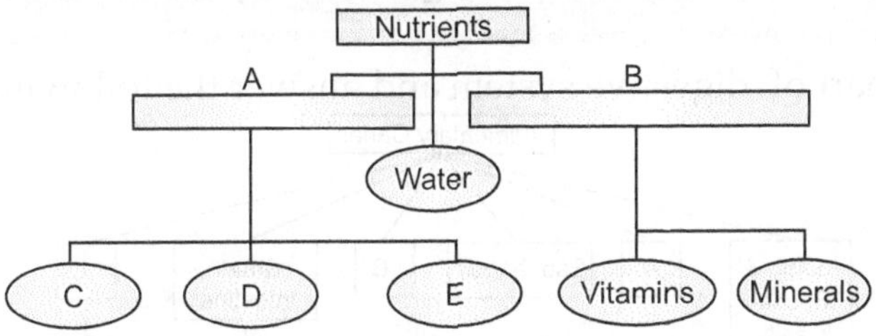

(i) Label the part A and B.
(ii) What is the common characteristic on basis of which they have been labelled as A and B differently?
(iii) Label the different components/ types of nutrients?
(iv) Why is water a nutrient?
(v) What will you call a diet which has all of these components?

Ans. (i) A- Macronutrient, B- Micronutrient
(ii) Amount in which the nutrients are required in diet is the common characteristic to classify nutrients.
(iii) C- Carbohydrates , D- Proteins , E- Fats
(iv) Water contains dissolved salts and it is important for proper functioning of body. Thus, water also acts as nutrient.
(v) A balanced diet.

3. Study the diagram of food pyramid given below and answer the following questions:

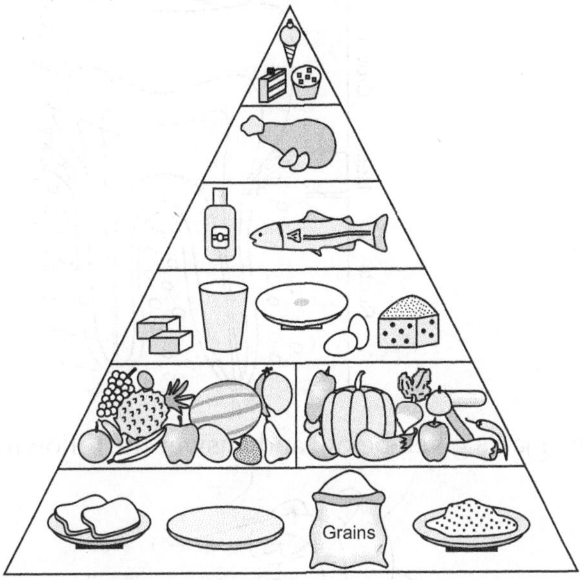

(i) What is shown in the diagram?
(ii) What kind of food is put in upper part of pyramid?
(iii) How is the energy derived from food measured?
(iv) How much energy per day is required by an adult man at rest?
(v) Why energy required by adult woman at rest is different from your answer in part (iv)?

Ans. (i) The pyramid shows various kinds of food that are required for a healthy life.
(ii) The upper part of pyramid contains food with high calorific value.
(iii) By measuring its calorific value.
(iv) 1650 Kcal
(v) Women have lower basal metabolic rate due to which they require lesser calories to do normal functions.

Chapter 11. Digestive System

1. Study the flowchart of digestive system and answer the following questions:

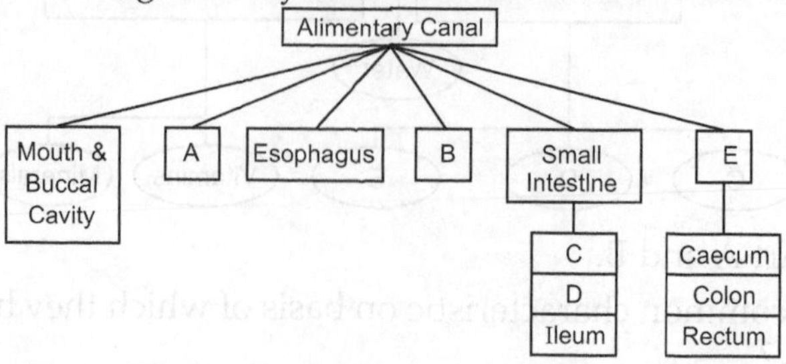

(i) Label the part A and give its one function .
(ii) Label the organ B. How it disinfects the food?
(iii) Label C. What is its shape?
(iv) What is D? Which glands secretion is (are) poured into this part?
(v) Name the organ E. What is absorbed in this part?

Ans. (i) A-(mouth and buccal cavity). Its functions are chewing, swallowing, digestion and absorption of simple sugars.
(ii) B-Stomach; By acid produced in stomach which disinfects the food.
(iii) C-Duodenum ; J shaped
(iv) D-Jejunum ; Liver and pancreas secretions are poured into Jejunum.
(v) D-Large intestine. Water is absorbed in large intenstine

2. Draw a labelled diagram to show the internal structure of a mammalian tooth with two roots:

Ans.

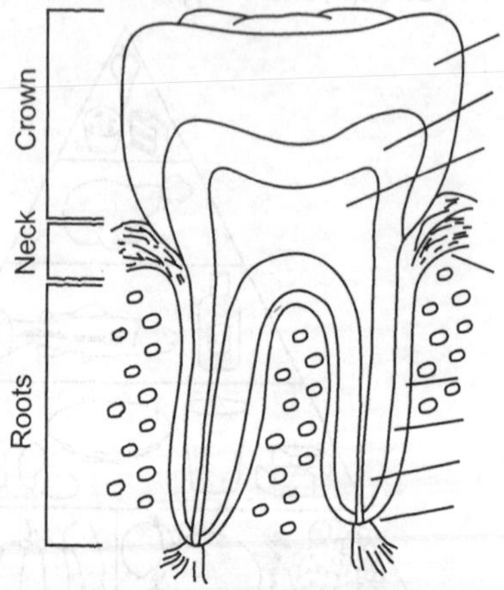

3. Study the diagram of jaws given below and answer the following questions:

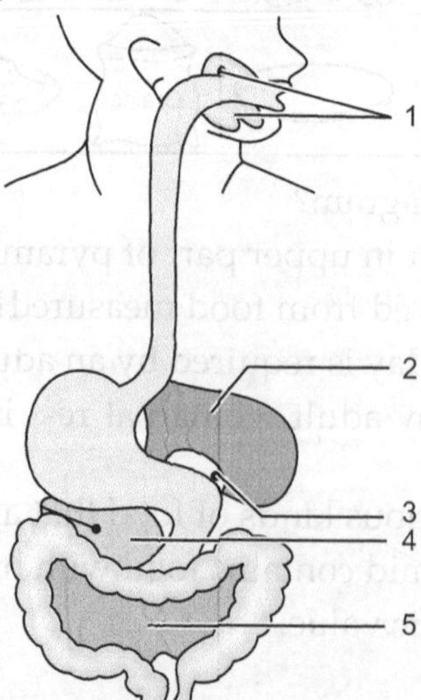

(i) Identify the part labelled 1 in above picture.
(ii) Label the organ 2 and give name of its secretion.
(iii) Label 3. Explain the reason for green colour of 3?
(iv) What is 4. how it pour its secretion in digestive tract?
(v) What glands are present in organ 5.

Ans. (i) 1- Salivary gland
(ii) 2- Liver , Bile is secreted from liver.
(iii) 3- Gall bladder; Bilirubin and Bileverdin pigment is responsible for green coloured gall bladder.
(iv) 4- Pancreas; pancreatic duct
(v) 5- Intestinal glands

4. Study the diagram given below and answer the following questions:

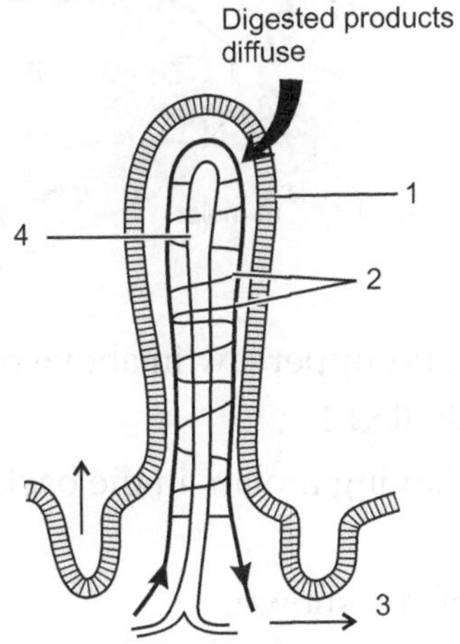

(i) What does this diagram show?
(ii) What is the function of this organ?
(iii) What type of lining is present on the structure?
(iv) What type of nutrients is absorbed by blood capillaries?
(v) Which part absorb fats?

Ans. (i) Villus
(ii) Absorption of nutrients
(iii) Single celled epithelium.
(iv) Carbohydrates and amino acids
(v) Lacteale

5. Study the diagram of incisor tooth and answer the follwing questions:

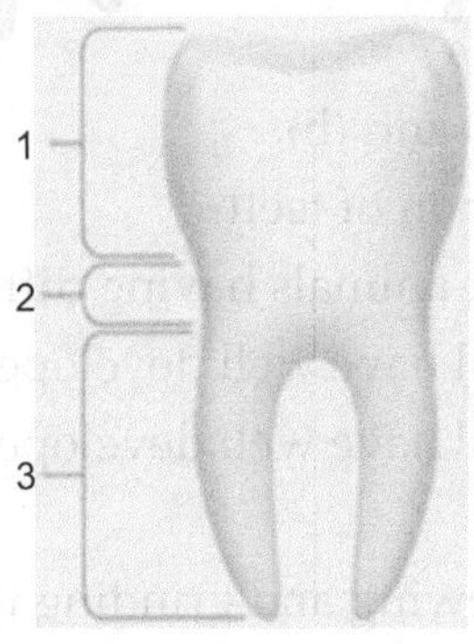

(i) Label the parts 1, 2, 3 ?
(ii) What is the function of part labelled 3?
(iii) Name the part of tooth which helps in cutting?

(iv) How do you recognize this type of tooth?

(v) Name the fleshy part which protects Part 3?

Ans. (i) 1- Crown , 2- Neck, 3- Root

(ii) Roots hold tooth in its pace

(iii) 1- Crown

(iv) Cutting edge and two roots

(v) Gum

6. Study the diagram of jaws given below and answer the questions that follow:

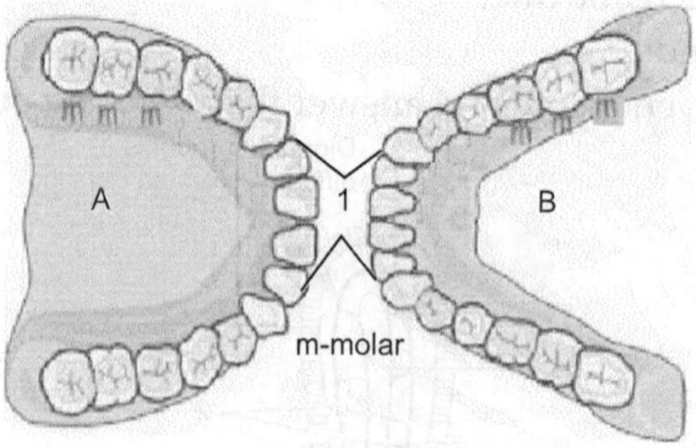

(i) Identify the lower jaw and upper jaw in above picture.

(ii) Label the front teeth labelled 1 .

(iii) Explain the reason for having molars in the back side of jaw.

(iv) What is dental formula?

(v) Write dental formula of jaw shown.

Ans. (i) A- Upper jaw and B- lower jaw

(ii) 1 – Incisors

(iii) The molars are in the back side of jaw because they help in grinding food.

(iv) Arrangement of teeth on half of upper and lower jaw is the dental formula.

(v) $\frac{2123}{2123}$ = 32 teeth

7. Study the diagram of different types of teeth given below and answer the questions that follow:

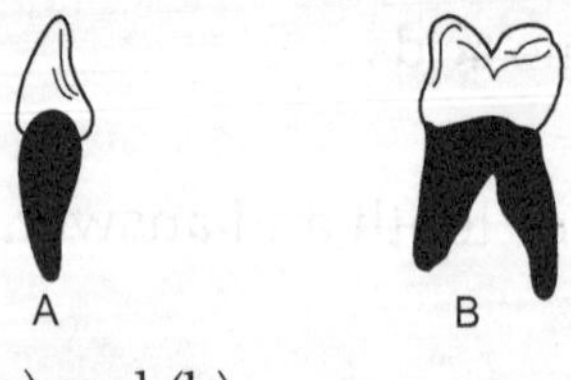

(i) Label the teeth named (a) and (b).

(ii) What is the general function of teeth?

(iii) What is the term used for animals having different type of teeth in their jaws?

(iv) What type of animals will have well developed Type (a) teeth?

(v) What type of animals will have well developed Type (b) teeth?

Ans. (i) (a)-Canine and (b)-molar

(ii) Teeth help in cutting, chewing and grinding the food materials.

(iii) Hetrodont

(iv) Flesh eating animals

(v) Grass eating animals

8. The diagram given below shows the internal structure of a tooth. **[November 2019]**
Answer the questions that follow:

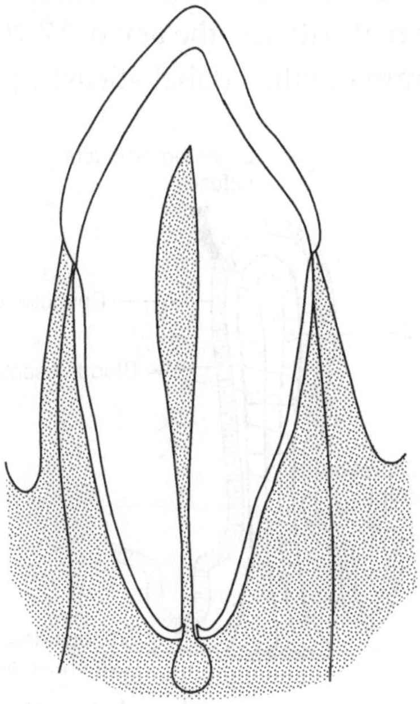

(i) Copy the diagram and label the following part:
Enamel, Cement, Dentine, Pulp cavity

(ii) What is the function of canines?

(iii) How many teeth does an adult human have?

(iv) Write the Dental Formula of an adult human.

(v) Why are the last molars in each jaw called the 'Wisdom tooth'

Ans. (i)

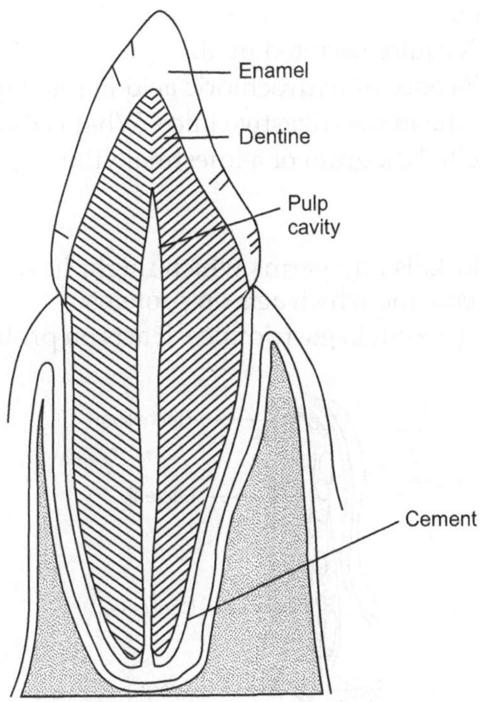

(ii) Canines help in holding and tearing the food.

(iii) 32

(iv) $\frac{2, 1, 2, 3}{2, 1, 2, 3}$

(v) The last molars in each jaw are called wisdom tooth because they appears when human body reaches maturity *i.e.*, the age of 17-20 years.

9. Draw a well-labelled diagram of villus (label atleast 5 parts).

Ans.

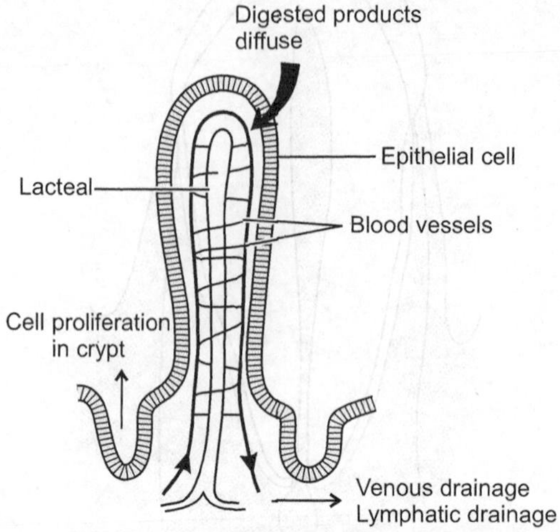

Microscopic structure of an interstinal villus

10. The picture given below is a part of the alimentary canal. **[February 2020]**
Answer the questions that follow:

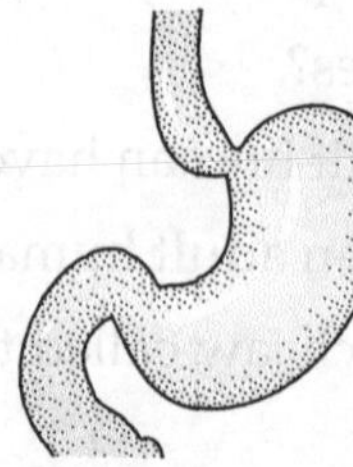

(i) Identify the organ.
(ii) Name the digestive juice secreted by it.
(iii) What is the significance of hydrochloric acid in the digestive juice?
(iv) Mention the enzyme in the digestive juice. What is its action in food?
(v) Draw a neat, labelled diagram of a intestinal villus.

Ans. (i) Stomach
(ii) Gastric juice
(iii) Hydrochloric acid kills any germs which might have entered along with food and it activates pepsin enzyme which acts on proteins.
(iv) Pepsin enzyme is present in gastric juice. It acts on proteins and converts into peptides.
(v)

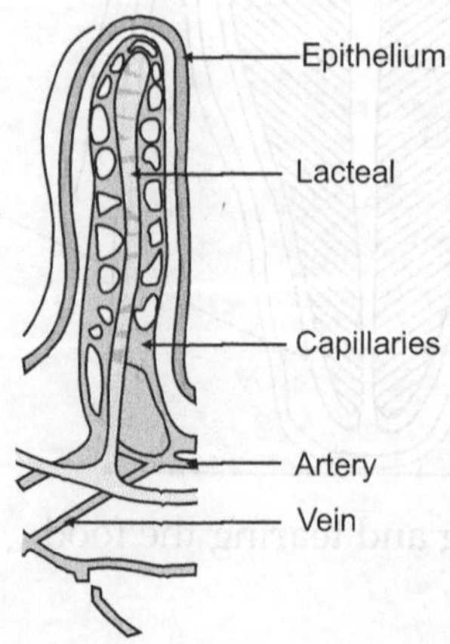

Chapter 12. Skeleton-Movement and Locomotion

1. Study the diagram of a joint and answer the questions that follow:

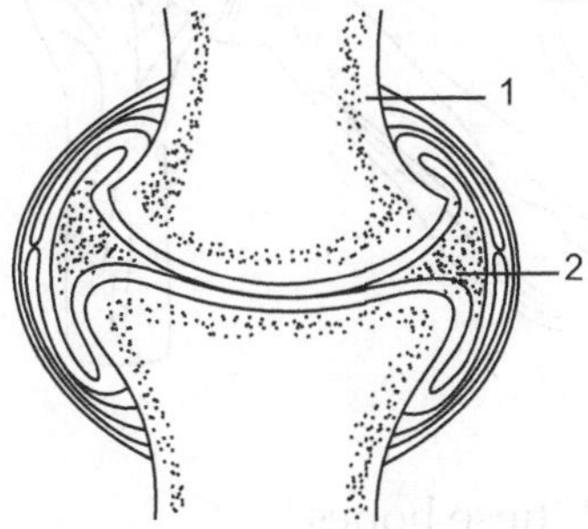

(i) Name the part -1 and 2.
(ii) What is the kind of joint shown?
(iii) Name the part of human body where it is located?
(iv) What fluid is contained in the cavity?
(v) What is the function of the fluid?

Ans. (i) 1-Epiphyseal bone , 2-Synovial cavity
(ii) Synovial joint
(iii) Knee
(iv) Synovial fluid
(v) Fluid lubricates the joint.

2. Study the diagram of pelvis given below and answer the questions that follow:

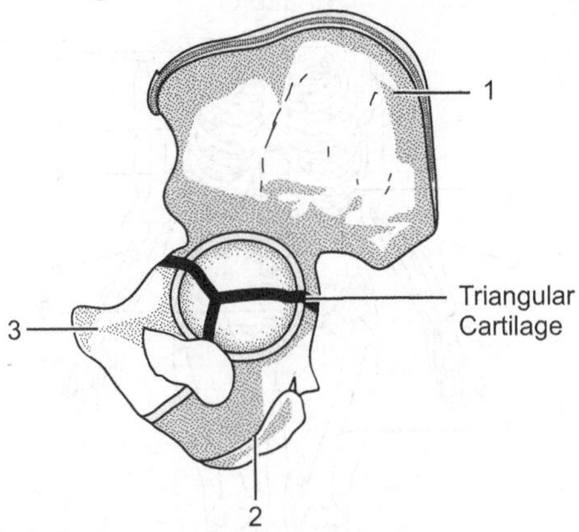

(i) Where is this bone located in human body?
(ii) Identify the bone labelled 1, 2 and 3 in above picture.
(iii) What joins these bones in centre?
(iv) What bone articulate with the cavity formed at the joining of three bones?
(v) Why women have wider pelvis than men?

Ans. (i) Hip
(ii) 1- Ileium; 2- Ischium ; 3- Pubis
(iii) Cartilage
(iv) Femur
(v) Wider pelvis helps in childbirth.

3. Study the figure related to types of synovial joint and answer the following questions:

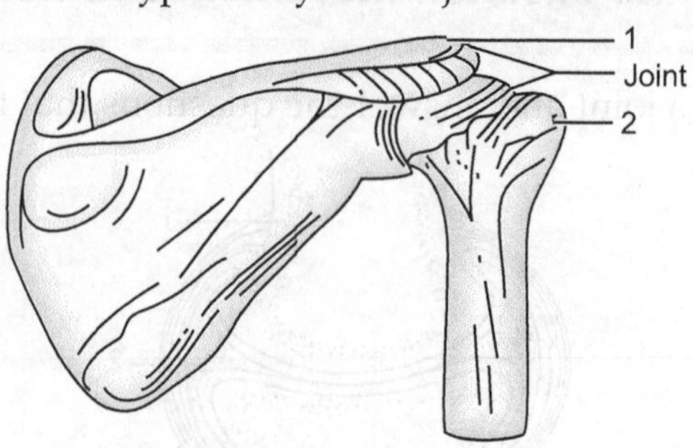

(i) Label the part 1 and 2.
(ii) What is the location of these bones?
(iii) What kind of joint is found here?
(iv) Why bone is covered with cartilage?
(v) What will be the kind of movement possible in this type of joint?

Ans. (i) 1- Collar bone or Clavicle, 2- Humerous
(ii) Shoulder
(iii) Ball and socket joint
(iv) A bone is covered with cartilage because it provides cushioning to moving end.
(v) Rotational movement.

4. Given below is a diagram of skeleton. Name the bones numbered 1-11:

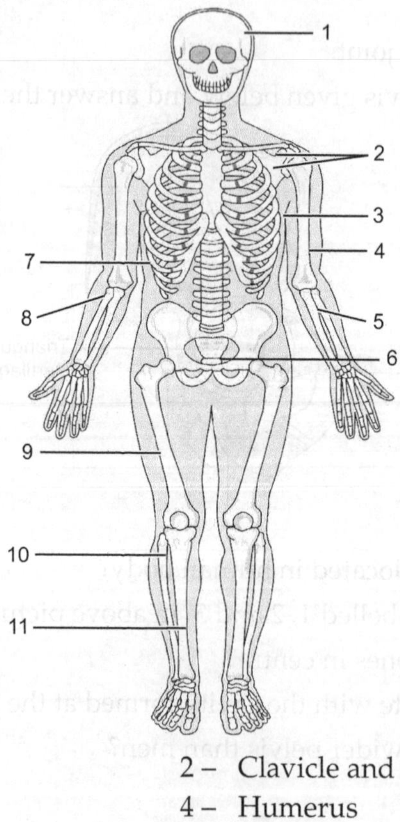

Ans.

1 –	Cranium	2 –	Clavicle and scapula
3 –	Sternum	4 –	Humerus
5 –	Ulna	6 –	Coccyx
7 –	Ribs	8 –	Radius
9 –	Femur	10 –	Fibula
11 –	Tibia		

5. Study the diagram of a long bone given below and answer the questions that follow:

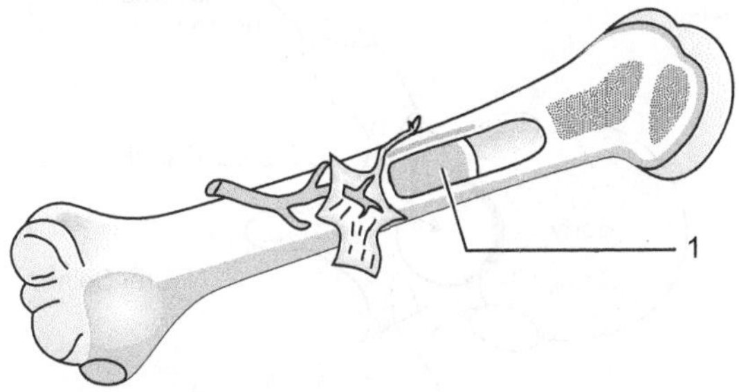

(i) What is shown in the diagram?
(ii) What kind of material named 1 is present inside bone?
(iii) What is function of this type of bone except support and protection?
(iv) Give example of any one other large bone having similar structure.
(v) Why this bone has ball like structure at the upper end?

Ans. (i) Femur
(ii) Bone marrow
(iii) Bone marrow produce blood cells.
(iv) Femur, hip bone , sternum, shoulder blade (any one).
(v) Because through this it will articulate with other bone to form joint.

6. The diagram given below shows the bones in hindlimb of a man. **[November 2019]**
Answer the questions that follow:

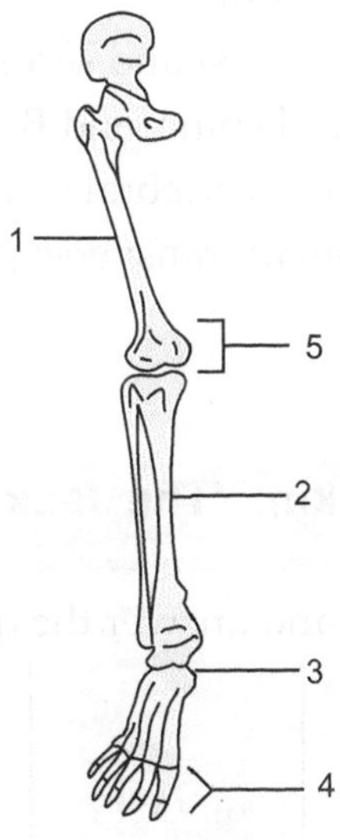

(i) Label the parts 3 and 4.
(ii) What is the fluid present in 5?
(iii) Name the corresponding bones of the parts 1 and 2 in the forelimb.
(iv) Mention any two functions of the human skeleton.
(v) Draw a neat, labelled diagram of a thoracic vertebra.

Ans. (i) 3- tarasal; 4- phalanges.
(ii) Synovial fluid.
(iii) Corresponding bone of part 1 [femur] is humerus and part 2 [tibia] is radius.
(iv) Two functions of human skeleton are:
(1) It provides a supporting framework and gives define shape to the body.
(2) It protects the delicate organs present in the body.

(v) Thoracic verebrae

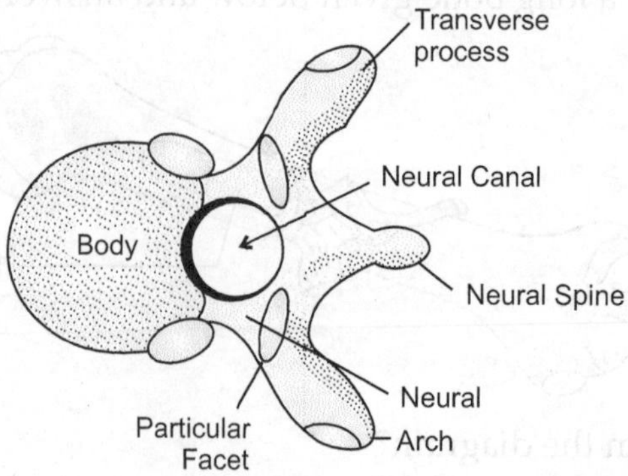

7. Study the diagram of vertebrae given below and answer the questions that follow:

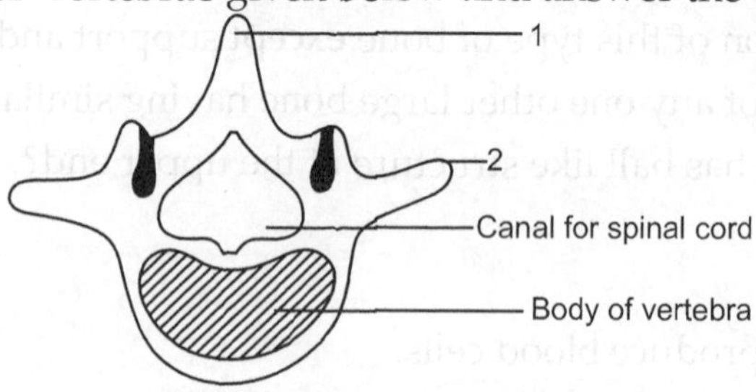

(i) Label the teeth named A and B.
(ii) What is the general function of cavity in vertebrae?
(iii) What is the major difference in first vertebra and other vertebrae?
(iv) What is name of second vertebra?
(v) What type of joint is found in first and second vertebra?

Ans. (i) A- spinous process Or Neural spine and B- transverse process
(ii) The spinal cavity cavity in the vertebral column consist of spinal cord lies in it.
(iii) First vertebra (Atlas) has small transverse processes
(iv) Axis
(v) Pivot joint

Chapter 13. Skin: "The Jack of All Trades"

1. Study the diagram of an artery and answer the questions that follow.

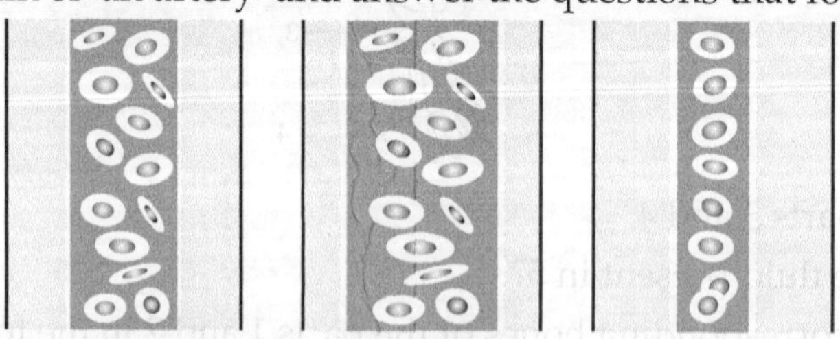

Normal **(a)** **(b)**

(i) Name the biological term used for state (a).
(ii) Under what condition it will happen?
(iii) Name the biological term used for state (b).
(iv) When a person is suffering from high fever or exercising extensively, he sweats profusely, what condition will occur?
(v) How bodies regulate these processes when body temperature fluctuates?

Ans.

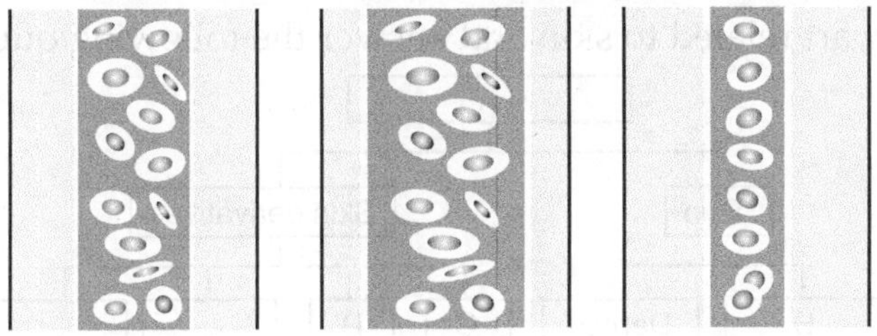

Normal (a) Vasodilatation (b) Vasoconstriction

(i) Vasodilation

(ii) Vasodilation will occur when body is heated.

(iii) Vasoconstriction

(iv) (a) Vasodilation

(v) Brain has a temperature regulation center located in hypothalamus. Normal body temperature is 37 degree. Its temperature fluctuates then body controls loss of heat from skin by vasodilation and vasoconstriction.

2. Study the diagram of a part of skin given below and answer the questions that follow:

(i) Identify the structure shown?

(ii) Identify the gland labelled as 1 in above picture.

(iii) What is the main ingredient of secretion this gland?

(iv) Which muscle is attached as 2 to this structure?

(v) Name any two conditions under which these muscles contract?

Ans. (i) Hair

(ii) 1- Sebaceous gland

(iii) Oils

(iv) Erector

(v) (1) Fear, (2) Cold

3. Study the flowchart related to skin and answer the following questions:

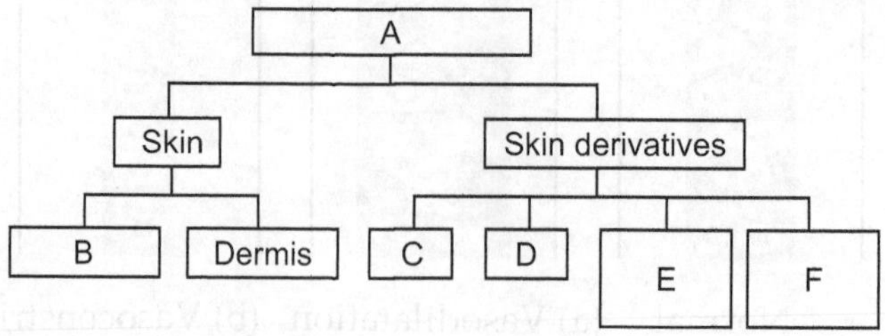

(i) Name the system A

(ii) Label B and give its function.

(iii) What kind of derivative is found as C and D?

(iv) Name the glands and their secretion labelled as E and F.

(v) What type of protein is present in nails and hair?

Ans. (i) A-Integumentary system

(ii) B-Epiderrmis ; protection

(iii) C- Hair ; D -Nail

(iv) E- Sebaceous gland – oil ; F - Sweat gland - sweat

(v) Keratin.

4. Study the diagram of a skin given below and answer the questions that follow:

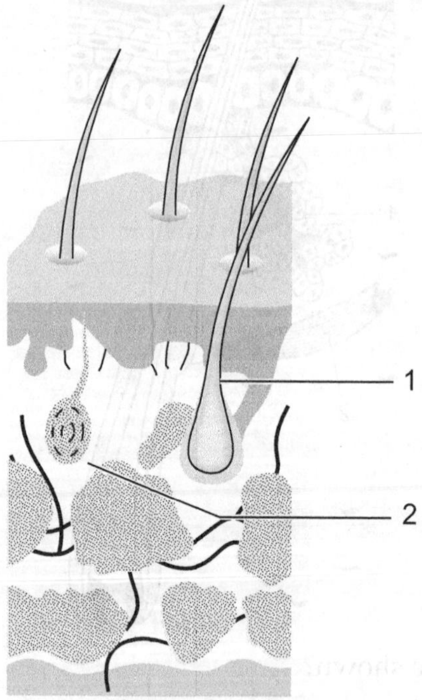

(i) What is shown in the diagram?

(ii) What kind of protein is present in labelled 1?

(iii) What is function of labelled 2 except cooling?

(iv) Give example of any one other large structure having modified structure of skin.

(v) Which important structure is missing in diagram? .

Ans. (i) Cross-section of skin

(ii) Keratin

(iii) Excretion of waste

(iv) Nail

(v) Blood vessels. .

5. Study the diagram of internal structure of skin given below and answer the questions that follow:

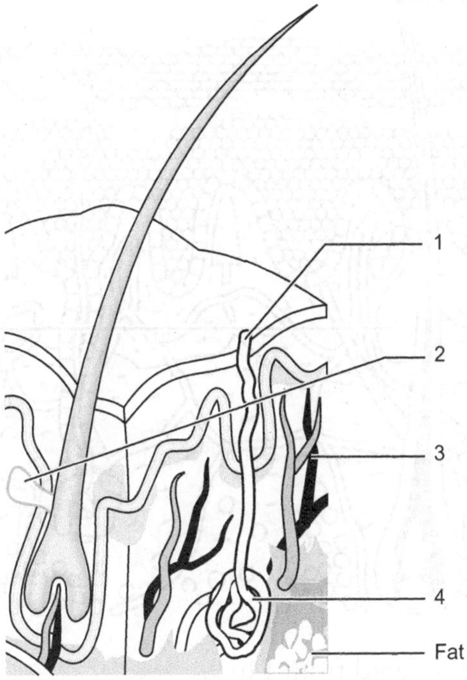

(i) Label the structure part 1 and give its function.

(ii) What secretion is produced by part 2?

(iii) Label and give the function of Part 3?

(iv) In which layer of skin part 4 is located?

(v) What is function of fat stored under skin?

Ans. (i) A sweat pore which help in removal of waste and help in keeping body cool.

(ii) Sebum

(iii) 3- Nerve ending. Nerve endings help in feeling sensations.

(iv) Dermis

(v) Fats stored under skin is for reserve food and protection.

6. Draw a labelled diagram of the generalised vertical section of the mammalian skin:

Ans.

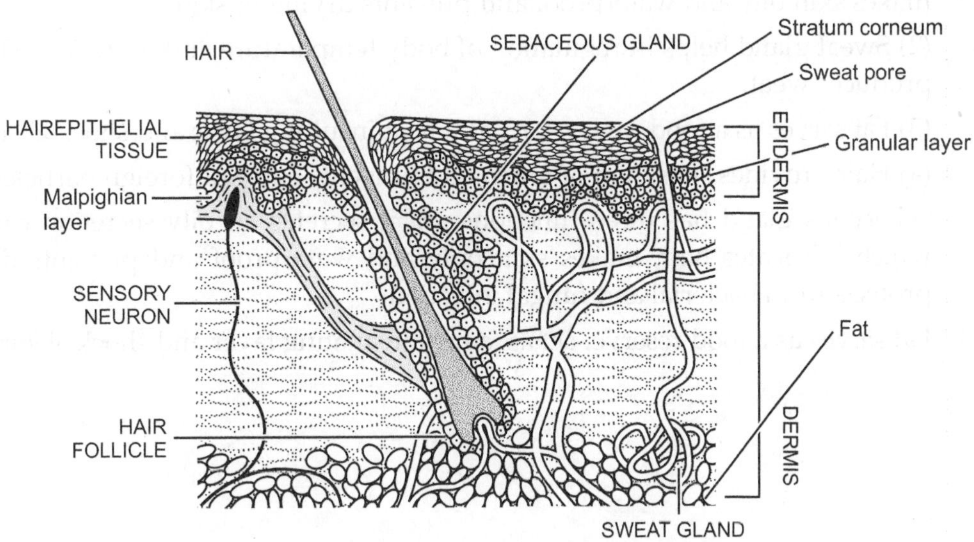

7. Given below is a diagrammatic sketch of the vertical section of the human skin.

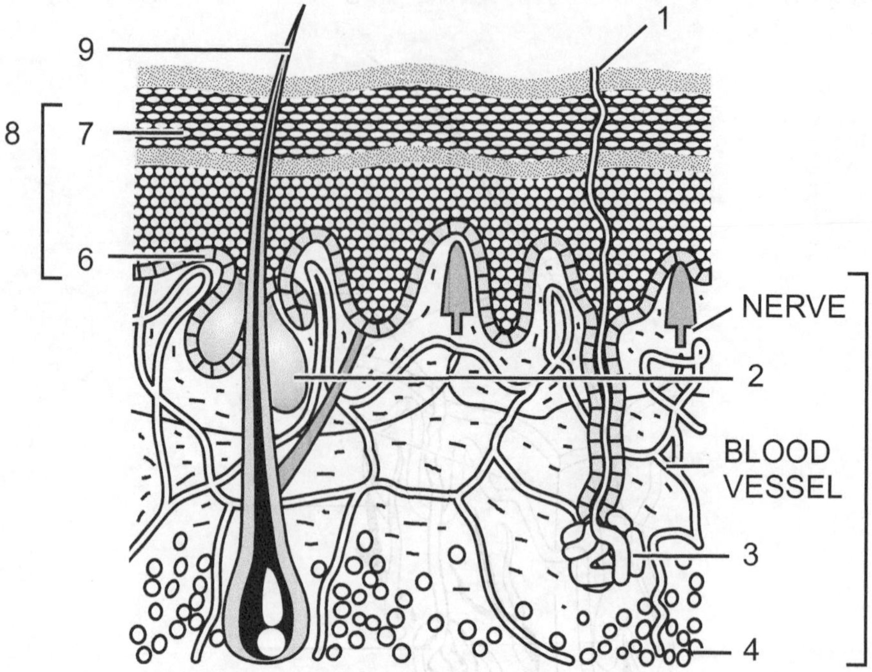

(i) Label the parts numbered from 1 to 9.

(ii) State one main function of each of the following parts:

(1) Part 2 (2) Part 3

(3) Part 4 (4) Part 9

Name any one of the above parts which has at least three functions.

(iii) Part 4 may add to one's good appearance or the figure. State one example of this function which may be comman to both men and women.

Ans. (i) 1. Sweat pore 2. Sebaceous gland

3. Sweat gland 4. Fat

5. Dermis 6. Stratum malpighi

7. Stratum corneum 8. Epidermis

9. Hair shaft

(ii) (1) Sebaceous gland secretes an oily secretion called sebum which lubricates hair, makes skin oily and waterproof and prevents drying of skin.

(2) Sweat gland helps in regulation of body temperature and excretion. These glands produce sweat.

(3) Fat serves as a food reserve, act as a heat insulting layer and shock absorber.

(4) Hair provides sensation of touch and prevents entry of foreign particles.

Sebaceous gland has three functions *i.e.,* it secretes an oily secretion called sebum which lubricates hair, makes skin only and waterproof and prevents drying skin, protects from bacterial and fungal infection.

(iii) Fat serves as a food reserve, act as a heat insulating layer and shock absorber.

Chapter 14. The Respiratory System

1. Observe the given diagram and answer the following question:

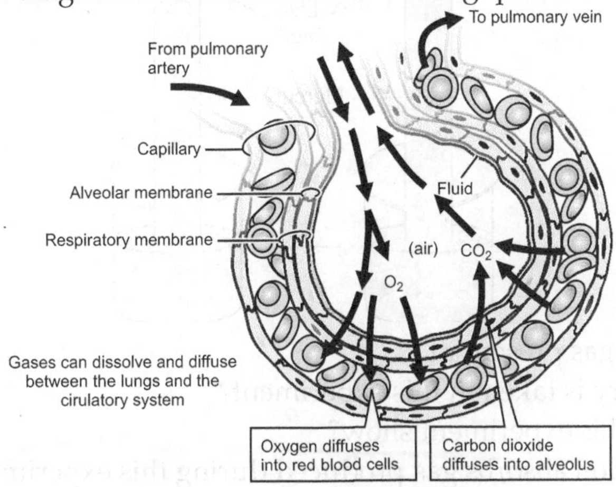

(i) Name the process explained in the given diagram.
(ii) What type of air is present in the bronchiole?
(iii) What type of blood does pulmonary artery contain?
(iv) What type of blood does pulmonary vein contain?
(v) Name the corresponding structure found in plants.

Ans. (i) The process explained is diagram is gaseous exchange in the alveolus..
(ii) Air rich in oxygen is present in the branchiole.
(iii) Deoxyygenated blood.
(iv) Oxygenated blood.
(v) Stomata

2 Study the diagram of an artery and answer the questions that follow:

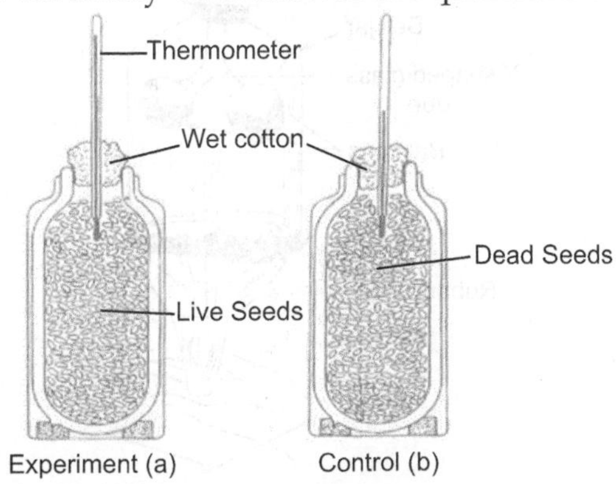

(i) Name the biological process taking place in (a).
(ii) Why temperature rises in (a)?
(iii) Why temperature does not rise in (b)?
(iv) Why control is used in the experiment?
(v) Why this experiment is done with thermos flask?

Ans. (i) Respiration
(ii) Respiration evolves heat.
(iii) Seeds in (b) are dead and do not respire. Thus temperature does not rises in (b).
(iv) Controls are used to check experimental errors.
(v) This experiment is done with thermos flask because if it is not done then temperature increase will be low due to loss of heat to environment.

3. Study the diagram given below and answer the following questions:

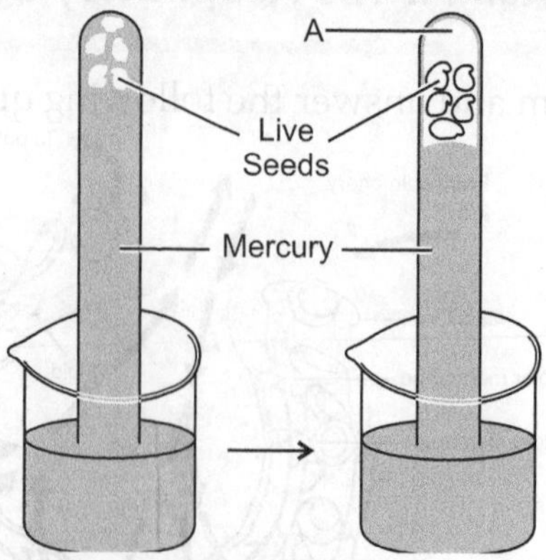

(i) Identify the gas present in 1.
(ii) Why mercury is taken in this experiment?
(iii) What does this experiment show?
(iv) Which solution absorbs gas produced during this experiment?
(v) What will happen if gas produced in this experiment is removed by some solution which can absorb it?

Ans. (i) CO_2 (Carbon dioxide)
(ii) Mercury will produce a totally inert environment.
(iii) This experiment shows that germinating seeds can respire without oxygen.
(iv) KOH (potassium hydroxide)
(v) Level of mercury will again rise up in tube if gas produced in this experiment is removed by KOH solution which can absorb it.

4. The figure alongside represents the apparatus which is used to demonstrate the process of breathing.

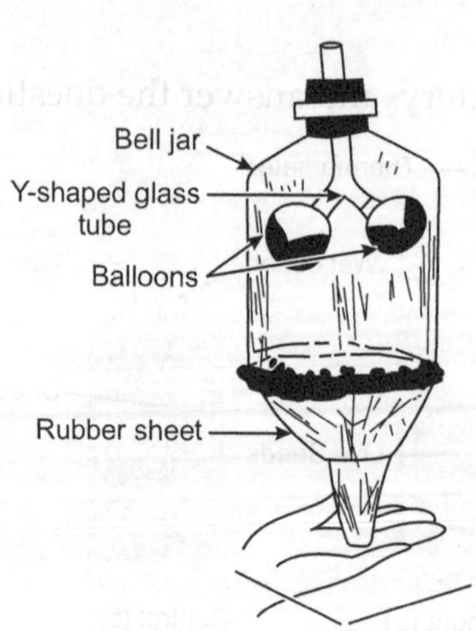

(i) What do the following parts in the diagram represent?
(1) Balloons (2) Rubber sheet (3) Bell jar
(ii) What will happen to the bolloons when rubber sheet is pulled down?
(iii) What is the role of diaphragm in breathing?
(iv) Which breathing movement is indicated when the rubber sheet is pulled down?

Ans. (i) Bolloons: Lungs
Rubber sheet: Diaphragm
Bell jar: Ribcage.
(ii) When the rubber sheet is pulled down, the balloons will inflate.
(iii) Diaphragm contracts and relaxes thereby decrease and increase the volume of thoracic cavity, for expiration and inspiration respectively.
(iv) Inspiration

5. The adjoining diagram represents the organs of human thorax. Study the diagram and answer these questions.

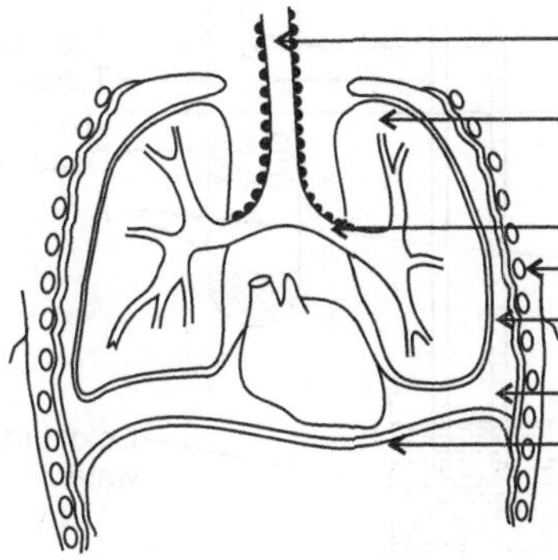

(i) What are the functions of partss 1, 5 and 7?

(ii) Why should the lining of alveoli be moist?

(iii) What are the positions of diaphragm at the time of insiration and expiration?

Ans. (i) **Part 1. Trachea:** The tube which helps to take the air in and out of the lungs. The ciliated epithelium of the inner wall of trachea prevents the entry of dust particles and other micro-organisms into the lungs.

Part 5. Pleural membrane: It protect the lungs against friction.

Part 7. Diaphragm: Helps in breathing by contracting and relaxing thereby increase and decrease the volume of thoracic cavity.

(ii) The lining of alveoli should be moist for the easy diffusion of respiratory gases.

(iii) During inspiration diaphragm flattens. During expiration diaphragm becomes dome-shaped

6. Study the diagram and answer the following questions:

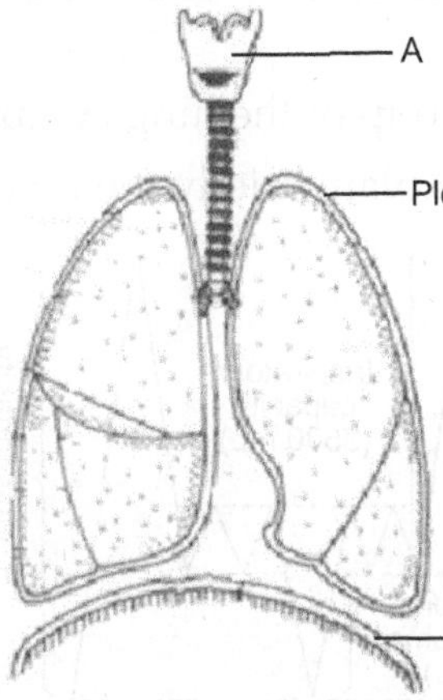

(i) Name the system.

(ii) Label part 1 and give its function.

(iii) Name the muscle present on underside of lungs.

(iv) What kind of fluid is found in pleural cavity and what is its function?

(v) Why there is difference in size of right and left lung?

Ans. (i) Respiratory system

(ii) 1 – Larynx. Larynx is responsible for production of voice.

(iii) Diaphragm

(iv) Pleural fluid is found in pleural cavity and its function is protection .

(v) There is a difference in size of right and left lung because of heart present on left side, due to which left lung is smaller.

7. Study the diagram given below and answer the questions that follow:

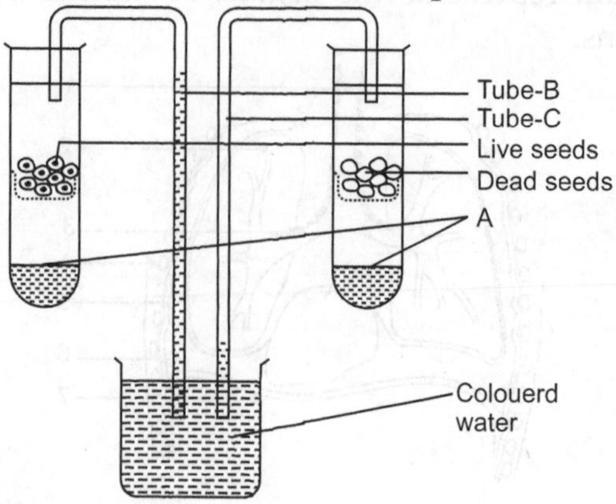

(i) What is contained as solution A?
(ii) What is function of solution A?
(iii) Why coloured water rises in tube B?
(iv) Why coloured water does not rises in tube C?
(v) What is inference of this experiment?

Ans. (i) Potassium hydroxide solution (KOH)
(ii) CO_2 is absorbed by KOH present in test tube.
(iii) The coloured water rises in tube B as space arises due to consumption of oxygen due to respiration.
(iv) The coloured water rises in tube C because boiled seeds do not respire.
(v) Germinating seeds use oxygen for respiration.

8. The volume of air in the lungs and rate at which it is exchanged during inspiration and expiration was measured:
The following diagram shows a group of the lungs volumes and capacities.
Study the diagram carefully and explain briefly the following:

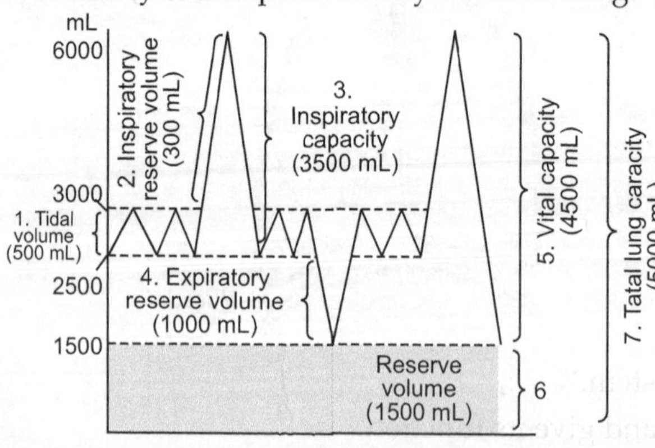

(i) Tidal volume
(ii) Inspiratory reserve volume
(iii) Expiratory reserve volume
(iv) Vital capacity
(v) Residual volume

Ans. (i) Tidal volume is the air breathed in and out in a normal quiet breathing and its volume is 500 ml.
(ii) Air that can be drawn in forcibly over the above the tidal air is called inspiratory reserve volume. Its volume is 3000 ml.

Diagram Based Questions

(iii) Air that can be forcibly expelled out after noral expiration is called expiratory reserve volume which is 1000 ml.

(iv) Vital capacity is the volume of air that can be taken in and expelled out by maximum inspiration and expiration. Its volume is 4500 ml.

(v) Residual volume is the air left in the lungs even after forcibly breathing out which is 1500 ml.

9. Study the diagram given below and answer the questions that follow:

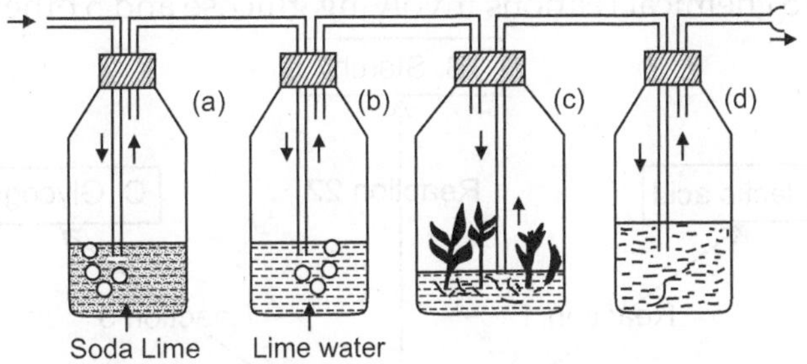

(i) Why soda lime is taken in (a)?

(ii) Why lime water is taken in (b)?

(iii) Label and process taking place in (c).

(iv) Which solution is taken in (d) and why?

(v) What can be done to further improve the results of this experiment?

Ans.

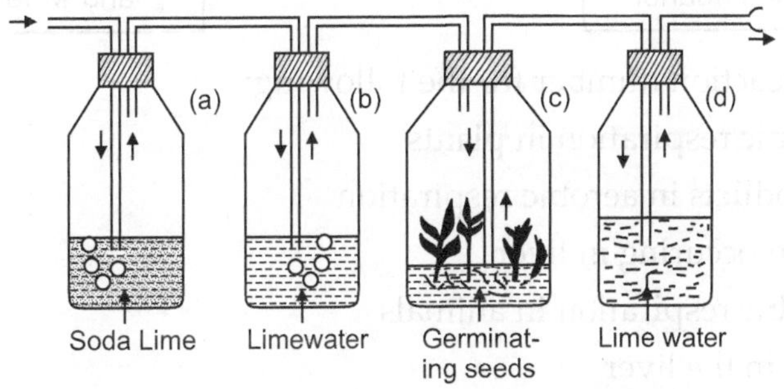

(i) Soda lime will absorb CO_2 present in air.

(ii) To check that no CO_2 is present in air.

(iii) Germinating seeds will respire and produce CO_2.

(iv) Lime water is taken to check presence of CO_2 produced by germination.

(v) Bottle may be kept in dark or covered with black paper so that no photosynthesis takes place.

10. Given alongside is a diagrammatic sketch of a part in human lungs:

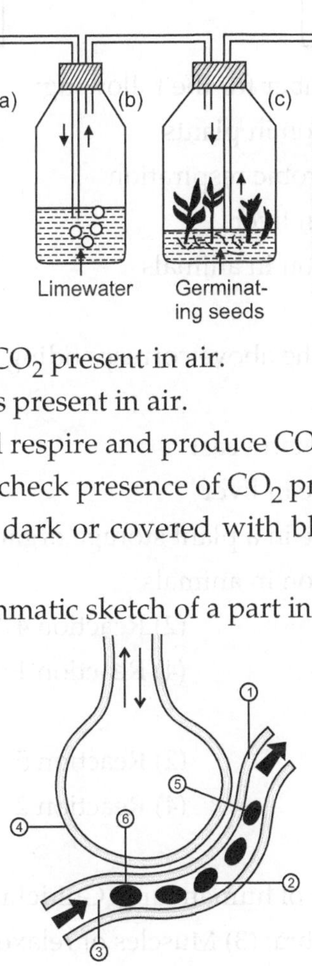

(i) Name the parts numbered 1-4.

(ii) What do the arrows 5 to 6 indicate?

Ans. (i) (1) Thin wall of capillary

(2) RBCs

(3) Plasma

(4) Thin wall of alveolus

(ii) Arrow 5 indicates carbon dioxide diffuses out and arrwo 6 indicates oxygen diffuses in.

11. Given below are 5 chemical reations involving glucose and 5 other chemical products (A-E).

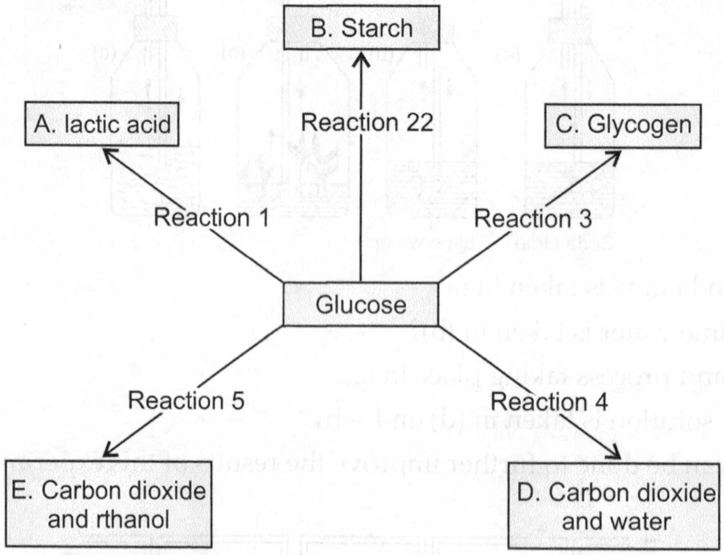

(i) Write the reaction number for the following:

(1) Anaerobic respiration in plants

(2) End products in aerobic respiration

(3) Reaction occuring in liver

(4) Anaerobic respiration in animals

(5) Storage in the liver

(ii) Which reaction 1-5 in the above corresponding to the following:

(1) Aerobic respiration

(2) Anarobic respiration in yeast

(3) Change taking place in liver

(4) Change taking place in a plant storage organ like potato

(5) Anaerobic respiration in animals

Ans. (i) (1) Reaction 5 (2) Reaction 4

(3) Reaction 3 (4) Reaction 1

(5) Reaction 3

(ii) (1) Reaction 4 (2) Reaction 5

(3) Reaction 3 (4) Reaction 2

(5) Reaction 1

12. Given alongside is the figure of human lung. Guidelines for 1 to 4 parts are indicated below: (1) Air rich in CO_2; (2) Vertebra; (3) Muscles of relaxed diaphragm; (4) Despressed ribs.

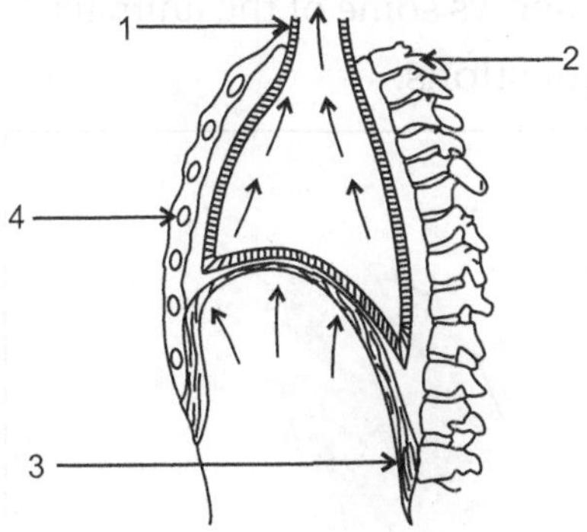

(i) Which state of breathing is represented in the figure?

(ii) What is dead airspace?

(iii) Name the muscles which relax in this process.

(iv) What is the composition of CO_2 in expired air?

Ans. (i) Expiration

(ii) The amount of air left in the respiratory tract that does not take part in the gaseous exchange.

(iii) Intercostal muscles.

(iv) 4%

Chapter 15. Hygiene

1. The diagram given below shows the structure of a housefly.

Answer the follwing questions:

(i) What is the economic importance of houseflies?

(ii) Why housefly is said as Public enemy no.1?

(iii) How does housefly contaminate food?

(iv) Mention some diseases that are caused by a housefly.

(v) What are the control measures to stay safe from houseflies?

Ans. (i) The housefly is a kind of natural scavenger cleaning and consuming the left-overs of food, wastes, etc.

(ii) Housefly is designated as Public enemy no. 1 as it spreads numerous diseases including dysentery, cholera, typhoid, etc.

(iii) Houseflies have spiny and hairy legs. They easily pick up filth containing germs from dirty places. They contaminate our food by sitting on it.

(iv) Dysentery, cholera, typhoid, etc.

(v) Elimination of breeding places, spraying insecticides and avoidance.

2. The diagram given below shows some of the animals.

Answer the following questions:

(a)

(b)

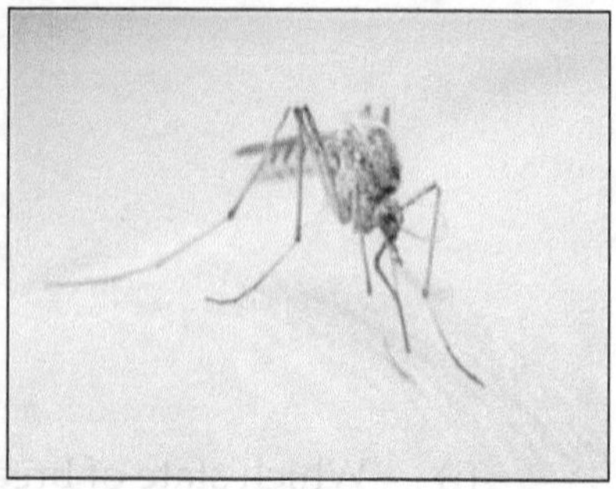
(c)

(i) Identify diseases caused by creatures shown in figures.
(ii) What is rat-flea?
(iii) Which of the following creature is the carrier of cancer-causing virus?
(iv) Where do they mostly exist?
(v) What are the control measures from all of the creatures?

Ans. (i) (a)- plague, (b)- viral disease, (c)- malaria, elephantiasis, etc.
(ii) Rate-flea is a parasite that lives on the skin of rats.
(iii) Cockroaches are suspected to be carrier of cancer-causing virus.
(iv) They mostly exist in connecting manholes and sewers.
(v) The control measures include elimination of breeding places, spraying DDT or rat poison on stagnant water, and keeping food covered.

3. The diagram given below shows the internal structure ofAmoeba. Answer the following questions:

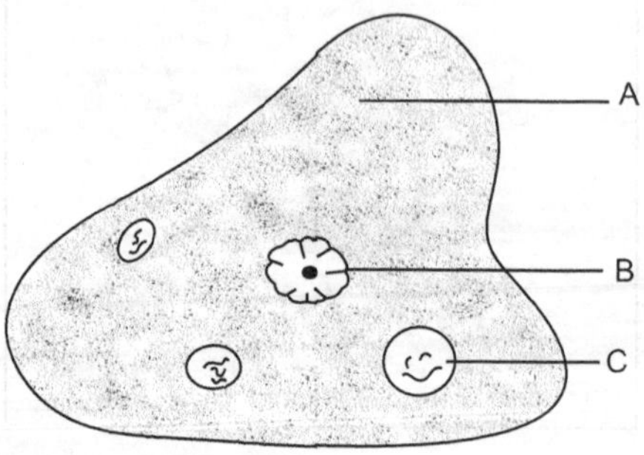

(i) Identify 1, 2, and 3.
(ii) What disease is caused by *Entamoeba histolytica*?
(iii) What is the function of Amoeba?
(iv) How is *Entamoeba* transmitted?
(v) Mention one symptom of Amoebic dysentery.

Ans. (i) 1-Pseudopodium, 2-nucleus, and 3-blood cell
(ii) Amoebic dysentery
(iii) The amoeba has no contractile vacuole and bears one large blunt pseudopodium for their movement.
(iv) Entamoeba transmitted through contaminated water and food.
(v) Bloody diarrhoe

Chapter 16. Disease: Cause and Control

1. Given below are four living organisms. Study and answer the following questions:

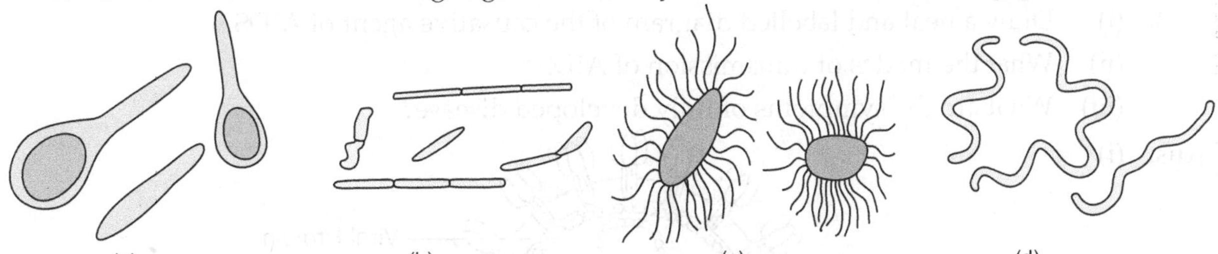

   (i) Name the disease caused by (a), (b), (c) and (d).
   (ii) Name the organism labelled (c).
   (iii) Which part of the body is affected by organism (b)?
   (iv) Which vaccine is useful against developing immunity for TB?
   (v) Is the disease caused by (d) communicable or non-communicable disease? State the reason.

Ans. (i) (a)- Tetanus, (b)- Tuberculosis, (c)- Typhoid, (d)- Syphilis
   (ii) *Salmonella typhi*
   (iii) Lungs
   (iv) BCG.
   (v) Syphilisis communicable disease because it is a sexually transmitted disease and spread from one person to another person.

2. Study the diagram and answer the following questions:

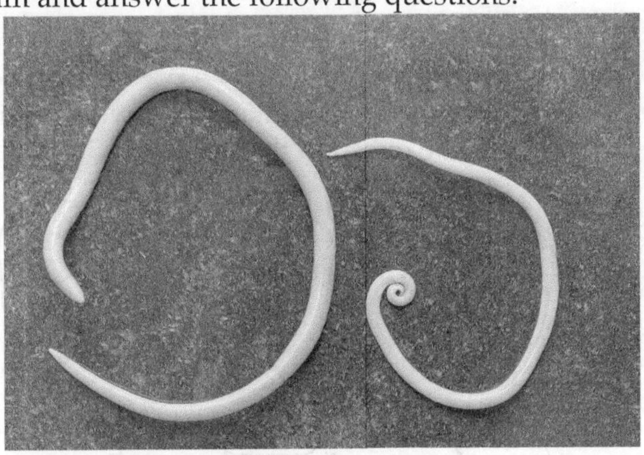

   (i) Identify the worms shown in picture.
   (ii) Which one is female? How will you identify it.
   (iii) Name the disease caused by it.
   (iv) Is it parasite or free-living?
   (v) What are the reasons for the disease caused by it.

Ans. (i) These worms are round worms.
   (ii) Longer worm: Female
   Smaller worm: Male
   Female worm is slightly longer in size and male worm is recognised by a slightly curved hind end.
   (iii) Ascariasis
   (iv) These are parasitic worms.

(v) The eggs of female worm which pass out along patient's faeces are scattered in soil and reaching inside the vegetables, children playing in ground may get eggs on their hands and without washing hands may enter into the body while eating food.

3. (i) Draw a neat and labelled diagram of the causative agent of AIDS.

(ii) What the modes of transmission of AIDS?

(iii) What are the symptoms of fully developed disease?

Ans. (i)

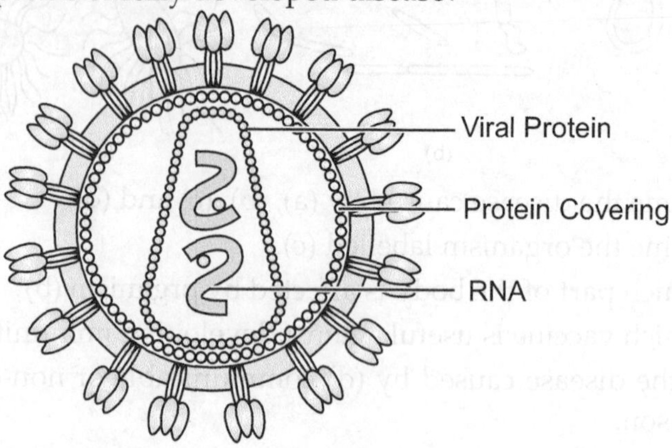

(ii) Modes of transmission of AIDS are sexual intercourse, contaminated blood transfusions, injection needles,

(iii) When AIDS is fully developed, most individuals die within 3 years from other infections or cancers. Symptoms during this period include swollen lymph nodes, fever, night sweats, and weight loss

Chapter 17. Aids to Health

1. The diagram given below is a process that takes place whenforeign bodies enter the body. Based on the diagram answer the question that follow:

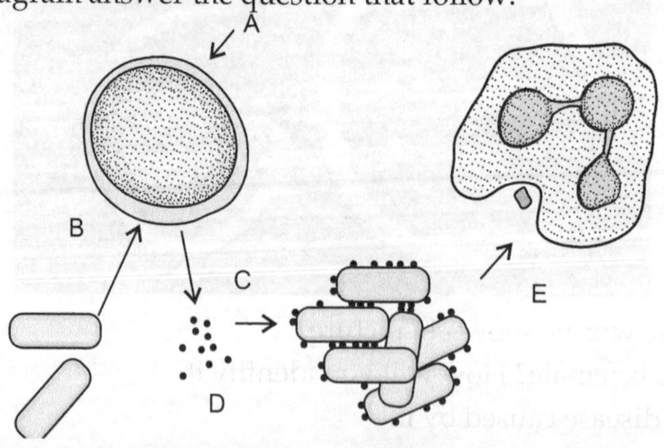

(i) Study the diagram given and label the diagram by mentioning the complete process.

(ii) Name the process that is represented in the image.

(iii) What happens to the bacteria that attack the body?

(iv) Name the cell involved in this process.

(v) Explain the step E in the image briefly.

Ans. (i) Lymphocyte recognises bacteria as foreign and secretes the appropriate antibody, bacteria attach to the receptor, antibodies make the bacteria clump together, Phagocytic white cell can then take in the clumped bacteria and digest them.

(ii) The given image represents the antibodies attacking bacteria.

(iii) The specialised lymphocytes recognise the bacteria and specific antibody is formed. This antibody when attached to the bacteria activates Phagocytosis and kills the bacteria.

(iv) The process involves specialised lymphocytes, antigen, antibody, and phagosome.

(v) In the image the step E represents the digestion of antibody-antigen clumps with the help of phagocytic white cells.

2. Study the diagrams and answer the following questions:

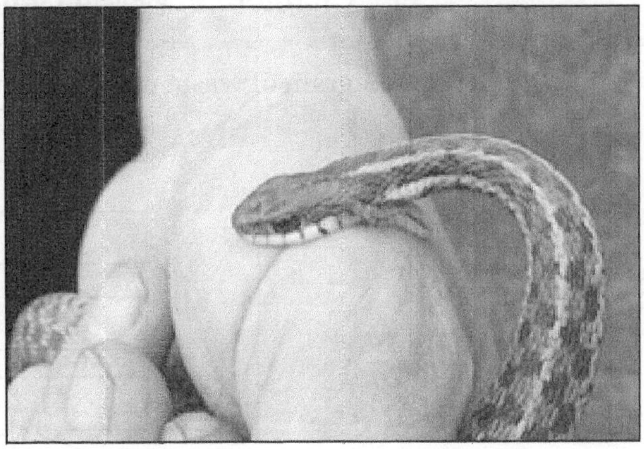

(i) Identify the type of treatments necessary in the situation shown in the diagram.

(ii) Categorise the type of immunity.

(iii) Name the remedy for it.

Ans. (i) By providing injections which contain antibodies for the poison injected in body.

(ii) Dog-bite: Innate immunity

Snake-bite: Acquired immunity

(iii) Antivenin injections for snake bite and vaccine for rabies in dog bite.

Chapter 18. Health Organisation

1. Study the diagram and answer the following questions:

(i) Identify the symbol.

(ii) State its main function.

(iii) Mention two activities of it.

(iv) Where is it used?

(v) On what colour background is this emblem painted?

Ans. (i) Red Cross

(ii) Its function is to prevent or remove human sufferings in peace time as well as at the time of war.

(iii) The two activities are-

(a) To extend relief and help to the victims of any calamity like flood, fire, etc.
(b) To extend all possible first-aid in any accident.

(vi) It is used by hospitals, ambulance and doctors.

(v) The emblem of the Red Cross Society is a red- coloured cross painted on a white background.

Chapter 19. Waste Generation and Management

1. Study the diagram of waste management and answer the following questions:

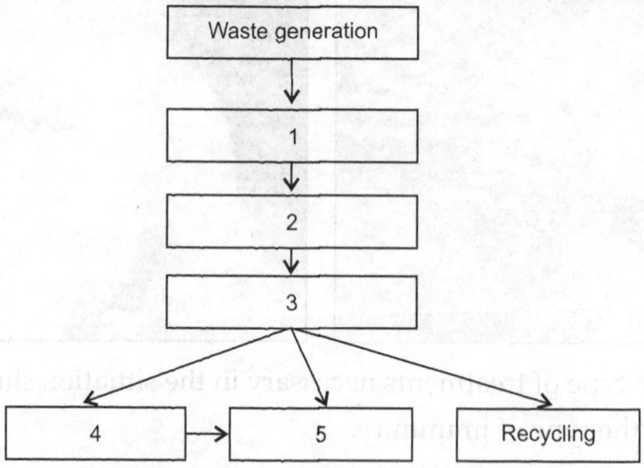

(i) Label the parts (1) what is done in this step?
(ii) What is done in part labelled (2)?
(iii) Name the function of step (3).
(iv) Name the step in which waste is burnt.
(v) Name the step labeled (5). What waste which is disposed off in it?

Ans. (i) 1- Segregation
(ii) 2-handling and transport
(iii) 3 treatment and disposal
(iv) 4- waste incineration
(v) 5- waste land filling , Ashes and organic waste

2. Study the diagram given below and answer the questions that follow:

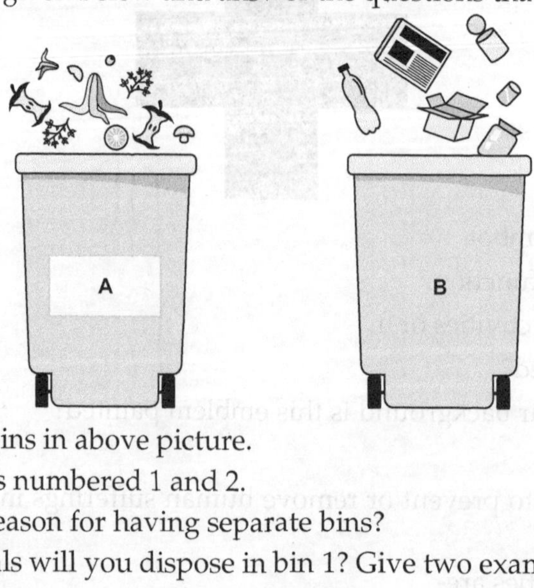

(i) Identify the bins in above picture.
(ii) Label the bins numbered 1 and 2.
(iii) Explain the reason for having separate bins?
(iv) What materials will you dispose in bin 1? Give two examples.

(v) What materials will you dispose in bin 2? Give two examples.

Ans. (i) Garbage bins

(ii) 1 –Wet waste ; 2- Dry waste

(iii) for segregation

(iv) vegetable waste and rotten food

(v) paper and plastic

3. Study the figure given below and answer the following questions:

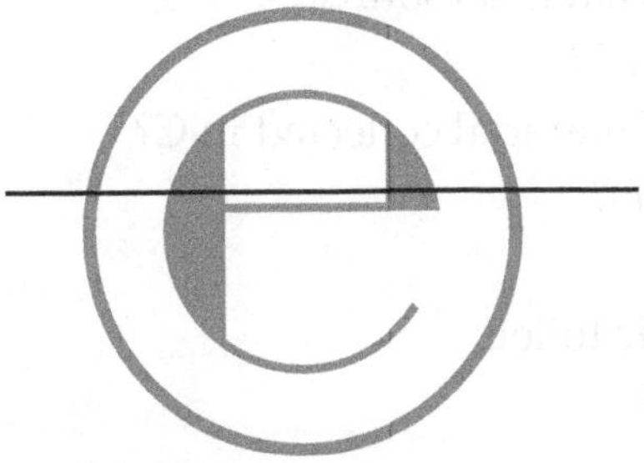

(i) What does the logo shown above tell?

(ii) How will you dispose-off material where this logo is seen?

(iii) Where will you see this logo in your home?

(iv) Why should we not dispose-off these materials in domestic garbage?

(v) What can you save on disposing-off waste properly?

Ans. (i) Electronic waste

(ii) It can be given to a kabadiwala.

(iii) On electrical appliances this logo is found at home.

(iv) We should not dispose-off these materials in domestic garbage because they contain re-useable material.

(v) We can save copper/ aluminum and other metals used in wires for reuse, recycling of paper and plastics can also reduce the excess use of plastics and reduce deforestation, etc.

4. Study the diagram given below and answer the following questions:

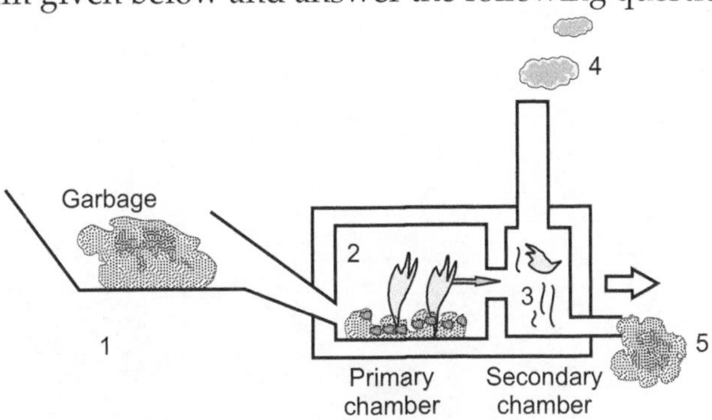

(i) Name the structure shown in the diagram.

(ii) What will happen to garbage in chamber 2?

(iii) What kind of garbage is disposed by this method?

(iv) Label the parts number 4 and 5.

(v) Give any one disadvantage of this process.

Ans. (i) Incinerator

(ii) Garbage is burnt

(iii) Organic waste

(iv) Smoke and ash

(v) Burning of garbage leads to air pollution.

5. The diagram shows flow chart of municipal sewage disposal

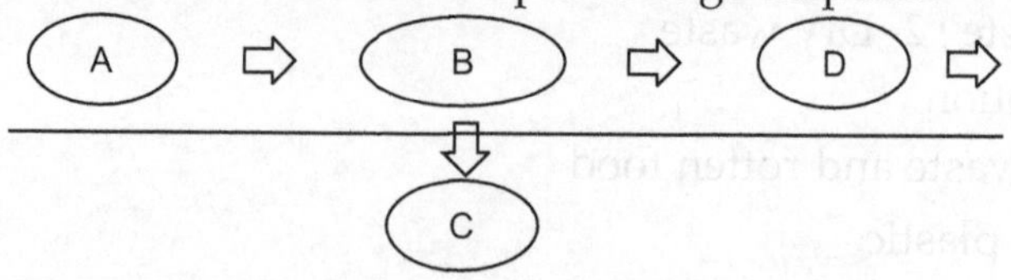

(i) A is Raw sewage. What is its source?
(ii) What will happen in B?
(iii) What is the name of material collected in C?
(iv) What is released as D?
(v) What is fate of D?

Ans. (i) Domestic and public toilets
(ii) Biological treatment
(iii) Sludge
(iv) Treated Sewage
(v) It can be used for irrigation.

Fill in the Blanks

Set 12

Chapter 1. Introducing Biology

1. Evolution deals with the study of (i) ....and (ii)............ of organisms.
2. The study of humans and their interaction with the society is called ....
3. (i) .... was probably the pioneers of an organised study of biology. Biology is divided into different ways like, based on (ii) .... of organisms (for e.g., botany, zoology, etc.), special groups of organisms (for e.g. (iii) ....and (iv) ...., etc.) and in terms of (v) .... (for e.g., morphology, biogeography, etc.)
4. Science of growing fruits and vegetables is called........................
5. Embryology: Embryo of plants and animals:: Biogeographic: ....
6. Bioengineering deal with making artifical ....
7. .... is the entire DNA sequence of an organ.
8. Silk is produced by rearing ....
9. Organisms which get food from the host body are called ....
10. Animals who give birth to young ones are called ....
11. .... is called the Father of Medicines.

Ans. 1. (i) origin, (ii) descent
2. Anthropology
3. (i) Greek (ii) Major kinds (iii) bacteriology (iv) entomology (v) approach on studying it
4. Horticulture
5. Geographical distribution of plants and animals.
6. (i) limbs, (ii) joints
7. Genome
8. silkworms
9. parasites.
10. *jeevaj* (viviparous)
11. Hippocrates

Chapter 2. Cell: The Unit of Life

1. Lysosome: Digestive enzymes: Mitochondria: ....
2. Chloroplast: Photosynthesis:: ....: Imparts colour to flowers and fruits
3. .... consists of membranous sacs and secretes 40 types of digestive enzymes.
4. The membrane of the (i) .... is called tonoplast. Tonoplast is a (ii) .... permeable membrane. The vacuole isfilled with a watery fluid called as (iii) ....
5. The (i) ....uses light which is bent by glass lenses to magnify the image while (ii) .... uses beam of electrons which are bent by magnets.
6. .... is surrounded by microtubules, located near the nucleus.
7. .... is the living substance of the cell.
8. Very thin flexible, living membrane which is differentially permeable is called ....

9. White blood cells are .... that can squeeze out through the capillary wall.
10. More than 1000 chromosomes are found in nucleus of certain ....
11. .... are hereditary units.
12. .... is a plastid which stores starch.

Ans. 1. Power house or ATP synthesis.
2. Chromoplasts.
3. Lysosome.
4. (i) Vacuole (ii) Semi permeable (iii) Cell sap
5. (i) Compound microscopes (ii) Electron microscope
6. Centriole
7. Protoplasm
8. palsma membrane
9. amoeboid
10. insects
11. Genes
12. Leucoplast

Chapter 3. Tissues: Plants and Animals Tissue

1. .... are long tube like structures meant for transporting water and other dissolved materials.
2. Skeletal, striped or voluntary muscles are called .... muscles.
3. .... epithelium are present in kidney tubules, salivary glands, etc.
4. Bone synthesizing cell with single nucleus is called ....
5. (i) .... and (ii) .... both are conducting or vascular tissues of the plants.
6. Cells of (i)...........epithelium are tall, thick, cylindrical, and pillar-like, commonly found in the digestive tract (in the lining of stomach, small intestine), kidney duct and thyroid gland. (ii)........... epithelium differentiate into secretory units.

Ans. 1. Tracheae (or xylem vessels) 2. Striated
3. Cuboidal 4. Osteoblasts
5. (i) Xylem (ii) phloem 6. (i) Columnar (ii) Glandular

Chapter 4. The Flower

1. .... protects the young flower bud and when green they also perform the process of photosynthesis.
2. When the perianth is green-like sepals it is called .... perianth.
3. .... constitutes the inner essential whorl of flower comprising carpels.
4. After fertilisation, the ovules develop into (i).................. and the (ii)............... matures into a fruit.
5. Calyx: Corolla:: Stamen: ....
6. Anther: ....: Ovary: Ovule
7. Male flower: Staminate flower:: Female flower: ....

Fill in the Blanks

8. The essential part of the flower is directly concerned with (i)................ Essential part consists of male parts called (ii)........ and female parts (iii)............ (iv)........... protect the young flower bud (v)...........attracts insects for pollination.

Ans. 1. Sepals **2.** Sepaloid
3. Gynoecium **4.** (i) Seeds, (ii) Ovary
5. Calyx: Corolla:: Stamen: Carpel **6.** Pollen
7. Pistillate flower
8. (i) Reproduction (ii) Stamens (iii) Carpels (iv) Sepals (v) Petals

Chapter 5. Pollination and Fertilisation

1. Transference of pollen grains from anthers to stigma of the same flower is called ....
2. Different timings for maturation of gynoecium and androecium are called ....
3. Pollination is the process of transfer of ....
4. The two kinds of pollination are (i) .... and (ii) ....
5. For self – pollination the flower need not be .... and showy.
6. Self- pollination does not yield .... varieties.
7. .... is a water-pollinated flower.
8. After fertilisation .... falls off or remains intact in dried and shrivelled form.
9. The stamens are (i).................. and hangout of the flower to be exposed to (ii)............... in anemophilous flower.
10. The stigma is a part of (i) .... It is covered with hair or with glandular papillae. It acts as a (ii) .... receptor during the process of pollination.
11. Maize: Anemophilous:: *Bignonia:* ....
12. Self-sterility: Orchids:: Heterostyly: ....

Ans. 1. Autogamy **2.** Dichogamy **3.** Pollen grains
4. (i) Cross–pollination, (ii) self–pollination **5.** Large **6.** New
7. *Vallisneria* **8.** calyx. **9.** (i) Long (ii) Wind
10. (i) Gynoeciumm (ii) Pollen **11.** Ornithophilous **12.** Prim rose

Chapter 6. Seeds-Structure and Germination

1. .... is a scar on the seed coat (tegmen/hilum).
2. .... are seed leaves (plumule/cotyledons).
3. .... part of flower forms seed (ovary/ovule).
4. In bean seeds, ... grows faster and the seeds are brought .... ground.
5. A seed is protected by .... and ....
6. Rice, wheat and maize are rich in .... food. (starch/protein)
7. Cotyledon: Bean seed:: ....: Maize grain
8. Pea seed: Hypogeal:: Bean seed: ....
9. Carbon dioxide: Soda lime:: oxygen: ....
10. Radicle: ....: Plumule: Coleoptile
11. ....: largest seed:: Orchid seed: smallest seed

12. The (i) .... in maize grain is not separate and is fused with (ii) .... (iii)..........andmicropylearenotclearlyvisibleinmaizegrain.Itcontainsasinglecotyledoncalled (iv) .... The (v) .... is the nutritive tissue of maize grain.
13. Bean seed has outermost hard brownish covering called (i) .... The inner protective layer of seed coat is called (ii) .... (iii) .... represents the spot where ovule was attached to ovary wall through placenta. A tiny pore called (iv) .... marks the opening through which pollen tube had entered the ovule. There are two (v) .... which protects and provide nourishment to the embryo.

Ans. 1. Hilum 2. Cotyledons 3. Ovule
4. Hypocotyl, above 5. Testa, tegmen 6. Starch 7. Endosperm
8. Epigeal 9. Pyrogallic acid 10. Coleorhiza 11. Double coconut
12. (i) Seed coat, (ii) fruit wall, (iii) Hilum, (iv) scutellum, (v) endosperm.
13. (i) Testa, (ii) tegmen, (iii) Hilum, (iv) micropyle, (v) cotyledons.

Chapter 7. Respiration in Plants

1. $C_6H_{12}O_6 + 6O_2 \rightarrow$ .... + ...+ 38 ATP
2. Rate of .... is more than the rate of .... in the day time in case of green plants.
3. .... is used to create vacuum to show anaerobic respiration.
4. During respiration (i) .... energy is converted into (ii) .... energy.
5. In anaerobic respiration in plants, the end product is (ii) .... and (ii) .... instead of lactic (iii) .... in animals.
6. .... are the openings found on older stems.
7. .... is the chemical which absorbs oxygen of the air.
8. .... is a respiratory substance.
9. Glycolysis occurs in the .... of the cells.
10. Number of ATP molecules released during anaerobic respiration is ....
11. Aerobic respiration: ....: Anaerobic respiration: Glycolysis
12. Aerobic respiration: Human beings:: Anaerobic respiration: ....
13. Cytoplasm: ....: TCA cycle: Mitochondria
14. Anoxybiotic: 2ATP:: Oxybiotic: ....
15. Plants: ethanol:: Animals: ....
16. (i) .... is a biochemical process that occurs in a series of steps whereas combustion is a (ii) .... process that occurs in a single step. Respiration is a/an (ii) .... reaction which occurs at body temperature, whereas (iii) .... is carried out by heat. (iv) .... is a non-cellular process which occurs at high temperature.

Ans. 1. $6CO_2 + 6H_2O$ 2. Photosynthesis, respiration 3. KOH
4. Chemical, heat 5. (i) Ethanol, (ii) carbon dioxide, (iii) acid 6. Lenticels
7. Pyrogallic acid 8. Glucose 9. Cytoplasm 10. 2
11. Kreb's cycle 12. Yeast 13. Glycolysis 14. 38 ATP
15. Lactic acid
16. (i) Respiration, (ii) physico-chemical, (iii) enzyme-catalysed, (iv) combustion, (v) Combustion

Chapter 8. Five kingdom classification

1. .... is a combination of algae and fungi.
2. Three kingdom classification was given by ....
3. .... is the process of casting off and regrowing of exoskeleton.
4. Arthropods have .... circulatory system.
5. The naked embryo present in Cryptogamae are called ....
6. The method of naming organisms is called (i) .... and was introduced first by (ii) .... The five kingdom classification was proposed by (iii) .... Kingdom Plantae is divided into (iv) ...., (v) ...., (vi) ...., (vii) ...., (viii) .... Kingdom Animalia is divided into (ix) .... and (x) ....
7. Arthropods: ....:: Annelids: Nephridia
8. Salamander: amphibia:: Heloderma: ....
9. Leech: ....:: Roundworm: Nematoda
10. Sea- cucumber: ....:: Sepia: mollusca

Ans. 1. Lichen 2. Ernst Haeckel 3. Moulting 4. Open

5. Spore

6. (i) Binomial Nomenclature, (ii) Carolus Linnaeus, (iii) R.H. Whittaker, (iv) Thallophya, (v) Bryophyta, (vi) Pteridophyta, (vii) Gymnosperms, (viii) Angiosperms, (ix) Vertebrates, (x) Invertebrates.

7. Malpighian tubules 8. Reptile 9. Annelida

10. Echinodermata

Chapter 9. Economic Importance of Bacteria and Fungi

1. .... is a common bread mould.
2. Penicillin is obtained from fungi ....
3. Causative agent of bacterial meningitis is ....
4. Rod shaped bacteria are called as ....
5. The process by which atmospheric nitrogen is fixed to be used by plants is called (i) .... The process by which organic matter of the dead remains are decomposed to ammonia is called (ii) .... Nitrification is the process where (iii) .... is converted into (iv) .... (v) .... is an example of denitrifying bacteria.
6. Penicillium: Blue mould:: ....: Green mould
7. White button mushroom: Agaricusbisporous:: Oyster mushroom: ....
8. Cocci: spherical bacteria:: Spirilla: ....

Ans. 1. Rhizopus 2. Penicillium notatum 3. Neisseria meningitidis

4. Bacilli

5. (i) Nitrogen fixation, (ii) Ammonification, (iii) ammonium compounds/ammonia, (iv) nitrites, (v) *Pseudomonas*

6. *Aspergillus* 7. Pleurotus 8. Twisted bacteria

Chapter 10. Nutrition

1. .... is not digested in man.
2. .... is the animals starch.

3. .... is required for transmission of impulses.
4. .... is an example of polysaccharides.
5. .... vitamin is called anit-rickets.
6. .... is caused by vitamin B_{12} deficiency.
7. Protein deficiency in children leads of ....
8. Element present in thyroxine is ....
9. .... is the simplest carbonhydrate.
10. A rich source of protein is ....
11. Night-blindness is caused by deficiency of ....
12. (i) .... is part of food which is a no nutrient yet important part of food. We get it mainly from (ii) .... and fruits which has a lot of cellulose. Cellulose in (iii) .... in our body, but it is important because it helps in proper (iv) .... of food in gut. Cellulose has capacity to absorb water thus it helps in avoiding (v) .... too. Due to roughage food becomes (vi) .... and this makes us feel that our stomach is full. This fullness gives us feeling of satisfaction. Cellulose help in removal of (vii) .... from the body. Thus, roughage though not a nutrient, is (viii) .... for us. That is why doctors advice green leafy vegetables and (xi) .... when we have constipation. Remember most food are (x).................... of nutrients.
13. (i).................... are carbohydrates made of single unit of sugar which provide (ii).......... .... energy.
14. (i) .... (ii) and .... are required for proper formation of teeth.
15. (i) .... and (ii) .... are deficiency diseases caused due to reduced intake of food in growing children.
16. (i) .... are proteins produced against (ii) .... during immunity process to protect our body.
17. (i) Vitamin A, ... and (ii) .... are fat soluble.

Ans. 1. Cellulose 2. glycogen 3. Sodium 4. Starch
5. Vitamin-D 6. Pernicious anaemia 7. Kwashiorkor 8. Iodine
9. Glucose 10. Egg 11. Vitamin-A
12. (i) Roughage, (ii) leafy vegetables, (iii) indigestible, (iv) movement, (v) constipation, (vi) Bulky, (vii) toxins, fats and cholesterol, (viii) important, (ix) fruits, (x) mixture,
13. (i) Monosaccharides (ii) instant 14. (i) Fluorine (ii) Calcium
15. (i) Kwashiorkor (ii) Marasmus 16. (i) Antibodies (ii) antigens
17. (i) D (ii) E

Chapter 11. Digestive System

1. .... is present in the stomach for proper digestion.
2. Pancreatic juice contains fat digesting enzyme called ....
3. Enzyme helps in .... of ingested food.
4. Through mastication, the food is mixed with saliva and shaped intto a soft, flexible mass called ....
5. In humans, the last three teeth are ....
6. Vitamin .... deficiency causes pellagra.
7. .... covers the crown of the teeth.
8. End product of carbohydrate digestion is ....
9. Removal of undigested food is called ....
10. Enzymes are (i) .... in nature, they are destroyed by heating. They work best in a narrow temperature range of (ii) .... °C to (iii) .... °C. They are

also sensitive to changes in pH and act best in (iv) .... or (v) .... medium. For example, the enzyme –(vi) .... requires an acidic medium, while (vii) .... requires an alkaline pH. Chemicals which accelerate the rate of chemical reactions are called (viii) .... As enzymes also do the same function in organism, they are called (ix) .... Enzymes act on specific substrate and produce specific (x) ....

11. (i) .... is shortest part of small intestine measuring only about (ii) .... inches in length.
12. (i) .... is middle part of the small intestine which help in absorption of (ii) ....
13. (i) .... opens into (ii) .... via ileocecal sphincter.
14. (i) .... juice secreted in stomach is highly (ii) .... in nature.
15. (i) .... is well developed in grass herbivores but is vestigial organ in (ii) ....
16. Complete digestion of food takes place in ....
17. .... helps in emulsifying fats.
18. .... teeth are well-developed in carnivorous animals.
19. The number of teeth in human are ....
20. Pancreas is both .... and .... in function.
21. .... helps in the curdling of milk.
22. The movement of food in the oseophagus is ....
23. In stomach food remains in the form of ....
24. (i) ...., (ii)..........................., and (iii) .... are the parts of large interstine.
25. (i) .... is the largest digestive gland in the human body and secretes (ii) ....

[February 2020]

Ans.1. HCl 2. Lipase 3. Digestion 4. Bolus
5. Molars 6. B_3 7. Enamel 8. Glucose
9. Egestion 10. (i) Proteins, (ii) 35°C, (iii) 40 °C, (iv) acidic, (v) alkaline, (vi) pepsin, (vii) trypsin, (viii) catalyst, (ix) biocatalyst, (x) end product.
11. (i) Duodenum, (ii) 10 inches 12. (i) Jejunum, (ii) nutrients
13. (i) Ileum, (ii) large intestine 14. (i) Gastric juice, (ii) acidic
15. (i) Caecum, (ii) humans 16. Small interstine
17. Bile 18. Canine
19. Thirty-two 20. (i) Exocrine (ii) endocrine
21. Renin 22. Peristalsis
23. Chyme 24. (i) Caecum, (ii) colon and (iii) retum\
25. (i) Liver, (ii) bile.

Chapter 12. Skeleton-Movement and Locomotion

1. Our skeleton consists of bones (i) .... and (ii) ....
2. (ii) .... is the chief componet of our skeleton.
3. Shape wise the bones can be classified as (i) ...., (ii) ...., (iii) .... and (iv) ....
4. The central hollow partt of the long bones is known as ....
5. A bone becomes soft and flexible when placed in dilute ....
6. The skeletal system includes (i) .... (ii) .... and (iii)....................................... . Skeletal system is set of all the bones in the human body. It

provides (iv) .... and (v) .... to our body so we have some shape. Tendons attach our (vi) .... to muscles so we can move around. Ligaments attach bones to other (vii) .... We have (viii) .... bones in our body. Bones are made up of (ix) .... , (x) .... , sodium, and other minerals, as well as the protein collagen.

7. There are two types of girdles- (i).................. girdle and (ii) .... girdle
8. Opening at the back of skull is called (i) ....through which (ii) .... passes out.
9. In ear (i) ....small bones present are called ear (ii) ....
10. Vertebral column has 33 ring shaped bones out of which (i) .... fuse to form sacrum and 4 fuse to form (ii) ....
11. There are 24 ribs out of which seven are called (i) .... ribs because they are attached with (ii) .... in the front.
12. Biceps and (i) .... are (ii) .... muscles causing opposing movements.

[February 2020]

Ans. 1. (i) Cartilages, (ii) ligaments
2. Bone
3. (i) Long, (ii) short, (iii) flat irregular
4. Bone marrow
5. Weak hydrochloric acid
6. (i) tendons, (ii) ligaments, (iii) cartilage, (iv) strength, (v) rigidity, (vi) bones, (vii) bones, (viii) 206, (ix) calcium, (x) phosphorus
7. (i) Pelvic (ii) pectoral
8. (i) foramen magnum (ii) spinal cord
9. (i) three (ii) ossicles.
10. (i) 5 fuse (ii) coccyx.
11. (i) true ribs (ii) breast bone or sternum
12. (i) triceps, (ii) antagonistic

Chapter 13. "Skin-The Jack of all Trades"

1. Hair keeps the body ....
2. Mammary glands are modified ....
3. The subcutaneous tissue of man is rich in ....
4. Blood vessels and never fibres are absent in the .... of the skin.
5. .... pigmentt gives colour to the hair.
6. The skin synthesises vitamin.
7. .... is a hereditary skin disease.
8. Melanin pigment is located in the cells of .... layer.
9. The part of the hair which projects from the skin is ....
10. Skin is the (i) .... part of body of the body. It forms a protective covering and also forms a barrier to the entry of (ii) .... Skin can be broadly classified in to two parts – (iii)......................... and (iv) .... Epidermis is (v) .... layer which is further divided into (vi) .... sub parts – outer (vii) .... , middle (viii) .... and inner (ix) .... Root of hair lies in dermis though its elongated part (x) .... protrude out of epidermis.
11. Cells in the cornified layer are (i) .... and are made up of (ii) ....
12. Sweat glands are (i) .... tubes which open to skin surface through sweat (ii) ....
13. Blood capillaries supplu (i) .... and (ii) ....to skin and collect waste materials.
14. The skin can store (i)...................... and (ii) .... in its tissues.
15. Fat stored provides (i) .... and (ii) .... the body from heat and cold.
16. (i) .... is heating milk to a temperature of (ii) .... °C and then chilling quickly.

[February 2020]

Ans. 1. Warm
2. Sweat glands
3. Fat
4. Epidermis
5. Melanin
6. D
7. Albinism
8. Malpighian

9. Shaft
10. (i) largest, (ii) Germs, (iii) Epidermis, (iv) Dermis, (v) Outer, (vi) Three, (vii) Stratum corneum, (viii) Malpighian layer, (ix) Stratum germinative, (x) Hair shaft
11. (i) dead (ii) kertin
12. (i) coiled (ii) pore
13. (i) food (ii) oxygen
14. (i) fats (ii) water
15. (i) energy (ii) insulates
16. (i) Pasteurization (ii) 60

Chapter 14. The Respiratory System

1. Respiration is controlled by ....
2. Respiration is a .... process.
3. The right lung of man has .... lobes.
4. The total lung capacity of a human is .... litres.
5. During inspiration diaphragm ....
6. The functional unit of lungs is ....
7. Adam's apple is ....
8. The part of trachea concerned with sound, ...
9. .... + $6O_2$ $\xrightarrow{\text{enzymes}}$ 6 .... + 6 .... + 686 kcal
10. Complete the location and function in the following table.

Structure	Location	Function
Epiglottis	(i)	(ii) ....

11. All living organisms need energy which they get from (i) .... by burning it through respiration. During this process (ii) .... is consumed to produce energy, (iii) .... and water. Plants do not have any particular organ for exchange of gases. They exchange gases with help of (iv).......... and (v) .... which are present on stems and leaves. Thus we can say that plants do not show (vi) .... but they also respire. During the day time as photosynthesis is also taking place, plants use oxygen produced by this process. At night there is no photosynthesis, but (vii) .... continues. Thus, plants take oxygen from air and remove carbon dioxide. Roots absorb (viii) .... from the air present in air pockets in soil. If plant is watered too much and for long periods, then roots will not be able to (ix) .... and plant may die. Germinating seeds respire (x) .... as seed coat is impermeable to oxygen.
12. A small amount of energy liberated in the breakdown of the glucose molecule is released as (i) .... energy. But a major part of it is converted into chemical energy in the form of (ii) ....
13. The Krebs cycle provides (i) .... and (ii) .... needed for ETCs.
14. Respiration takes place in every cell of the body where (i) .... combines with (ii) ....
15. Inhaled air is rich in (i) .... and exhaled air has more of (ii) ....
16. Venous blood carries oxygen at (i) .... into the capillaries in lung whereas the pressure of carbon dioxide is (ii) ....
17. Glucose is incompletely brokendown to (i) .... alcohol in plants during (ii) .... respiration.

Ans. 1. enzymes 2. biochemical/catabolic 3. Three
4. 5-6 litres 5. Flattens 6. Alveoli 7. Larynx

8. Larynx
9. (i) $C_6H_{12}O_6$ (ii) $6O_2$ (iii) $6H_2O$
10. (i) Entrance of the trachea of windpipe.
(ii) To prevent food from entering the trachea while swallowing.
11. (i) food, (ii) glucose, (iii) carbon dioxide, (iv) lenticels, (v) stomata, (vi) breathing, (vii) respiration, (viii) oxygen, (ix) respire, (x) anaerobically
12. (i) heat (ii) ATP.
13. (i) hydrogen (ii) electrons
14. (i) oxygen (ii) food / nutrients.
15. (i) oxygen (ii) carbon dioxide.
16. (i) 40 mm (ii) 46 mm.
17. (i) Ethyl (ii) anaerobic.

Chapter 15. Hygiene

1. Hepatitis A is the infectious hepatitis transmitted through ....
2. Too much eating of (i) .... and (ii) .... spoils the teeth causing caries and gum infections.
3. .... improves blood circulation.
4. ....is injurious to health and may even lead to cancer.
5. For adults ....hours continuous undisturbed sleep is sufficient.
6. .... should be cleared everyday preferably in the mornings.
7. .... eat up mosquito larvae.
8. .... is a parasite which lives on rat's skin.
9. Tannery wastes contain the pathogen .... and are the most water serious water pollutants.
10. .... is a kind of natural scavenger cleaning and consuming the left-overs of the food, etc.
11. The municipal bodies must ensure the supply of clean germ-free (i) .... and an effecient (ii) .... disposal.
12. Trachoma: Eye:: Hepatitis: ....
13. Cholera:Vibrio cholera:: Amoebic dysentery: ....
14. Anopheles: .... :: Aedes: Dengue

Ans. 1. Contaminated water, 2. (i) Sweets (ii) chocolates 3. Physical exercise 4. Tobacco 5. 6-7 6. Bowels 7. *Gambusia* 8. Rat- flea 9. *Anthrax bacilli* 10. The housefly 11. (i) Drinking water, (ii) sewage 12. Liver 13. *Entamoeba histolytica* 14. Malaria

Chapter 16. Disease: Cause and Control

1. (i)............................. is a disease which is also known as 'Elephantiasis' which is caused by parasitic worm. It is said as 'Elephantiasis' because (ii) .... become swollen. The swelling is due to the blockage of (iii) .... by the worms. It is treated by eradication of the (iv) ....
2. One must always breathe through............
3. .... is a common disease of the eye.
4. Hepatitis spread through.........
5. Rats spread........... disease.
6. Yellow fever is caused by........
7. .... pairs of legs are present in housefly.

8. *Entamoeba histolytica* causes.......
9. *Mycobacterium tuberculosis* $\xrightarrow[\text{(disease)}]{\text{cause}}$ (i) .... $\xrightarrow[\text{(spread)}]{\text{mode of transfussion}}$ sputum of infected person.

   (ii) .... $\xrightarrow[\text{(appear)}]{\text{symptom}}$ Reddish eruptions on the chest $\xrightarrow[\text{(pathogen)}]{\text{causative germ}}$ (iii) ....

Ans. 1. (i) Filariasis, (ii) Legs, (iii) Lymph circulation, (iv) Vector
2. nose,
3. Trachoma,
4. Contaminated water,
5. Plague,
6. Aedes mosquito,
7. Three,
8. Amoebiasis.
9. (i) Tuberculosis, (ii) Typhoid, (iii) *Salmonella typhi*

Chapter 17. Aids to Health

1. .... is a blood serum protein produced in response to the injection of antigen.
2. Streptomycin is obtained from bacterium ....
3. Polio is a .... and given to the person of age ....
4. Toxoids are made up of .... which is secreted from ....
5. *Salvarson* was used to destroy germs caused by .... and ....
6. The diluted form Lyson, carbolic acid, benzoic acid, etc are used as ....
7. In T-cells, the T stands for ....
8. The process in which the killer cells invade the pathogen or even kill an infected body cell is termed as ....
9. When the walls of the blood capillaries get squeezed the process is termed as ....
10. World health day is celebrated in ....

Ans. 1. Antibody
2. Streptomyces
3. Oral vaccine, age 3-12 years.
4. Diluted formalin
5. Syphilis and sleeping sickness.
6. Antiseptic
7. T stands for thymus. 8.Local defence mechanism
9. Diapedesis
10. 7th April

Chapter 18. Health organisation

1. The World Health Organisation established in (i) ...., is a specialised agency of the (ii) ....
2. The two main organisations (i) .... and (ii) .... take steps individually at the national levels as well as collectively at the global level to safeguards the (iii) .... interests of their people.
3. The activity of WHO is to (i) .... and (ii) .... about the occurrence of (iii) .... diseases.
4. WHO has (i).......................... regional offices in the world and its headquarters are located in (ii) ....

Ans. 1. (i) 1948, (ii) UNO.
2. (i) Red Cross, (ii) WHO, (iii) health.
3. (i) collect, (ii) supply information, (iii) epidemic diseases.
4. (i) six, (ii) Geneva

Chapter 19. Waste Generation and Management

1. Waste is generated from (i) ....of our houses as well as all of the industries which produce various useful materials for human consumption. Waste may be categorised into (ii) .... and (iii) .... on the basis of possibility of them being degraded in nature by biological process. Waste (iv) .... is solution to tackle a complex problem of waste handling and help in keeping our (v) .... clean. It consist of (vi) ...., (vii) .... and (viii) .... of garbage, (ix) .... and other (x) ....products.
2. (i)............................ should be treated before (ii)................ in rivers.
3. (i) .... matter is left in open (ii) .... to decompose.
4. Electrostatic (i) .... and (ii) .... are used to reduce industrial pollution.
5. (i) .... is most commonly used method of disposing organic waste but it causes (ii) ....
6. Reuse, (i) .... and (ii) ....

Ans. 1. (i) kitchen, (ii) biodegradable, (iii) non biodegradable, (iv) management, (v) environment, (vi) collection, (vii) transportation, (viii) disposal, (ix) sewage, (x) waste

2. (i)Affluent, (ii) discharge
3. (i) Organic, (ii) land
4. (i) precipitator, (ii) (scrubber
5. (i) Incineration, (ii) pollution
6. (i) recycle, (ii) reduce

State Main Functions

Set 13

Q. State the main function of the following terms:

Chapter 1. Introducing Biology

1. Parasitology

Ans. Help to know about the parasites, their types and their effects on the body.

2. Eugenics

Ans. To improve the human species by selectively mating people with specific desirable hereditary traits, to reduces human suffering by "breeding out" disease, disabilities and so-called undesirable characteristics from the human population.

3. Ichthyology

Ans. Helps to study the behaviour and habits of fishes, their history, and their developmental patterns.

4. Horticulture

Ans. Helps to improve the ways how we use plants for food and other human purposes, as well as repairing the environment and personal aesthetics.

5. Palaeontology

Ans. Reconstructs the evolution of life on earth, support and help to understand the evolution with the help of fossils, foot prints, past climatic events, etc.

Chapter 2. Cell: The Unit of Life

1. Ribosome

Ans. Helps in the synthesis of protein.

2. Centrosome

Ans. Form asters and organise the formation of spindle fibres during cell division.

3. Glycogen granule

Ans. Serves as food for the cell.

4. Endoplasmic reticulum

Ans. A supportive framework for the cell and responsible for the synthesis and transport of proteins and fats.

5. Mitochondria

Ans. Responsible for the synthesis of respiratory enzymes and release of energy from pyruvic acid produced in cytoplasm in the form of ATP.

6. Plasma membrane

Ans. Separates contents of the cell form its surroundings. Regulates the entry of certain solutes and ions. Maintain the shape of animal cell.

7. Lysosomes

Ans. Intracellular digestion, Destroy foreign substances. When the cell is old or injured, lysosomes rapidly destroy cell organelles and hence, are called suicide bags.

8. Golgi apparatus

Ans. Synthesis and secretion of enzymes, hormones, etc. Formation of acrosome of sperm.

9. Cytoplasm

Ans. Different organelles contained in it perform different functions. All metabolic activities occur in it.

10. Asters of centrosomes

Ans. Initiates and regulates cell division. Forms spindle fibres.

11. Chromosomes

Ans. Carry genetic characters form parents to offspring.

12. Glycon granule

Ans. Serves af food for the cell.

13. Vacuole

Ans. Gives turidity to the cells. Storage of water and other substances, food, pigments and waste products.

Chapter 3. Tissues: Plants and Animal Tissue

1. Apical meristem

Ans. Tissue, that is responsible for the increase in the length of stem and the root.

2. Xylem parenchyma

Ans. It takes part in sideways transport of water and stores food

3. Ligament

Ans. Short bands of tough, flexible tissue which connect the bones of the body together.

4. Cardiac muscles

Ans. Involuntary in functioning and contract without getting tired soon.

5. Neuron

Ans. Highly specialised for receiving and conducting the messages or information from one part of the body to another

6. Adipose tissue **[November, 2019]**

Ans. Adipose tissue stones fat which acts as an insulator for retaining body heat.

7. Xylem **[November, 2019]**

Ans. A conducting tissue which helps in transport of water and dissolved salts from roots upwards to other parts of the plant.

Chapter 4. The Flower

1. Sepals

Ans. Protect the developing flower and keep it from drying out.

2. Pollen grains

Ans. Carry the male reproductive cell of plants and transfer it to the female reproductive cells or the ovules to facilitate the sexual reproduction to takes place.

3. Pistil

Ans. Helps transfer pollen and fertilises the seeds in the ovaries

4. Corolla

Ans. Bright in colours which attract insects for pollination.

5. Bracts

Ans. Attracts insects for pollination and protects inflorescences (flower structures).

Chapter 5. Pollination and Fertilisation

1. Pollen tube

Ans. Transports the male gamete cells from the pollen graineither from the stigma (in flowering plants) to the ovules at the base of the pistil or directly through ovule tissue in some gymnosperm.

2. Antipodal cells

Ans. Nutritive in function; it nourishes the embryo sac. Substances produced by the antipodals helps in the growth and development of the endosperm.

3. Synergids

Ans. Act as the producing centre of the signals which guide the pollen tube.

4. Micropyle

Ans. An opening in the ovule which remains as a small pore in the seed coat. This facilitates the entry of pollen tube into the ovule and oxygen and water into the seed during germination.

5. Embryo sac

Ans. Site of fertilisation of the egg and development of the embryo.

Chapter 6. Seeds: Structure and Germination

1. Micropyle present in seed

Ans. Through micropyle water enters into the seed and diffusion of respiratory gases also occurs through it.

2. Testa

Ans. Testa protects the delicate inner parts of seed from injury and from the attack of bacteria, fungi and insects.

3. Endosperm

Ans. Provides nourishment to the embryo.

4. Plumule

Ans. Develops into shoot.

5. Seed

Ans. Seed contains embryo that develops into a new plant.

6. Pyrogallic acid

Ans. A chemical that absorbs oxygen.

7. Seed coat **[November, 2019]**

Ans. Seed coat protects the embryo from mechanical damage.

Chapter 7. Respiration in Plants

1. Lime water

Ans. Used to test the presence of carbon dioxide gas which is evolved by respiration process in plants because carbon dioxide turns lime water milky.

2. Soda lime

Ans. Absorbs carbon dioxide gas so that it may not interfere with the results of respiration.

3. ATP

Ans. The energy currency of the cell, which provides energy to perform various vital activities.

4. Pneumatophores

Ans. The aerial roots found in mangrove plants that helps in respiration.

5. Carbolic acid

Ans. An antiseptic which prevents bacterial or fungal growth in dead seeds.

6. Enzyme

Ans. A protein which increases the rate of a chemical reaction.

Chapter 8. Five Kingdom Classification

1. Flagella

Ans. Locomotory organs found in protists which help in locomotion.

2. Mesosome

Ans. Mesosome increases the surface area of plasma membrane so that the cell can carry out various cellular functions like respiration, enzymes secretion, DNA replication, etc. more efficiently.

3. Rhizoids

Ans. Hair-like projections found in Bryophytes which provides anchorage to the substratum and also helps in absorption of water and nutrients from the substratum.

4. Osculum

Ans. Large opening present in sponges through which water containing wastes is excreted out from the body.

5. Clitellum

Ans. Helps in sexual reproduction in annelids.

Chapter 9. Economic Importance of Bacteria and Fungi

1. Antibiotics

Ans. Chemical substances which kills or inhibit the growth of other microbes.

2. Hyphae

Ans. Secretes some enzymes that digest the food in substratum and absorbs it *i.e.,* it helps in extracellular digestion.

3. Spore

Ans. Spores when fall into suitable substratum and under requisite heat and moisture germinates and form new mycelium, it is a mode of asexual reproduction of some fungi and also helps the microbes to withstand unfavourable conditions.

4. Capsule

Ans. Slimy outermost layer of bacterial cell that provides protection to bacteria from mechanical injury, drying, etc. and it enhances the ability of bacteria to cause disease. It also helps bacteria to attach to a surface.

5. Plasmid

Ans. Facilitate the replication process in bacteria, it also enhances the survival of an organism.

Chapter 10. Nutrition

1. Carbohydrates

Ans. The principal energy sources in the body. For example, sugars, starch.

2. Proteins

Ans. Proteins are needed for growth, repair and building cells of the body. For example, milk, meat, fish, etc.

3. Fats

Ans. Provides more energy that carbohydrates. It has less oxygen than carbohydrates. Common foods rich in fats are butter, cream, vegetable oils, etc.

4. Cellulose

Ans. A kind of carbohydrates which contributes in providing roughage for proper functioning of the gut.

5. Calcium

Ans. Contributes in clotting blood, require for muscle contraction. It is the constituent of bones and enamel.

6. Phosphorus

Ans. Required in bones and teeth for synthesis of nucleic acid, for muscle contraction and conduction of nerve impulse.

7. Vitamins

Ans. The chemical substances which help in maintaining good health. Some vitamins are fat soluble and some are water soluble.

8. Water

Ans. Water act as solvent in body to produce digestive juices, helps in transportation of digestive food and oxygen throughout the body and involve in maintaining temperature.

Chapter 11. Digestive System

1. Incisor

Ans. For cutting and biting the food.

2. Canine

Ans. For tearing the food.

3. Pepsin

Ans. converts proteins into proteoses and peptones.

4. Pancreas

Ans. It produces digestive juice for digestion and insulin a hormone.

5. HCl

Ans. It kills the bacteria that enter the stomach along with the food/activates propepsin of pepsingram into pepsin.

6. Oesophagus

Ans. It passes the food into the stomach by peristaltic movements.

7. Lacteal

Ans. Lacteal is present in the intestinal villi, help to absorb fatty acids and glycerol.

8. Epiglottis

Ans. It prevents the food to enter the trachea while swallowing.

9. Trypsin

Ans. Hydrolyses proteins into peptones and polypeptides.

10. Liver

Ans. Deamination, synthesis of RBC's in foetus, fat metabolism.

11. Lipase

Ans. Hydrolyses fats into fatty acids and glycerol.

12. Rennin

Ans. It converts milk protein caseinogen into insoluble casein.

13. Sucrase

Ans. Hydrolyses sucrose into monosaccharides.

14. Amylase

Ans. Hydrolyses starch into maltose.

16. Lactase

Ans. Hydrolyses lactose into glucose and galactose.

15. Maltase

Ans. Hydrolyses maltose into glucose.

Chapter 12. Skeleton: Movement and Locomotion

1. Pelvic girdle

Ans. 1. Articulation of femur.

2. Thoracic cage

Ans. 2. Protection of heart and lungs.

3. Synovial fluid

Ans. 3. Acts as a lubricant to absorb mechanical shock.

4. Cranium

Ans. 4. Protection of brain.

5. Skull **[November, 2019]**

Ans. 5. The skeleton of head which protects our brain (cranium) and forms our facial bones.

Chapter 13. Skin-"The Jack of all Trades"

1. Malpighian layer

Ans. Its cells divide to produce new cells of replace the worn out cells.

2. Sweat glands

Ans. They help in sweating thereby control temperature.

3. Sebaceous glands

Ans. To secrete sebum that makes the skin only.

4. Subcutaneous fat layer

Ans. To store fat and act as heat-insulating layer.

5. Hair

Ans. These are sensory and warm the body.

Chapter 14. The Respiratory System

1. Epiglottis

Ans. It closes the glottis during swallowing of food.

2. Diaphragm

Ans. The contraction and relaxation of diaphragm increases and decreases the volume of thoracic cavity during breathing.

3. Cartilaginous rings of trachea

Ans. It prevents the collapsing of trachea.

4. Mucus membrane inside the nasal passage

Ans. To make the air moist and traps the dust particles and micro-organisms.

5. Larynx

Ans. To produce voice.

6. Pleura

Ans. To protect the lungs

7. Pleural fluid **[November, 2019]**

Ans. To lubricate the lungs for its easy movements (expansion and contraction).

8. Inter-costal muscle

Ans. The relaxation of inter-costal muscles move the ribs inward and downward which decreases the volume of thoracic cavity.

9. Alveoli

Ans. To increase the surface area for exchange of gases.

10. Ciliated epithelium lining of respiratory tract.

Ans. It drives away/removes dust and any other foreign particles entering the respiratory tract.

11. Mitochondria

Ans. It is the seat of cellular respiration where pyruvate gets oxidised to release energy in form of ATP.

12. Ribs

Ans. Ribs during inspiration moves upwards and outwards thus chest cavity enlarges all around which forces the air to come inside. During expiration ribs move inwards and downwards decreasing the chest cavity and forceing the air to expels out. The movement of ribs is brought about by intercostal muscles.

13. Abdominal muscles

Ans. Abdominal muscles increases the space of thoracic cavity during inspiration by pushing the diaphragm towards abdominal cavity and allowing the air to enters into thoracic cavity. Similarly during expiration abdominal muscles pushes the diaphragm back to its original position, thus decreaseing the chest cavity and air expels out.

Chapter 15. Hygiene

1. Vectors

Ans. An agent that acts as an intermediate carrier of the pathogen.

2. DDT

Ans. A pesticide which is a serious water pollutant.

3. Mosquito

Ans. An insect that act as a vector and causes a number of diseases like malaria, dengue, filariasis, etc.

4. Pesticides

Ans. Substances that kill inescts and pests.

5. Pseudopodium

Ans. An extension in the cell membrane of organisms like *Amoeba* which is used to engulf food particles.

Chapter 16. Disease: Cause and Control

1. Anti-typhoid vaccination

Ans. Used for preventing typhoid and it should be repeated every year.

2. BCG

Ans. A vaccine is useful in developing immunity for tuberculosis.

3. Hepatitis B vaccine

Ans. A vaccine is used to help prevent infection from the hepatitis B virus.

Chapter 17. Aids to health

1. Phagocytosis

Ans. A process by which certain living cells called phagocytes ingest or engulf other foreign cell.

2. Immunoglobulin

Ans. The antibodies produced by plasma cells. Which act as critical part of the immune response by specifically recognised cell and adding in their destruction.

3. T-lymphocytes

Ans. Immune cells that protect the body from pathogens and cancer cells. T-cells originate from bone marrow and mature in the thymus. They are important for cell mediated immunity and the activation of the immune cells to fight infection.

4. Vaccination

Ans. The vaccine stimulates an immune response by the body that will fight off that type of illness.

5. Passive immunisation

Ans. Used when there is a high risk of infection and insufficient time for the body to develop its own immune response, or to reduce the symptoms of ongoing disease.

6. Disinfectants

Ans. Strong chemical substances that used to clean the floor to kill the germs. Cresol, phenol, Lysol, 40% formalin, lime, etc is common disinfectants.

7. Antibiotics

Ans. Chemical substances or in other words protein that found in the blood that works against the germs.

8. Vaccine

Ans. A vaccine is injected e.g., polio drops, it involve in the stimulation of WBCs to produce antibodies against the particular disease.

9. Sulpha drugs

Ans. Used as antibiotics against the germs of syphilis that causes the disease named as sleeping sickness.

Chapter 18. Health Organisation

1. UNO

Ans. UNO helps in creating the need ofan international body to look after the health problems of people of the world.

2. Red cross

Ans. Red Cross helps the victim of any calamity and supply blood for the needy victim.

3. WHO

Ans. WHO looks after the health problems of people of the world.

4. Health Organisation

Ans. Health organisation safeguards the health interests of the people by taking steps individually at the national level as well as collectively at the global level.

5. International bodies

Ans. International bodies look after the people suffering from calamities and provide their requirement and thus look after the health problems of people of the world

❑

Short Answer Type Questions | Set 14 |

Chapter 1. Introducing Biology

1. Define the following:

   (i) Biology (ii) Applied biology (iii) Human Biology

Ans. (i) Biology may be defined as the study of living organisms. It is divided into many specialised fields that cover their morphology, physiology, anatomy, behaviour, origin and distribution.

(ii) Applied biology is the understanding of how organisms work from subcellular to whole organism level.

(iii) Human biology is the study of humans as a living organism and his relationship with other living organisms.

2. How genetics help people to understand their characteristics features transferred to their Offsprings?

Ans. Genetics is a branch of biology which is related to the study of genes, genetic variation, and heredity in organisms. All the genetic information is transferred from parent to their offsprings with the help of DNA.

3. How biology becomes important for human life?

Ans. The study of biology becomes important for human life as it helps to study the:

(i) Importance of interdependence and independence.

(ii) Diversity of life forms and their importance to humans.

(iii) Workings and difficulties of the living forms and then develop respect for all life forms.

(iv) Need for nutrition, health care, animal husbandry and natural resources.

(v) Cause, effect, and principles of treatment of various diseases and improve the quality of human life.

(vi) Help to solve problems of population, pollution, and conservation

4. How does Biology help us in leading a healthier life.

Ans. Biology is a study of living things; hence it helps in understanding the behavioural characteristics of the living. The knowledge and study of biology has immensely helped in human health and advancement. Basic knowledge can help inself-analysing causatives agents of diseases, different microbes causing diseases, etc. Thereby examining and studying their mode of living and propagating diseases. This study has paved the way for production of new drugs synthesised naturally or synthetically killing causatives of diseases.

5. Identify the branch which deals with the study of the following animals:

   (i) Lizard (ii) Silkworm (iii) Fishes (iv) Parasites

   (v) Insects

Ans. (i) Herpetology (ii) Sericulture (iii) Ichthyology (iv) Parasitology

(v) Entomology

Chapter 2. Cell: The Unit of Life

1. Define plasma membrane and explain its functions.

Ans. **Definition**: Plasma membrane (also called cell membrane) is a network of lipids and proteins that forms the boundary between a cell's contents and the outside of the cell.

Function: It allows the outward and inward movement of molecules across it. The movement of molecules across the plasma membrane takes place by diffusion, osmosis, active transport,phagocytosis (cell eating) and pinocytosis (cell drinking).

2. What is the cell theory? Who propounded it and when?

Ans. Postulates of cell theory:

(i) Cell is the smallest unit of structure of all living things.

(ii) Cell is the unit of function of all living things.

(iii) All cells arise form pre-existing cells.

Cell theory was propounded by Theodor Schwann and Matthlas Scheiden in the year 1839 and was modifed by Rudolf Virchow in 1858.

3. Explain the function of chloroplast.

Ans. Chloroplast is a double membranous cell organelle. It contains a green colour photosynthetic pigment called chlorophyll. Chlorophyll is important for photosynthesis as it helps to channel the energy of sunlight into chemical energy. With photosynthesis, chlorophyll absorbs energy and then transforms water and carbon dioxide into oxygen and carbohydrates.

4. Name the part of the cell concerned with the following:

(i) It comprises RNA and protein and is found in both prokaryotes and eukaryotes

(ii) Centre for synthesis of ribosomal RNA.

(iii) It bears ribosomes on their surface and actively involved in the synthesis and storage of protein.

(iv) It is found only in the animal cell and contains short bundles of microfilaments arranged at right angles to each other.

(v) These are the sites where cell respirations occur to release energy in the form of ATP.

Ans. (i) Ribosome (ii) Nucleolus (iii) Rough endoplasmic reticulum (RER)
(iv) Centrosome (v) Mitochondria

5. Mention three features found only in plant cells and one found only in animal cells:

Ans. Features found only in plant cells:

(i) Presence of cell wall

(ii) Presence of large vacoles. The liquid contained in vacuoles is called cell sap

(iii) Presence of plastids

Features found only in animal cells:

(i) Presence of centrosome

6. Explain the characteristic feature of eukaryotic cell and identify the cell organelle which is involved in forming complex sugar from simple sugars.

Ans. Eukaryotic cells are cells that contain a nucleus and organelles enclosed by a plasma membrane. Eukaryotic cells are larger and more complex than prokaryotic cells, which are found in Archaea and Bacteria, the other two domains of life.

Golgi apparatus is involved in forming complex sugar from simple sugars. It gathers simple molecules and combines them to make more complex molecules. It then takes those big molecules and packages them into vesicles and either stores them for later use or sends them out of the cell.

7. Define the following:

(i) Cell theory (ii) Protoplasm (iii) Plastid (iv) Chromatin fibres
(v) Centrioles

Ans. (i) Cell theory is the scientific theory which explains that living organisms are made up of cells and the cells are the basic structural and organisational unit of all organisms. Also according to this theory all cells come from pre-existing cells.

(ii) Protoplasm is the colourless material comprising the living part of a cell, including the cytoplasm, nucleus and other organelles.

(iii) Plastids are doubled membrane 'sac-like' organelles. It is generally involved in either the manufacture or storage of food. They include chloroplasts, chromoplasts and leucoplasts.

(iv) Chromatin fibre are coiled and condensed to form chromosomes.

(v) Centriole is a small structure which is made of microtubules. Microtubules exist as part of the centrosome which helps to organise microtubules in the body.

8. Give abbreviations

(i) SER (ii) RER

(iii) ATP (iv) DNA

(v) RNA

Ans. (i) Smooth endoplasmic reticulum

(ii) Roughendoplasmic reticulum

(iii) Adenoisne triphophate

(iv) Deoxyribonucleic acid

(v) Ribonucleic acid

9. Name the plastid and pigment likely to be found in the cells of:

(i) petals of sunflower (ii) ripe tomato

(iii) skin of green mango (iv) cells of potato

Ans.

Cells	Plastid	Pigment
(i) petals of sunflower	Chromplasts	Xanthophyll
(ii) ripe tomato	Chromplasts	Carotene
(iii) skin of green mango	Chloroplasts	Chlorophyll
(iv) cells of potato	Leucoplasts	No pigment

Chapter 3. Tissues: Plants and Animal tissue

1. Can you consider a cluster of eggs as a tissue? Why?

Ans. A tissue is a group of similar cells from the same origin that together carry out a specific function. An egg is a zygote or a cell but a cluster of eggs cannot be considered as a tissue as it does not form a organ like a tissue. Instead it gives rise to a new individual organism if gets fertilised.

2. Write down the characteristic feature of muscle tissue found in the wall of heart.

Ans. Cardiac muscle tissue is found in the wall of heart.

Characteristic feature of the cardiac muscle tissue:

(i) It is a striated, involuntary muscle.

(ii) The cells of these muscles are cylindrical, branched and uninucleate.

(iii) Fibres of these muscles contain large number of mitochondria and glycogen granules.

3. Write abbreviations of the following:

(i) RBC (ii) WBC

Ans. (i) Red blood cells (i) White blood cells

4. (i) Make a flow chart on the classification of connective tissues.

(ii) Write a short note on lymph.

Ans. (i)

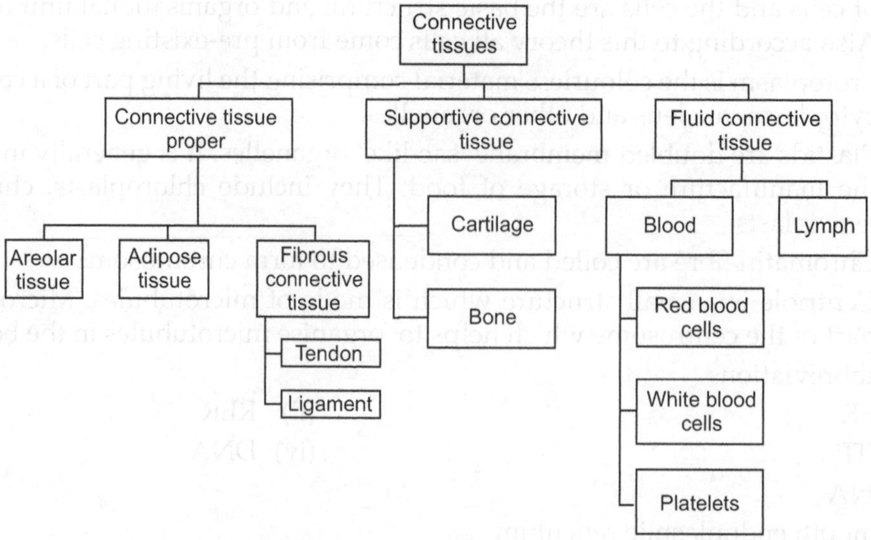

(ii) Lymph is a fluid containing blood plasma with infection-fighting white blood cells, throughout the body. It is a type of fluid connective tissue which is mainly concerned with transportation and immunity.

5. Name the three kinds of muscles found in the human body. In each case, name one region in the body where they are found.

Ans. (i) **Striated muscles:** Provide the force for locomotion and all voluntary movements of the body. These muscle are found in the limbs.

(ii) **Unstriated muscles:** Provide movements for the passage of food in the intestines. These muscles are found in iris of the eye, lining of blood vessels, urinary bladder, etc.

(iii) **Cardiac muscles:** Provide rhythmic contration and relaxation movements. These muscles are found only in the heart.

6. Explain the following:

(i) Companion cells (ii) Dendrite (iii) Meristematic tissues

Ans. (i) Companion cells are the type of living cells found within the phloem of flowering plants. Each companion cell is usually closely associated with a sieve element. It regulates the activities of a sieve cell.

(ii) Dendrites are tree-like extensions at the beginning of a neuron that help to increase the surface area of the cell body. They receive stimulation in order for the cell to become active. They conduct electrical messages to the neuron cell body for the cell to function.

(iii) Meristematic tissues are those tissues in which the cells remain young and immature with the power of continuous division. They have ability to enlarge, stretch and differentiate into other types of cells as they mature. They are found at all the growing points of a plant such as tips of roots, stems and branches.

7. What are the functions of areolar tissue and adipose tissue?

Ans. **Function of areolar tissues:**

(i) It fixes skin to underlying muscles.

(ii) It helps in repairing of tissue after an injury.

(iii) It acts as supporting and packing tissue between organs lying in the body cavity.

Function of adipose tissues:

(i) It serves as a fat reservoir.

(ii) Adipose tissue acts as food reservoir by storing the fat.

(iii) It acts as an insulator and regulates the body temperature.

8. What are involuntary muscles? Explain with the help of diagram. Where are they found?

Ans. The muscles which do not move on our will are called involuntary muscles. These muscles are also called as smooth muscles. They are made up of spindle shaped tapering cells which are uninucleate.

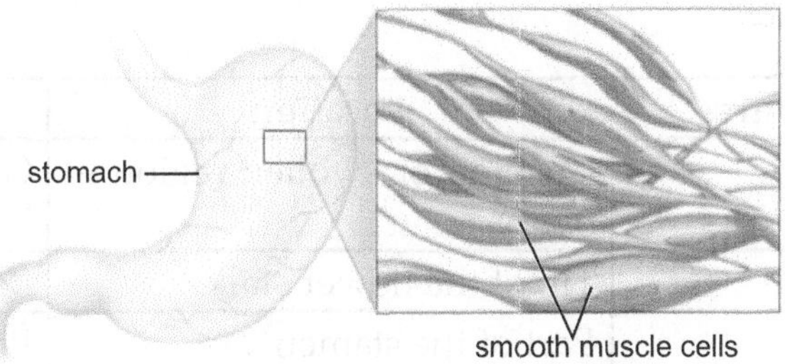

Smooth muscles (involuntary muscles)

The movement of food in the alimentary canal or the contraction and relaxation of blood vessels are involuntary movements. They are also found in the iris of the eye, in ureters and in bronchi of the lungs.

9. List the tissues found in the human heart.

Ans. (i) Muscular tissue (Cardiac muscles)

(ii) Epithelial tissue (Lining of blood vessels of the heart)

(iii) Connective tissue (Fluid connective tissue in the form of red-blood corpuscles)

10. Name the kind of cells found in the following places:

(i) Surface of the human skin

(ii) Salivary gland

(iii) Brain

(iv) Inner lining of the wind pipe

Ans. (i) Squamous epithelium

(ii) Cuboidal epithelium

(iii) Neuron

(iv) Cuboidal epithelium

Chapter 4. The Flower

1. Define the following terms:

(i) Perianth (ii) Staminate flower (iii) Perigynous flower

Ans. (i) Perianth is the floral structure consists of the calyx (all sepals) and the corolla (all petals) especially when the two whorls are fused.

(ii) Staminate flower is a male flower, bearing only stamens.

(iii) In perigynous flower, the gynoecium is present in the centre and the rest of the floral parts are arranged at the rim of the thalamus at the same level, e.g. plum and rose.

2. Write a short note on calyx.

Ans. Calyx (also called sepals) is the outermost, lowermost, non-essential but protective whorl of the flower. These are typically green and enclose the rest of the flower in the bud stage; however, they can be absent or prominent and petal-like in some species. Sepals are capable of photosynthesis due to the presence of chlorophyll. They protect the bud of the flower.

Sepals may persist when flower blooms or they may fall off. Two patterns can be seen in calyx as follows:

(i) Gamosepalous: Sepals fused to form a cup-like structure.

(ii) Polysepalous: Sepals are separate.

3. Where the following structures/parts located and what are their functions?
   (i) Placenta (ii) Thalamus
   (iii) Anther (iv) Stigma

Ans.

	Structures	Location	Function
(a)	Placenta	Cushion or swollen region in ovary	Given origin to ovules
(b)	Thalamus	Tip of the flower stalk	Bears all the parts of the flower
(c)	Anther	Part of the stamen	Produces male gametes or pollen grains
(d)	Stigma	Terminal knob-like part	Serves as the landing place for pollen grains during pollination

4. Explain the sexuality in flowers.

Ans. A flower can be divided based on the presence of reproductive structure. These are as follows:

(i) A flower that contains both male and female reproductive structures is called a bisexual flower. Bisexual flowers are also called hermaphrodites. They are perfect flowers.

(ii) A flower that has only one reproductive structure, either male or female, is called a unisexual flower.

(iii) A unisexual flower, which contains only androecium, is called a male or staminate flower.

(iv) A unisexual flower, which contains only gynoecium, is called a female or pistillate flower.

(v) A flower in which both male and female reproductive organs are absent is called a neuter flower.

5. Explain the part of gynoecium that determines the compatible nature of pollen grain.

Ans. Gynoecium (also called pistil) is the female organs of a flower, comprising the stigma, style and ovary. Stigma determines the compatible nature of pollen grain. It has the ability to recognise, the pollen, whether it is of the right type (compatible) or of the wrong type (incompatible). If it is of the right type, it accepts the pollen for pollination that lead to fertilisation. If the pollen is of wrong type, the pistil rejects the pollen.

6. Explain the types of flower based on the fusion of stamens.

Ans. Based on the fusion of stamen, flowers are of following types:

(i) Polyandrous: In this, the stamens in a flower remain free, e.g., cruciferae.

(ii) Monadelphous: In these, the stamens are united into one bundle, e.g., china rose.

(iii) Diadelphous: These have the stamens united into two bundles, e.g., pea.

(iv) Polyadelphous: In polyadelphous, the stamens are united into more than two bundles, e.g., citrus. There may be a variation in the length of filaments within a flower, as in Salvia and mustard.

Chapter 5. Pollination and Fertilisatino

1. Define the following:
   (i) Triple fusion
   (ii) Anemophilous flower
   (iii) Double fertilisation

Ans. (i) Triple fusion is a fusion that involves a sperm nucleus and two polar nuclei which occurs in the double fertilisation in a seed-bearing plant that results in the endosperm formation.

(ii) The flowers pollinated by wind are known as the anemophilous flower. These flowers are usually characterised by small size, inconspicuous and also nectar or fragrance, etc. It produces large number of pollen grains which are smooth, light and dry.

(iii) Double fertilisation is characteristic of seed plants. It is process in which one sperm nucleus fuses with the egg nucleus to form an embryo and another fuses with polar nuclei to form endosperm.

2. Explain the following terms:

(i) Ornithophily (ii) Elephophily

(iii) Artificial pollination.

Ans. (i) Ornithophily: The process of pollination affected by birds.

(ii) Elephophily: The process of pollination affected by elephants.

(iii) Artifical pollination: The process of pollination affected by men through artificial modes.

3. What happens to the following after fertilisation?

(i) Ovules (ii) Calyx (iii) Petals (iv) Stamens

Ans. (i) Ovules become the seed after fertilisation.

(ii) After fertilisation, the calyx sheds or falls or remains intact in a shriveled and dried form.

(iii) Post by fertilisation, the petals shed and fall.

(iv) Once fertilisation takes place, the stamens fall off.

4. Mention any two contrivances in flowers which favour cross-pollination.

Ans. The contrivances in flowers that favour cross-pollination are:

(i) The property of self-fertilisation

(ii) Unisexuality

(iii) Dichogamy: The differing maturation timings of gynoecium and androecium

(iv) The various structural barriers.

5. Name the parts of the ovary which give rise to:

(i) Seed............. (ii) Fruit............. (iii) Fruit wall ....

Ans. (i) Ovules (ii) Ovary (iii) Ovarian wall

6. Give three advantages and disadvantages of self-pollination.

Ans. **Advantages of self-pollination** are:

(i) It ensures continuity of the race.

(ii) It helps to preserve the parental characters as the gametes from the same flower are involved.

(iii) Flowers need not be showy nor need them to produce nectar or scent.

Disadvantages of self-pollination are:

(i) The genetic defects cannot be eliminated from the breed.

(ii) Repeated self-pollination leads to loss of vigour and the vitality of the breed. The seeds produced are smaller and weaker.

(iii) It does not yield new varieties.

7. State the name of the chief pollinating agent against the corresponding plant by choosing from those given in brackets.

(i) Dahila .... (Crow, butterflies, mosquito)
(ii) Maize .... (Bess, locusts, rain, wind).
(iii) Vallisneria .... (Wind, water, ants, rabbits).

Ans. (i) Butterflies (ii) Wind (iii) Water

8. What are the advantages of the following in the flower to the plant concerned?
(i) Long and feathery stigma
(ii) Brightly coloured petals
(iii) Smooth and light pollen
(iv) Protruding and easily movable anthers
(v) Fragrant nectar

Ans. (i) In wind-pollination, long and feathery stigma assists in trapping the pollen grains
(ii) The bright colour of the petals causes the insects to get attracted for cross-pollination.
(iii) Smooth and light pollen are easily carried away by the wind facilitating cross-pollination.
(iv) Protruding and easily movable anthers is the factor which brings about their movement even with the slightest wind.
(v) Fragrant nectar attracts insects for the process of pollination.

9. Give two examples of the following:
(i) Wind pollinated flower (ii) Insect pollinated flower
(iii) Heterostyly (iv) Protogyny

Ans. (i) Maize and Pine (ii) Sweet pea and *Salvia*
(iii) Prim rose and *Oxalis* (iv) Custard and Apple

10. Name the devices found in the flowers which favour cross-pollination. Explain any two.

Ans. The devices found in flowers which favour cross-pollination are unisexuality, dichogamy, self-sterility, herkogamy and heterostyly.
(i) Unisexuality is a condition in which plants possesses either stamen or carpels but not both. The plants may be monoecious which bear both male and female flowers. For example, maize, pumpkin, cucumber, castor, etc.or dioecious in which male and female flowers are borne by two different plants, *i.e.,* the staminate flowers are born by male plants and the pistillate flowers by the female plants. For example, mulberry, betel, papaya, cannabis, *Vallisneria,* etc.
(ii) Dichogamy a process in which stamens and carpels of a bisexual flower mature at different times. It may be of two types:
(1) Protandry is a condition in which the male part (*i.e.,* anther) mature earlier than female part (*i.e.,* stigma) thereby ensuring that self-fertilization does not occur.
(2) Protogyny is a condition in which the stigma matures earlier than anthers.

11. What happens in the following after the fertilisation?
(i) Placenta (ii) Ovary wall (iii) Secondary nucleus
(iv) Antipodal cells (v) Inner integument

Ans. (i) Stalk of the seed (ii) Pericarp (iii) Endosperm
(iv) Disorganised (v) Tegmen

12. Describe the advantages and disadvantages of cross-pollination to the plant.

Ans. The advantages of cross-pollination are.
(i) Nature favours cross-pollination
(ii) The offspring produced are healthier
(iii) The seeds produced are viable and in abundance.
(iv) Cross-pollination can be used to produce new varieties using two different varieties of the same species or even two species.

The disadvantages of cross-pollination:
(i) As some pollinating agents are always required which may not be available at the

required time, polination is not always certain.

(ii) Wastage of pollen takes place as the pollen has to be generated in abundance to ensure pollination

(iii) It is uneconomical for the plant as the flowers are required to be fragrant, vibrant, large and producing nectars to attract the agents causing pollination.

13. What is the function of the pollen tube? Explain it with the help of a diagram.

Ans.

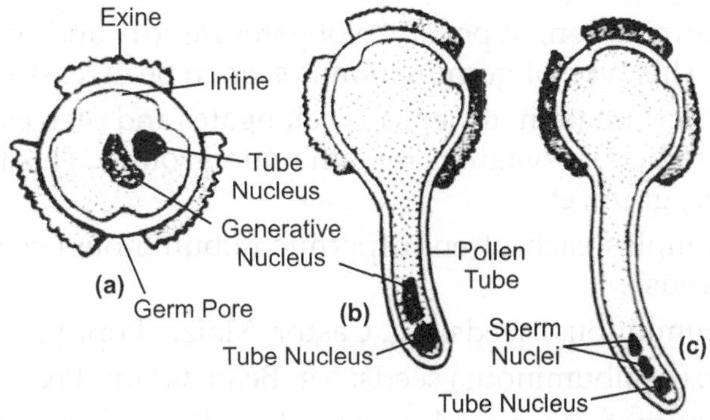

The pollen tube emerges out of the grains by breaking through its exine, growing from the style and stigma as a result of dissolving the tissues brought about by the enzymes, finally reaching the ovary. Here, it makes way to the ovule via the micropyle which is a small pore.

25. State two advantages of cross pollination. **[November 2019]**

Ans. Tow advantages of cross pollination are:

(i) Offspring's produced are healthier.

(ii) New varieties are produced.

Chapter 6. Seeds-Structure and Germination

1. (i) What do you mean by germination of seed?

(ii) Give a brief description of the conditions necessary for seed germination.

Ans. (i) Germination is the process of formation of a seedling developed from the embryo.

(ii) The conditions necessary for seed germination are-

(1) **Water**- Seed absorbs water from soil through micropyle so that the seed coat ruptures and radicle and plumule can come out to form root and shoot system respectively. Water also activates enzymes that acts on stored food to convert into diffusible form, which can be utilised by growing embryo.

(2) **Suitable temperature**- An optimum temperature of about 25-35^{0}C is required which favours germination process.

(3) **Oxygen-** Oxygen is required for oxidation of food which will provide energy for rapid cell division and cell growth during germination of seed.

2. It is commonly observed that potatoes kept in a basket during late rainy season start giving out small shoots. Would you consider it germination? Explain to support your answer.

Ans. Yes, this process is called germination. Potato is a tuber and it has auxillary buds which under suitable temperature, water and availability of oxygen germinates and give rise to small shoots.

3. Classify the seeds based on the storage of food materials in them and give suitable examples.

Ans. On the basis of number of cotyledons seeds are classified into following types:

(i) Monocotyledonous: Only one cotyledon is present. Example: maize, wheat, etc.

(ii) Dicotyledonous: Two cotyledons are present. Example: pea, gram, etc.

On the basis of presence or absence of endosperm, seeds are:

(i) Albuminous: Endosperm present. Example- custard apple, cereals, millets etc

(ii) Exalbuminous: Endosperm absent and cotyledons stores food and are thick and fleshy. Examples- gram, mango, pea, etc.

4. What is the difference between an embryo and a seed?

Ans. A seed is a mature ovule after fertilisation. It contains a tiny living-plant called the embryo. The embryo with in the seed remains in an inactive or dormant state, until it is exposed to favourable conditions, when it germinates.

5. What is the role played by:
   (i) hypocotyl in epigeal germination
   (ii) epicotyl in hypogeal germination

Ans. (i) In epigeal germination, hypocotyl elongates rapidly and cotyledons are pushed above the ground. This type of germination is seen in bean, castor, etc.

(ii) In hypogeal germination, the epicotyl elongates and pierces out of the seed in opposite direction to radicle, cotyledons remain underground. This type of germination is seen in maize, pea, gram, etc.

6. Give any two examples each of endospermic (albuminous) seeds, and non-endospermic (exalbuminous) seeds.

Ans. Endospermic (albuminuous) seeds: e.g. Castor, Maize, Poppy.
Non-endospermic (Exalbuminous) seeds: e.g. Bean, Gram, Pea

7. With regard to germinate bean seed, answer the following questions:
   (i) State the function of the micropyle.
   (ii) Name the parts of seed which grows into seedling
   (iii) Draw a labelled diagram of structure named above.
   (iv) Name the part of seed which provides nutrition for growing seedling.

Ans. (i) The 'micropyle' serves two important functions:
Allows absorption of water and makes it available to embryo for germination.
Enables diffusion of respiratory gases for the growing embryo.

(ii) The embryo of the seed grows into seedling.

(iii) Structure of embryo

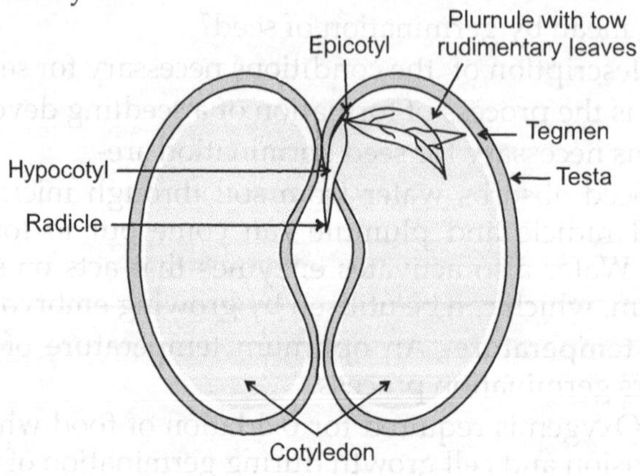

(iv) The cotyledons of the seed provide nutrition for the growing seedling or the embryo.

Chapter 7. Respiration in Plants

1. Explain respiration along with chemical equation? What are the characteristics of respiration?

Ans. Respiration is a catabolic process where glucose breaks down in a series of steps to release carbon dioxide, water and energy. The energy released is used to carry out various life activities.

$$C_6H_{12}O_6 + 6O_2 \rightarrow 6CO_2 + 6H_2O + \text{Energy}$$

The characteristics of respiration are:
(i) It is a catabolic process which occurs at body temperature.
(ii) It is an enzyme catalyzed reaction.

(iii) It is a biochemical process where energy is liberated in form of ATP and heat.

2. Why should anyone not sleep under the tree during the night?

Ans During night, plants only respire *i.e.*, they take oxygen in and release carbon dioxide. No photosynthesis occurs at night so carbon dioxide is not utilised by plants. If a person sleeps unders the tree during night he will inhale the carbon dioxide released by the tree which may prove harmful to him.

3. What is the importance of fermentation? Name the organism which play an important role in this process.

Ans. Fermentation process mainly occurs in yeast. It is used to produce beer, wine, cheese, etc. It is a type of anaerobic respiration where glucose is converted into ethyl alcohol and carbon dioxide.

4. Give abbreviations:

(i) ATP (ii) ADP (iii) TCA (iv) PYR

Ans. (i) Adenosine Triphosphate (ii) Adenosine diphosphate

(iii) Tricarboxylic acid cycle (iv) Pyurvic acid

5. What do you mean by respiration? List different ways of gaseous exchange in plants.

Ans. Respiration is a biochemical process which occurs in a series of steps where glucose is oxidised to release energy that is used to carry out various metabolic activities in living organisms.

The different ways of gaseous exchange in plants are:

(i) Through stomata in leaves

(ii) Lenticels in old woody stems

(iii) General surface of roots

(iv) Pneumatophores present in mangroves plants.

6. When a substrate is being metabolised, why does not all the energy that is produced get released in one step? It is released in multiple steps. What is the advantages of step-wise release?

Ans. The step-wise release of energy is advantageous because if all the energy would be released at once most of the energy would be lost in the form of heat energy. So there would be wastage of maximum energy. If it is released in step-wise, energy can be properly utilised by cells to carry out their metabolic activities and excess energy can be stored with minimum or no wastage.

7. Give the overall chemical reaction of the following-

(i) Anaerobic respiration in plants

(ii) Anaerobic respiration in animals.

Give two organisms which respire anaerobically only, throughout their lives.

Ans. (i) $C_6H_{12}O_6 \rightarrow 2C_2H_5OH + 2CO_2 + 2ATP$

(ii) Lactic acid + 2ATP + heat energy

Yeasts, *Escherichia coli*, fungi respire anaerobically throughout their lives.

8. What are the three important aspects about the overall chemical equation of respiration, pertaining to the following:

(i) Single or several steps

(ii) Direct or enzyme catalysed

(iii) Forms of energy liberated.

Ans. (i) Respiration occurs in a series of chemical steps.

(ii) It is carried out by enzymes.

(iii) Energy is liberated as ATP and heat energy.

9. What happens to the energy liberated in respiration?

Ans. The energy which is liberated in respiration is used to carry out various life processes like digestion, excretion, reproduction, etc. Some energy is stored in form of ATP. A part of the energy is also lost in form of heat to the surroundings.

10. Can cell respiration occur in any organism at a temperature of about 65°C? Give reason.

Ans. Cell respiration cannot occur in an organism at a temperature above 65°C because the enzymes which are involved in respiration process gets inactivated at higher temperature.

11. The following two chemical reactions are supposed to indicate a certain process occurring in the green plants under two different conditions.

(a) $C_6H_{12}O_6 + 6O_2 \rightarrow 6CO_2 +$ ______ + 38ATP

(b) $C_6H_{12}O_6 \rightarrow$ ______ $+ 2CO_2 + 2ATP$

(i) Fill in the blanks of each reaction.

(ii) Name the process represented by the two chemical equations.

(iii) What are the conditions under which the two reactions 'a' and 'b' are occurring respectively?

Ans. (i) (a) $6H_2O$ (b) $2C_2H_5OH$

(ii) Reaction 'a' represents aerobic respiration
Reaction 'b' represents anaerobic respiration.

(iii) Reaction 'a' occurs in presence of excess supply of oxygen.
Reaction 'b' occurs in absence of oxygen.

11. What is respiration? How are respiration and burning similar and how are they different?

Ans. Respiration is a catabolic breakdown of glucose in a series of chemical steps catalysed by enzymes to release energy in form of ATP to carry out various life processes.

Respiration and burning are similar as they both require oxygen, produce energy, produce carbon dioxide and water.

Differences between respiration and combustion.

Respiration	Burning
It is a biochemical process which occurs in a series of steps.	It is a physio-chemical process which occurs in a single step.
It is a cellular process carried out by enzymes.	It is a non-cellular process carried out by heat.
It occurs at body temperature.	It occurs at very high temperature.
Energy is liberated as ATP and heat.	Energy is liberated as heat and light.

12. Describe one experiment each you would perform to demonstrate the following phenomena for the germinating seeds:

(i) produce heat

(ii) give out carbon dioxide

(iii) can respire even in total absence of air.

Ans. (i) Experiment to show germinating seeds produce heat:

Two thermos flasks were taken marked as A and B. Soaked bean seeds were taken in flask A and boiled bean seedswashed with formalin were taken in flask B.

A thermometer was inserted in each of the flasks and their mouths were plugged with cotton wool as shown in figure below. The initial thermometer readings of both were noted.

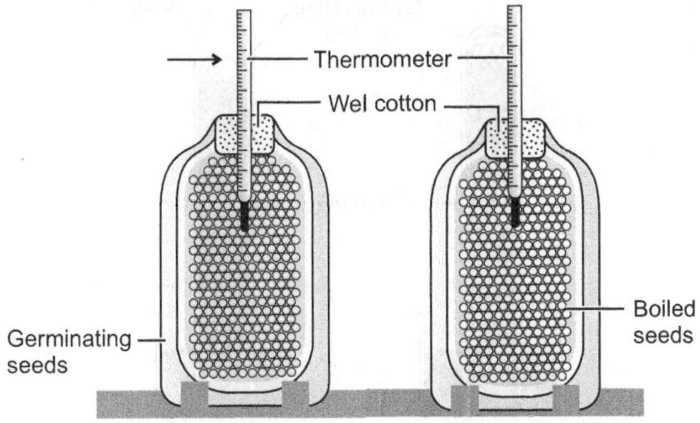

Flask A **Flask B**

After few hours it was seen that thermometer of flask A shows higher readings which indicates heat is liberated by germinating seeds due to respiration. Thermometer in flask B did not show any increase in readings as seeds were dead.

(ii) Experiment to show germinating seeds give out carbon dioxide:

Two flasks A and B were taken and some wet cotton wools were placed at the bottom of both the flasks. Soaked bean seeds were taken in flask A and boiled bean seeds washed with formalin were taken in flask B.

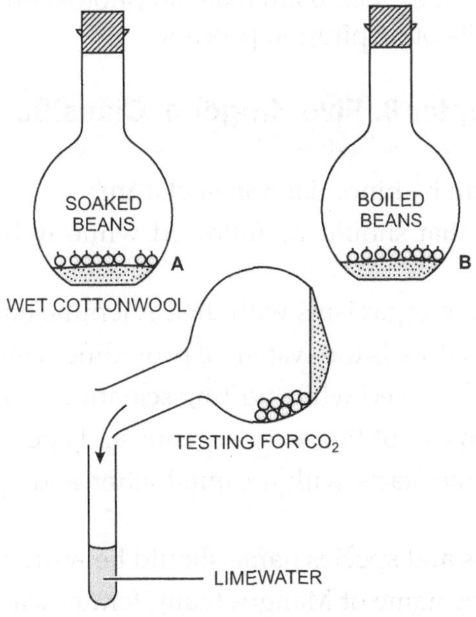

The flasks were corked and kept undisturbed for few days. Seeds in flask A show germination whereas seeds in flask B were not germinated.

Gases in each flasks were tested by removing corks from mouth of the flasks. Flask A was first tilted over a test tube containing lime water followed by flask B. When flask A was tilted over the test tube containing lime water, it turns milky showing the presence of carbon dioxide gas. But flask B did not show any effect. This shows carbon dioxide is liberated by germinating seeds due to respiration.

(iii) Germinating seeds can respire even in total absence of air:

Few soaked and peeled off seeds were pushed into the mouth of a test tube filled with mercury and inverted in a beaker of mercury. Seeds float on top surrounded by mercury.

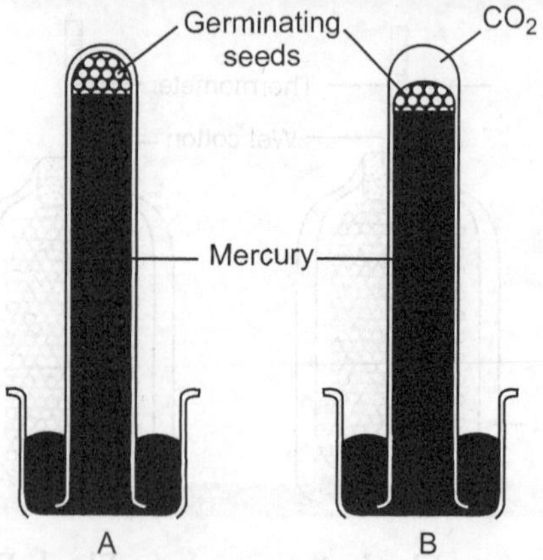

After few days it would be seen that mercury level had fallen down and presence of some gas could be detected. If some KOH would be introduced to the tube it would be seen that evolved gas would be absorbed and mercury level again rises up. This confirms gas evolved was carbon dioxide.

This experiment proves that germinating seeds can respire even in absence of air.

13. Suppose we conduct an experiment to demonstrate respiration in a green plant. What special precaution is required to be taken in it?

Ans. The apparatus containing the green plant should be covered with a black cloth, so that sunlight may not fall on the green plant to avoid photosynthesis process as this process may interfere with the results of respiration process.

Chapter 8. Five Kingdom Classification

1. (i) What do you mean by binomial nomenclature?

(ii) Give some rules that should be followed while writing scientific names. Explain it with an example.

(iii) Give some common organisms with their scientific names.

Ans. (i) Binomial nomenclature is the system of providing scientific names to living organisms.

(ii) Some rules to be followed while writing scientific names are:

(1) Each name consists of two words, genus and species.

(2) The genus name starts with a capital letter and species name starts with a small letter.

(3) Both the genus and species name should be written in italics or underlined.

Example: Scientific name of Mango- Mangiferaindica

(iii)

Common name	Scientific name
Man	*Homo sapiens*
Domestic cat	*Felisdomesticus*
Peepal tree	*Ficus religiosa*
Maize	*Zea mays*
Silk moth	*Bombyx mori*

2. (i) Define classification.

(ii) Write any two drawbacks in classifying organisms under the old two kingdom classification.

(iii) Give names of five kingdoms proposed by R.H. Whittaker.

Ans. (i) Classification is a method of categorising organisms on the basis of their similarities and differences.

(ii) Two drawbacks in classifying organisms under the old two kingdom classification are:

(a) According to this classification system, all living organisms are classified into 2 Kingdoms- Plantae and Animalia but Bacteria and Fungi have no chlorophyll and do not carry out photosynthesis still they are placed under Kingdom Plantae.

(b) Similarly Euglena which is an unicellular organism cannot be exclusively called plant or animal. They have chloroplasts like the plants and have cell mouth to feed like animals, possess contractile vacuole for excretion.

(iii) The five kingdoms areKingdom Monera, Kingdom Protista, Kingdom Fungi, Kingdom Plantae, Kingdom Animalia.

3. Complete the blanks in below figure:

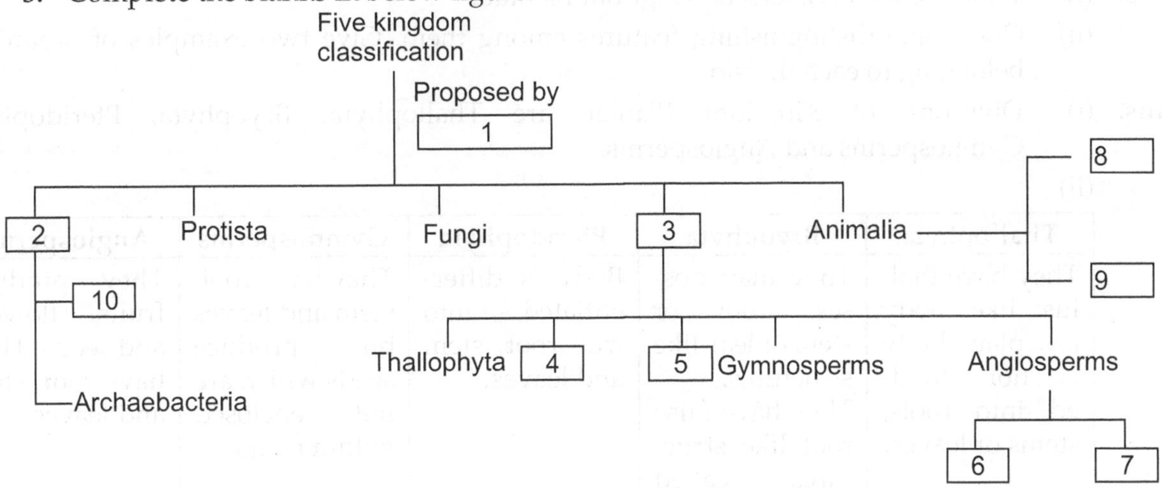

Ans. 1. R.H. Whittaker 2. Monera 3. Plantae 4. Bryophyta
5. Pteridophyta 6. Monocots 7. Dicots 8. Vertebrates
9. Invertebrates 10. Eubacteria

4. (i) Draw a neat labelled diagram of Spirogyra and label the following parts:
(1) Outermost layer of the cell
(2) Small, round spherical structures which stores starch
(3) Spirally coiled, ribbon shaped structure which perform photosynthesis.
(4) Darkly coloured and dot like structure generally present in the centre of the cell.

(ii) Why do we classify organisms? Who proposed 5 kingdom classification and on what basis the classification has been done?

Ans. (i) (1) Cell wall (2) Pyrenoids (3) Chloroplasts (4) Nucleus

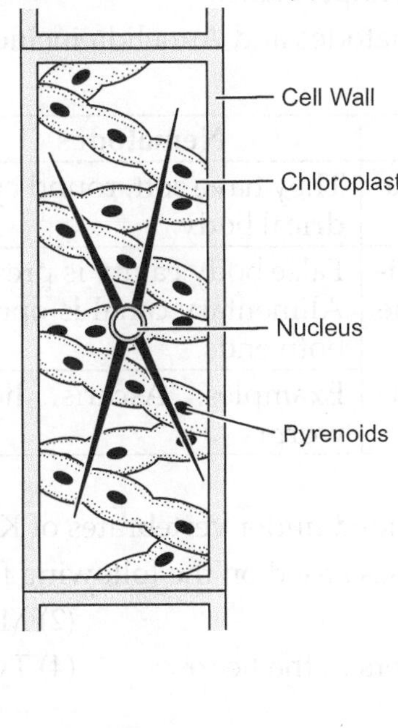

Short Answer Type Questions

(ii) We classify organisms because-

(1) Wide variety of organisms can be studied conveniently and quickly.

(2) Pattern of evolution can be easily understood.

(3) Information regarding existing diversity of plants and animals can be known.

R.H. Whittaker proposed 5 kingdom classification. This classification is based on following criteria:

(I) Cell structure *i.e.*, prokaryotic or eukaryotic.

(2) Mode and source of nutrition *i.e.*, autotrophic or heterotrophic.

(3) Body organisation whether unicellular or multicellular.

5. (i) What are the divisions of Kingdom Plantae?

(ii) Give some distinguishing features among them. Give two examples of organisms belonging to each division.

Ans. (i) Divisions of Kingdom Plantae are Thallophyta, Bryophyta, Pteridophyta, Gymnosperms and Angiosperms.

(ii)

Thallophyta	Bryophyta	Pteridophyta	Gymnosperms	Angiosperms
They have thallus like body *i.e.*, plant body is not divided into roots, stems or leaves.	They may possess root or stem or leaf-like structures. They have false root like structures called rhizoids.	Body is differentiated into true root, stem and leaves.	They have root, stem and leaves but produce seeds which are not enclosed within fruits.	They produce fruits, flowers and seeds. They have root, stem and leaves.
They reproduce vegetatively or asexually or sexually.	They reproduce by spore formation.	They reproduce usually by spore formation.	They reproduce usually by spore formation.	They reproduce sexually.
Examples: Volvox, spirogyra	Examples: Funaria, liverwort	Examples: Dryopteris, Marsilea	Examples: Cycas, Pinus	Examples: Cereals, pulses

6. (i) Name the phylum of Kingdom Animalia which includes all the worms.

(ii) Give some characteristics features of these phylum with suitable examples.

(iii) Give two example of Amphibians. **[February, 2020]**

Ans. (i) Platyhelminthes, Nematodes and Annelida includes all the worms.

(ii)

Platyhelminthes	Nematodes	Annelida
They have flat, ribbon like unsegmented body.	They have soft, round cylindrical body.	They have soft, cylindrical ring like segmented body.
Body cavity absent and alimentary canal has only one opening.	False body cavity is present. Alimentary canal is open at both ends.	True body cavity is present and alimentary canal is open at both ends.
Examples: Planaria, liver-fluke	Examples: Ascaris, hook-worm	Examples: Earthworm, leech

(iii) Frog, toad.

7. (i) Name the classes included under vertebrates of Kingdom Animalia?

(ii) Distinguish these classes based on the following features-

(1) Body temperature (2) Mode of breathing

(3) Presence of chambers in the heart (4) Type of skin

Ans. (i) Pisces, Amphibia, Reptilia, Aves and Mammalia are the classes included under vertebrates of Kingdom Animalia.

(ii)

Pisces	Amphibians	Reptilia	Aves	Mammalia
(1) They are cold blooded *i.e.,* body temperature fluctuates with surrounding temperature.	They are cold blooded	They are cold blooded	They are warm blooded *i.e.,* body temperature remains constant.	They are warm blooded.
(2) They breathe by means of gills.	Adults breathe by lungs but larvae by gills.	They breathe by lungs.	They breathe by lungs.	They breathe by lungs.
(3) They have two-chambered heart.	They have three-chambered heart.	They have three-chambered heart except crocodile.	They have four-chambered heart.	They have four-chambered heart.
(4) Body is covered with scales.	Smooth, non-scaly, moist, slimy skin is present.	Rough, horny scales on skin.	Body is covered with feathers.	They have hairy skin.

8. Name the Phylum /Class of each of the following invertebrates.

(i) Leech
(ii) Hydra
(iii) Scorpion
(iv) Slug
(v) Butterfly
(vi) Centipeda
(vii) Crocodile
(viii) Toad
(ix) Sea-cucumber
(x) Sponge
(xi) Whale
(xii) House lizard
(xiii) Shark
(xiv) Salamander
(xv) Heloderma
(xvi) Brittle star

Ans (i) Annelida
(ii) Cnidaria
(iii) Arthropoda/Arachnida
(iv) Mollusca
(v) Arthropoda/Insecta
(vi) Arthropoda/ Myriapoda
(vii) Reptilia
(viii) Amphibia
(ix) Echinodermata
(x) Porifera
(xi) Mammalia
(xii) Reptilia
(xiii) Pisces
(xiv) Amphibia
(xv) Reptilia
(xvi) Echinodermata

9. Listed below are eleven characteristics in column A and 5 classes of chordata in column B. Match at least two characteristics with each class.

Column A	Column B
Three chambered heart Feathers Two chambered heart Eardrum in a pit	(i) Pisces
Scales External ears Gills	(ii) Amphibia

Moist skin Horny scales on skin Homeothermal	(iii) Reptilia
Testes in scrotum	(iv) Aves (v) Mammalia

Ans.

Column A	Column B
(i) Pisces	(a) Two chambered heart (b) Gills (c) Scales
(ii) Amphibia	(a) Three chambered heart (b) Moist skin
(iii) Reptilia	(a) Horny scales on skin (b) Ear drum in a pit
(iv) Aves	(a) Feathers (b) Homeothermal
(v) Mammalia	(a) External ears (b) Testes in scrotum

10. Given below is a list of ten animals each followed by three terms or features. Underline the term which does not match with the animal.

(i) Amoeba – Nucleus, tentacle, food vacuole
(ii) Hydra- Invertebrata, Cnidaria, Crustacea
(iii) Fish- Gills, paired fins, ear drum
(iv) Earthworm - Invertebrata, Annelida, Insecta
(v) Grasshopper - Wings, trachea, proboscis
(vi) Butterfly - Insecta, Invertebrata, Mollusca
(vii) Whale - Gills, mammary glands, fat under the skin
(viii) Pigeon- Feathers, wings, hair
(ix) Monkey – External ear, sweat glands, lateral line
(x) Bat – Aves, Mammalia, Chordata

Ans. (i) Amoeba – Nucleus, <u>tentacle</u>, food vacuole
(ii) Hydra- Invertebrata, Cnidaria, <u>Crustacea</u>
(iii) Fish- Gills, paired fins, <u>ear drum</u>
(iv) Earthworm - Invertebrata, Annelida, <u>Insecta</u>
(v) Grasshopper- Wings, <u>trachea</u>, proboscis
(vi) Butterfly – Insecta, Invertebrata, <u>Mollusca</u>
(vii) Whale- <u>Gills</u>, mammary glands, fat under the skin
(viii) Pigeon- Feathers, wings, <u>hair</u>
(ix) Monkey – External ear, sweat glands, <u>lateral line</u>
(x) Bat – <u>Aves</u>, Mammalia, Chordata

11. Complete the following table:

Main feature	group
1. Animal with exoskeleton	....
2. Animal with rough scales	....
3. Animals which are fixed	....

4.	Mollusca
5.	Fishes
6. Diploblastic animals	....
7.	Birds

Ans.

Main feature	Group
1. Animal with exoskeleton	Arthropoda
2. Animal with rough scales	Reptilia
3. Animals which are fixed	Porifera
4. Animal with soft body and mantle	Mollusca
5. Animals with streamlined body and have soft scales.	Fishes
6. Diploblastic animals	Coelenterata
7. Animals with feathers	Birds

12. Give any one difference between each of the following-
(i) Protozoa and Metazoa
(ii) Vertebrate and invertebrate
(iii) Insecta and Arachnida
(iv) Flatworm and roundworm

Ans. (i) Protozoa are single celled organisms whereas Metazoa are multicelled organisms.
(ii) Vertebrates have backbone with internal skeleton whereas invertebrates do not possess backbone and internal skeleton.
(iii) Insecta body is divided into head, thorax and abdomen, have three paired jointed legs with two pairs of wings. But Arachnida body is divided into two parts, have 4 pairs of jointed legs and no wings.
(iv) Flatworms have soft, flattened, unsegmented body and body cavity is absent. Roundworms are long, cylindrical, unsegmented with false body cavity.

13. Name three animals (belonging to different classes) which breathe by means of lungs but have no external ears (pinnae).

Ans. Lizard of class Reptilia, Frog of class Amphibia and Pigeon of class Aves breathe by means of lungs but have no external ears (pinnae).

14. Give one major similarity and one major difference in the following pairs of animals.
(i) Insects and birds
(ii) Whales and fishes
(iii) Snakes and earthworms
(iv) Bat and pigeon
(v) Cuttlefish and dogfish
(vi) Wall lizard and frog

Ans. (i) Similarity - Both insects and birds have wings.
Difference - Insects are invertebrates whereas birds are vertebrates.
(ii) Similarity - Both are aquatic.
Difference - Whales have lungs for breathing and is a mammal. Fishes breathe through gills and belongs to class Pisces.
(iii) Similarity - Both do not have limbs.
Difference - Earthworm belongs to phylum Annelida, is an invertebrate whereas snake is a Reptile that is a vertebrate.
(iv) Similarity - Both have wings to fly, warm blooded.
Difference - Bat is a mammal and have external ears. Pigeon is a bird with no external ears.
(v) Similarity - Both are marine animals.
Difference - Cuttlefish is a mollusc whereas dogfish belongs to class Pisces.

(vi) Similarity - Both are cold-blooded animals and have three chambered heart.
Difference - Wall lizard is a terrestrial animal but frog lives partly on land and partly in water

Chapter 9. Economic Importance of Bacteria and Fungi

1. What is the black powder we see on stale bread?

Ans. When a piece of bread is kept in warm place, spores of moulds like *Rhizopus* fall on it and germinate to form mycelium that looks black powder.

2. Describe the mode of nutrition in fungi.

Ans. The nutrition of fungi is saprophytic in nature. They obtain food from dead or decayed organisms. The hyphae secrete some enzymes which digest the food in the substratum and then it is absorbed. This type of digestion is called extracellular digestion.

3. Explain the various types of bacteria along with diagram.

Ans. Based on the shapes, bacteria are of various types-

(i) Cocci – These are spherical shaped.
(ii) Bacilli – These are rod shaped bacteria.
(iii) Spirilla – These are spiral shaped or twisted bacteria.
(iv) Vibrio – These are comma shaped.

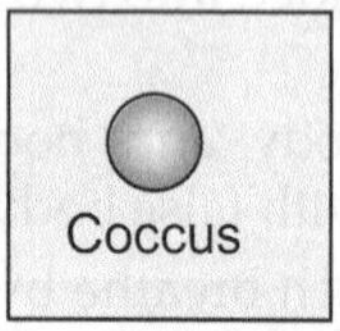

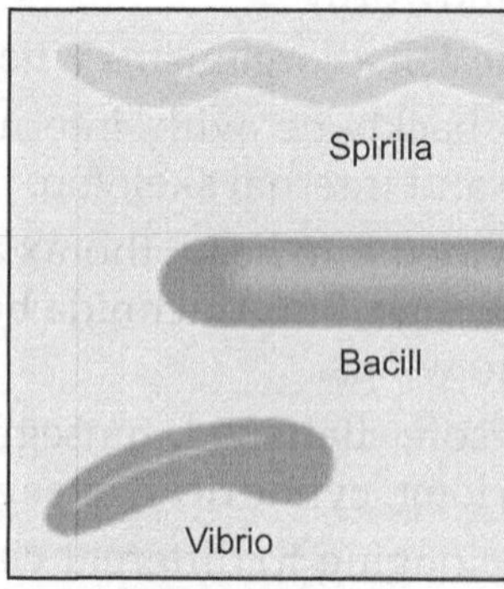

4. Give various methods to preserve our food.

Ans. The various methods to preserve our food are sterilisation, salting, pasteurisation, dehydration, refrigeration, irradiation, etc.

(i) Sterilisation is the process of heating at very high temperatures which kills the microbes as well as their spores.
(ii) Salting is the process where salt is added to food items like pickles, fishes, etc. which causes plasmolysis of bacterial cells and kills them.
(iii) Pasteurisation is heating and then suddenly cooling and is mainly applied to preserve milk.
(iv) Dehydration is the process of removal of water or moisture from food substances.
(v) Refrigeration means storing food at very low temperature.
(vi) Radioactive radiation is sometimes use to sterilize foods.

5. Mention two useful and harmful effects of wine.

Ans. Useful effects of wine:

(i) In mild quantities, wine acts as a stimulant.
(ii) It can be used to stimulate hunger, reduce anxiety and improve digestion.

Harmful effects of wine:

(i) In large quantities, wine can trigger asthma attack and increase blood pressure.
(ii) It can casue body ailments, particularly liver cirrhosis.

6. (i) Give two bacterial disease in
(1) Plants (2) Animals (3) Humans
(ii) Describe the harmful activities of bacteria.

Ans. (i) (1) Black rot of mustard and bacterial blight of cowpea.
(2) Anthrax, Black leg
(3) Cholera, Typhoid

(ii) Some harmful activities of bacteria are:
(1) They causes various diseases in plants, animals and human beings.
(2) They spoil our food and cause food poisoning or botulism.
(3) They are used as bioweapons which is a great threat to mankind.

7. (i) What is serum? Give an example.
(ii) Explain the useful activities of fungi to mankind.

Ans. (i) Serum is the plasma preparation from which fibrinogen has been removed and it contains specific antibodies or antitoxins which is obtained from blood of horses, monkeys, etc.
Example: Blood clotting Factor IX is used for the treatment of Haemophilia B, which is produced by genetically modified bacteria.

(ii) Useful activities of fungi to mankind are:
(1) It used in bakeries and brewing industries.
(2) It is used in cheese processing and obtaining some chemicals.
(3) It is used in various medicinal preparations like antibiotics, vitamins, enzymes, etc.
(4) Fungi increase the fertility of soil.

8. Describe the method of mushroom cultivation.

Ans. The methods of mushroom cultivation are
(i) Composting: Compost is prepared by mixing wheat or paddy straw with organic and inorganic fertilizers.
(ii) Spawning: Mushroom seeds of suitable mushrooms are spread on the compost.
(iii) Casing: A thin layer of soil is spread over compost keeping in mind that there would be proper circulation of air and temperature should be around 20-25°C.
(iv) Cropping and Harvesting: Within 2-6 weeks mushroom grows. Then the full grown mushroom are taken out.
(v) Preservation: They are preserved by vacuum cooling or giving gamma radiation, storing at 15°C, freeze drying in solution of brine, citric acid, etc.

9. Explain the economic importance of bacteria in terms of their utility.

Ans. Bacteria are useful in the following ways:
(i) They play an important role in nitrogen cycle.
(ii) Different flavours of tea are produced by certain bacteria.
(iii) They help in leather tanning.
(iv) They help in production of antibiotics, vitamins, serums, vaccines, etc.
(v) They are natural scavengers.
(vi) They are used in manufacture of various acids, vinegar, enzymes, dairy products, etc.

10. What are antibiotics? Name any two examples.

Ans. Antibiotics are chemical substances produced by a living organism that kill or stop the growth of disease-causing micro-organisms such as fungi and bacteria.
Example of antibiotics:
(1) Penicillin, (2) Streptomycin.

11. Mention useful effects and harmful effects of wine.

Ans. Wine is an example of alcoholic beverage produced by fermentation.
Harmful effects of wine is it may develop body ailments like liver cirrhosis, when consumed in large quantities. It can trigger asthma attack and increase blood pressure.

Useful effects of wine is when consumed in very mild quantities it may be a stimulant. It can be used to stimulate hunger, improve digestion and reduce anxiety.

12. What do you mean by antibiotics? Tabulate the different types of antibiotics used in various kinds of diseases.

Ans. An antibiotic is a chemical substance produced from microbes to kill or inhibit the growth of other microbes.

Antibiotic	Diseases treated
Penicillin	Tetanus, Diphtheria
Streptomycin	Tuberculosis
Chlorotetracycline	Typhoid
Erythromycin	Rickettsial fever

13. How do bacteria obtain their nourishment?

Ans. Modes of nutrition in bacteria:

(i) Autotrophic bacteria: Able to synthesize their own food.

(1) Photoautotrophs: Contain chlorophyll and use light energy for the synthesis of food.

(2) Chemoautotrophs: Obtain energy by oxidising inorganic compounds

(ii) Heterotrophic bacteria: Require organic compounds as their main source:

(1) Saprophytes: Grow on dead and decaying organic matter.

(2) Parasites: Obtain their food for living host on which they grow.

14. (i) How is nitrogen fixed by bacteria for use by plants? What is this process called?

(ii) Name the steps involved in the cycling of nitrogen in the air. Describe in brief.

Ans. (i) Plants cannot use free nitrogen present in atmosphere so nitrogen fixing bacteria like *Rhizobium, Azotobacter,* etc. convert free nitrogen into soluble nitrates to be used by plants. This process is called nitrogen fixation.

(ii) The steps involved in the cycling of nitrogen in the air are:

(1) Ammonification: Organic matter of dead remains of plants and animals are decomposed into ammonia by bacteria like *Clostridium*.

(2) Nitrification: In this process ammonia is first converted to nitrites by bacteria like *Nitrosomonas* and then into nitrates by bacteria such as *Nitrobacter*.

(3) Denitrification: In this process nitrates is break down to release nitrogen gas by denitrifying bacteria like *Pseudomonas*.

(4) Nitrogen fixation: The process by which free nitrogen of atmosphere is converted into soluble nitrates by nitrogen fixing bacteria like Rhizobium.

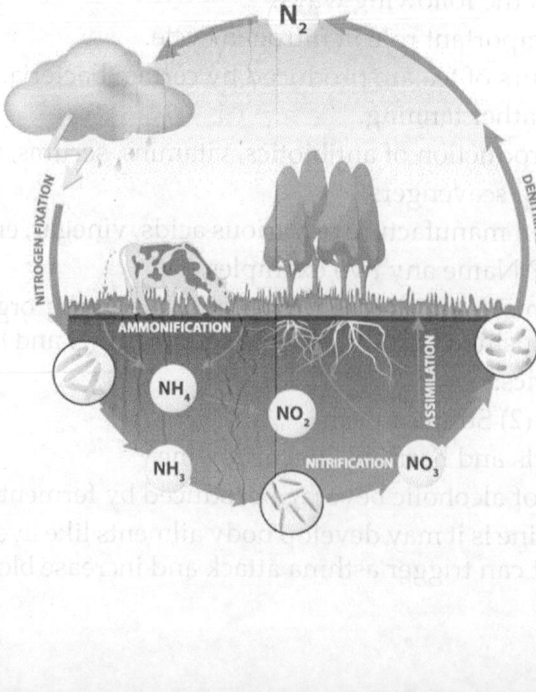

15. Complete the table:

Disease	Scientific name of causal agent
Cholera	(i) ....
(ii)	*Salmonella typhi*
Anthrax	(iii) ....
Tuberculosis	(iv) ....
(v)	*Clostridium tetani*
Penicillosis	(vi) ....

Ans.

Disease	Scientific name of causal agent
Cholera	*Vibrio cholerae*
Typhoid	*Salmonella typhi*
Anthrax	*Bacillus anthracis*
Tuberculosis	*Mycobacterium tuberculosis*
Tetanus	*Clostridium tetani*
Penicillosis	*Penicillium*

16. Mention any two ways by which food can be preserved? **[November, 2019]**

Ans. Food can be preserved by following ways:

(i) Pasteurization: This technique is mainly applied to milk.

(ii) Salting: It is used to preserve pickles, fish, etc.

17. Name the bacteria present in the roots nodules of leguminous plants. What is their role? **[November, 2019]**

Ans. The bacteria present in the root nodules of leguminous plants in Rhizobium. They convert free nitrogen form the soil atmosphere and convert into soluble nitrates which are used by leguminous plants.

18. Both bacteria and yeast reproduce by asexual method, but how does this method differ in them?

Ans. Bacteria reproduce asexually by binary fission or cell division. DNA duplicates followed by expansion of cell and two DNA are pulled apart. A constriction appears at the middle of bacterial cell separating the two cells. Two cells may remain attached or they get separated from each other and live as independent cells.

Yeast reproduce asexually by budding. An outgrowth called bud develops from the parent cell. Some of the nuclear material is transferred to the bud. The bud grows and may later get separated from parent or it may remain attached to the parent and form a chain of buds.

19. Would there be any bacteria in an aquarium?

Ans. Bacteria are present everywhere in air, water, soil, food stuffs, inside and outside the body of living organisms, etc. So bacteria are also present in an aquarium.

20. Give the technical terms for the following patterns of occurrence of bacteria:

(i) In pairs (ii) In long chains (iii) In clusters

Ans. (i) Diplococci (ii) Streptococci (iii) Staphylococci

Chapter 10. Nutrition

1. What will happen if a child does not take any vegetables and fruits in his/her diet?

Ans. He will develop deficiency disease due to shortage of vitamins and may get constipation.

2. Why cellulose is important part of food?

Ans. Cellulose it help in proper movement of food in food pipe and produce feeling of fullness of stomach by providing bulk to food.

3. Why people in hilly are need extra supplement of iodine?

Ans. Water in hilly areas contains less iodine because as water flows on surface iodine from minerals gets dissolved in it and iodine requirement in human is fulfilled from iodine dissolved in water.

4. Why growing children need different diet from adults?

Ans. Growing children need more nutrients for growth and development.

5. A doctor advises a bone patient to include more of milk and milk products in his everyday food. Why so?

Ans. Milk and milk products are good source of calcium. Calcium is an important component of bones and it makes bones stronger. So a doctor advises a bone patient to include more of milk and milk products in his everyday food so that he could get maximum calcium from it.

6. What are main functions of food?

Ans. Growth ,repair, energy, maintenance and protection.

7. How does our body get energy if we fast for too many days?

Ans. Firstly, all the carbohydrates in our body will be consumed and then proteins will be used to provide energy.

8. Why we must include water in our diet? Explain.

Ans. Our body consists of 70% water, which helps in digestion, regulate body temperature and essential for excretion also.

9. "If teeth are lost, taste is lost." Comment.

Ans. True, because without teeth we cannot digest food properly. Thus, food loses its charm.

10. State two reasons why whole grain is preferred over fine flour.

Ans. (i) Whole grain has more fibre (cellulose / roughage) which makes it easily digestible.
(ii) Grains have carbohydrates which are rich source of energy.

11. (i) Name the major nutrients present in the food.
(ii) Why are proteins necessary in our food? Name one protein deficiency disease of young children.

Ans. (i) Major nutrients in food are carbohydrate, fat, protein vitamins and minerals.
(ii) Proteins provide materials for the growth and repair of body cells and tissues. During emergency it gets oxidised to release enery. So proteins are necessary in our food. Kwashiorkor is a protein deficiency disease of young children.

12. Why human cannot digest cellulose?

Ans. Cellulose digesting enzyme (cellulase) is absent in humans. So, cellulose remains undigested in human alimentary canal.

13. (i) Define malnutrition
(ii) Categorise the following items into carbohydrates and proteins:
Cellulose, starch, egg albumin, milk, pea, glycogen, cane-sugar
(iii) Fats are considered as the main energy producers.

Ans. (i) Malnutrition is the condition in which a person suffers due to lack or deficiency of one or more essential elements of food.
(ii) Carbodydrates – Cellulose, strach, glycogen, cane-sugar
Proteins – Egg albumin, milk, pea
(iii) Oxygen is present in less amount in fats than in carbohydrates. So, the energy produced is more than carbohydrates. It produces approximately 9.4 kcal of energy.

14. Which mineral is assential for haemoglobin formation?

Ans. Iron is essential for haemoglobin formation.

15. (i) What happens if roughage is not included in the diet?

(ii) Taking the example of whole grain atta, fruit and green leafy vegetables, describe how roughage in our diet is useful.

Ans. (i) If roughage is not included in diet, it may lead to constipation.

(ii) Whole grain atta, fruit and geen leafy vegetables are main sources of roughage, They contain cellulose, which is not digested in our body. As it is fibrous in nature so it absorbs and retain a log of water which keeps our faecal matter soft, thus preventing constipation. It makes easy movement of undigested food through the intestine by stimulating contraction of muscles of wall of intestine.

16. (i) Name the mineral elements that is needed for the following respectively:

(1) Strong teeth (2) Proper working of thyroid
(3) Synthesis of haemoglobin

(ii) Name the nutrients whose deficiencies cause the followng disease in humans:

(1) Pernicious anaemia (2) Pellagra
(3) Night blindness (4) Goitre
(5) Kwashiorkor

Ans. (i) (1) Calcium (2) Iodine
(3) Iron

(ii) (1) Vitamin B_{12} (2) Vitamin B_3
(3) Vitamin A (4) Iodine
(5) Protein.

17. State the uses of:

(i) Iodine (ii) Iron
(iii) Vitamin C (iv) Carbohydrates
(v) Water (vi) Sodium
(vii) Phosphorous

Ans. (i) Iodine is an important componet of thyroxine hormone. Thus, for the proper functioning of thyroid gland, iodine is required.

(ii) For the synthesis of haemoglobin in blood.

(iii) Capillary wall's functioning is promoted.

(iv) Carbohydrates providde energy.

(v) Water is required for biochemical activities, tansportation of substances, etc.

(vi) To maintain fluid balance for nerve and muscle.

(vii) For healthy bones and teeth, absorption of minerals in the body.

18. Write the disease caused due to the deficiency of the following:

(i) Vitamin A (ii) Vitamin B_1
(iii) Vitamin B_{12} (iv) Vitamin D
(v) Vitamin C (vi) Vitamin K
(vii) Sodium (vii) Iron
(ix) Iodine (viii) Calcium

Ans.

(i) Night-blindness (ii) Beriberi
(iii) Pernicious anaemia (iv) Rickets
(v) Scurby (vi) Slow clotting of blood.
(vii) Muscular cramps (viii) Anaemia
(ix) Goitre (x) Rickets

19. Complete the table:

Vitamins/Minerals	Source	Deficiency disease
Calcium	Milk	(i) ....
(ii)	Citrus fruits	Scurvy
Calciferol	(iii)	(iv) ....
Riboflavin	Eggs	(v) ....

Ans.

Vitamins/Minerals	Source	Deficiency disease
Calcium	Milk	(i) Rickets
(ii) Vitamin C	Citrus fruits	Scurvy
Calciferol	(iii) fish liver oil	(iv) Rickets (in children) and osteomalacia (in adults)
Riboflavin	Eggs	(v) Inflammation of the tongue

20. Complete the missing information in the box

Nutrient	Source	Function	Deficiency disease
Thiamine (B1)	(i)	Glucose metabolism, muscle and heart function	(ii) ....
Niacin (B3)	Mushroom, whole grain, meat fish	(iii)	(iv) ....
Folic acid (B9)	(v)	Growth, protein metabolism, prevent anemia	(vi) ....
(vii) ...... (mineral)	Fish, milk product, iodized salt	(viii)	Goiter
(ix) .... (mineral)	(x)	Haemoglobin	Anemia

Ans.

Nutrient	Source	Function	Deficiency disease
Thiamine (B1)	(i) Whole grains, meat, fish	Glucose metabolism, muscle and heart function	(ii) Beri beri
Niacin (B3)	Mushroom, whole grain, meat fish	(iii) Help in metabolism and nervous system	(iv) Pellagra
Folic acid (B9)	(v) Green vegetables, lentils, beans	Growth, protein metabolism, prevent anemia	(vi) Improper brain function, anemia
(vii) Iodine (mineral)	Fish, milk product, iodised salt	(viii) Thyroid hormone, brain development	Goiter
(ix) Iron (mineral)	(x) Red meat, green vegetables	Haemoglobin	Anemia

21. Give two examples eac of the following and their usefulness if any in our body.

(i) Monoasccharides

(ii) Disaccharides
(iii) Polyaccharides

Ans. (i) Examples: Glucose and fructose
These monosaccharides need no digestion and are directly absorbed by our body. They are the instant source of energy for our body.
(ii) Examples: Sucrose, lactose
They need digestion and they are oxidised in the cells to release energy.
(iii) Examples: Cellulose, glycogen
Cellulose acts as a roughage and help in proper functioning of gut.
(1) Glycogen is stored in liver and muscles and gets converted into glucose to provide energy during emergency situations.
(2) A balanced diet is one which contains all the principal constituents of food in proper quantity. Example: A balanced diet consists of reuisite quantities of carbohydrates, protein, fats, vitamins and minerals.

22. List the six main purposes for which food is required by the body.

Ans. The size main purposes for which food is requried by the body are:
(i) Growth: It helps in building up of new cells for growth.
(ii) Repair: It provides materials to repair worn out cells.
(iii) Energy: It provides energy to carry out various life processes.
(iv) It maintains the chemical composition of cells.
(v) It provides raw materials for secretion of enzymes, hormones, milk, etc.
(vi) It protects us from diseases and infections.

Chapter 11. Digestive System

1. Why food should be thoroughly chewed before ingesting it?

Ans. Chewing produce saliva which contains digestive enzymes for breakdown of carbohydrates.

2. Why it is difficult to swallow dry milk powder?

Ans. Dry powders are not turned into balls (bolus) easily as they need a lot of mucus to moisten them. Thus, it is hard to swallow dry powders.

3. How ileum is adapted for better absorption of digested food?

Ans. Large surface area is obtained due to length and presence of villi which help ileum for better absorption of food.

4. Why stomach has elastic and muscular walls?

Ans. (i) Elastic –walls expand to accommodate food for storage
(ii) Muscular – walls to help churn the food

5. (i) Why potato turns blue black when dipped in iodine solution?
(ii) You have been supplied with a sample of food. How will you perform tests for the presence of starch and proteins in it?

Ans. (i) Potato has starch which turns blue black with iodine solution.
(ii) Test for starch:
(1) A little amount of food sample is taken along with water in a test tube and shaken. Then it is boiled to make a solution.
(2) Few drops of iodine solution is then added to the mixture after cooling.
(3) If the food contain starch it would give a blue-black colour.
Test for proteins:
(1) A small amount of food sample is taken in a test tube and few drops of dilute nitric acid is added to cover the food sample.

(2) Test tube is heated gently and then the acid is rinsed off with water, followed by addition of ammonium hydroxide.

(3) If the colour first changes from white to yellow and then after adding ammonium hydroxide yellow colour changes to orange, it gives the positive test for proteins.

6. Name the disorders of the human digestive system:

(i) Liver gets affected; skin and eyes turn yellow due to the deposit of bile pigment.
(ii) Ejection of stomach contents through the mouth.
(iii) Faecal discharge with more liquidity.
(iv) Faeces are clutched within the rectum.
(v) Food is not digested properly.

Ans. (i) Jaundice (ii) Vomiting
(iii) Diarrhoea (iv) Constipation
(v) Indigestion

7. Complete the missing information in the box:

Part of alimentary canal	Substrate	End product	Enzyme
Mouth	(i)	Maltose –to – glucose	(ii) ....
Stomach	Protein	(iii)	(iv) ....
Stomach	(v)	Fatty acids + glycerol	(vi) ....
Pancreatic juice	Protein	(vii)	(viii) ....
	Nucleic acids	(xi)	(x) ....

Ans.

Part of alimentary canal	Substrate	End product	Enzyme
Mouth	Starch	Maltose –to – glucose	(ii) Amylase or ptyalin
stomach	Protein	(iii) Peptones	(iv) Pepsin
stomach	(v) fats	Fatty acids + glycerol	(vi) Gastric lipase
Pancreatic juice	Protein	(vii) Amino acids	(viii) Peptidases
	Nucleic acids	(xi) nucleotides	(x) Nuclease

8. What are the functions of liver?

Ans. Produce : bile, fibrinogen, RBC's in foetus.
Control : blood sugar level, amino acid level
Regulate : blood volume
Destroy- : dead RBC's
Excrete : toxics
Store : vitamins and minerals

9. (i) What happens to the proteins and carbohydrates present in food?
(ii) What happens to the fats present in food?

Ans. (i) Proteins and carbohydrates → Amino acids and simple sugars → absorbed through thin epithelium of villi → reach capillaries → blood circulation → liver through hepatic portal vein.

(ii) Fats → Fatty acids and glycerol → absorbed into lymph vessel → lymphatic system → enters blood stream.

10. (i) The stomach secretes gastric juice which contains hydrochloric acid. What is its function?

(ii) What is the function of the rectum

(iii) What is roughage? Give two examples.

Ans. (i) HCl kills the germs which may have entered along with the food and it activates pepsin enzyme which digests proteins.

(ii) Rectum stores the semi-solid- faeces, the undigested food and expelled them to outside at intervals through anus.

(iii) Roughage is mainly the cellulose material which is not digested in our body but helps in keeping our faecal matter soft by retaining water preventing constipation and helps in easy movement of undigested food in intestine. Two example are green leafy vegetables, half crushed wheat.

11. What is digestion? Why do only animals require a digestive system?

Ans. Digestion is the break down of naturally occurring foodstuffs into soluble form so that they can be absorbed through living membranes.

Animals require a digestive system because they take food comprising of large and complex molecules like carbohydrates, proteins and fats which need to be broken down into simple and diffusible form so that they can be absorbed in our body. Our digestive system consists of different digestive glands which secretes various enzymes that degrades these complex molecules into simpler forms in different regions of digestive tract.

12. Give two reflexes related to digestive system which occur in humans.

Ans. (i) Production of saliva;

(ii) Tongue pressing food towards palate to swallow food.

13. (i) What are the end products of the digetion of starch, proteins and fats respectively?

(ii) Why is there no enzymes to digest vitamins?

(iii) What are the main characteristics of an enzyme?

Ans. (i) The end productrs of strach is maltose, proteins is amino acids, fats is fatty acids and glycerol.

(ii) Vitamins are utilised by cells in their original form and requires no digestion. They are either fat soluble or water soluble so they are directly absorbed by the cells as such. So there are no enzymes to digest vitamins.

(iii) The main characteristics of an enzyme are:

(1) It is a protein so it is destroyed by heating.

(2) It is substrate specific.

(3) It acts as a catalyst and can be used again and again.

(4) It acts at a particular pH.

(5) It works the best at optimum temperature *i.e.,* between 35-40°C.

(6) It increases the speed of the reaction.

14. State the difference between pepsin and trypin.

Ans. Differences between pepsin and trypsin are as follows: **[November, 2019]**

Pepsin	Trypsin
It acts in acidic medium.	It accts in alkaline medium.
It converts proteins into proteoses and peptones (peptides).	It converts proteins into petones (peptides) and polypeptides.

15. Describe four main purposes of food requirement of our body.

Ans. The food is required by the body for the following purposes.

(i) **Growth:** Food provides raw materials for building up new prtoplasm or cells for growth.

(ii) Repair: Food provides raw materials for repair of worn-out cells and for replacement of dead cells.

(iii) **Energy**: Oxidation of food during respiration provides energy which is utilised by the body for various activities.

(iv) **Reproduction:** Food provides raw materials for building new cells which are required for developing a new organism.

16. Distinguish between:

(i) Lipases and peptideses

(ii) Villi and microvilli

(iii) Sucrose and maltose

(iv) Extracellular and intracellular digestion

Ans. (i) Lipases are the type of enzymes found in digestion tract, which break fats into diglycerides and monoglycerides while peptidases break proteins into their respective peptides.

(ii) Villi are the foldings of mucosa layer, whereas present on the cells in mucosa layer. Both of them are meant to increase the area of absorption in small intestine.

(iii) Both enzymes are used in digestive tract to convert carbohydrates into its smaller units. Sucrose is the enzyme, which converts sucrose into glucose and froundefinedctose, which maltase enzyme converts maltos into glucose in small intestine.

(iv) Extracellulalr digestion is the type of digestion (breaking down of macromolecules) which takes place outside the cell, intracellular digestion is the type of digestion that takes place inside the cell.

17. Identify the enzymes based on their respective functions given below.

(i) Hydrolyses lactose into glucose and galactose.

(ii) Hydrolyses proteins into peptones and pclypetides.

(iii) Converts proteins into proteins and peptones.

Ans. (i) Lactase (ii) Trypsin (iii) Pepsin

18. Study the reaction related to nutrients and answer the following questions:
Clear blue solution of copper and sodium salts changes to green , yellow, and brick red solution with a type of food nutrient on heating

(i) Name the blue reagent.

(ii) What is the common substrate on which reagent react differently?

(iii) Reaction mixture should be heated at what temperature?

(iv) What does colour of reagent after reaction signifies?

(v) Why this reaction is used as food test?

Ans. (i) Benedict's solution

(ii) Simple sugars

(iii) 40 -50 °C.

(iv) Sugar concentration.

(v) This reaction is used as food test because all reducing sugar give positive response to this test.

19. (i) Mention two ways in which the ileum of a mammal is adapted for the absorption of digested food?

(ii) Complete the table by filling in the blanks (i) to (x):

Organ	Secretion	Enzyme	Nutrient	Product of digestion
Mount	(a)	(b)	Starch	(c)
Stomach	Gastric juice	(d)	(e)	Polypeptide
Pancreas	(f)	Steapsin	(g)	Fatty acids and glycerol
(h)	Intestinal Juice	(i)	Maltose	(j)

Short Answer Type Questions

Ans. (i) The two ways are:

(1) Inner lining of ileum is thrown into tiny finger like projections called villi which increases thè surface area thus facilitating the absorption of digested food.

(2) lleum is very long providing more surface area and is narrow so slow movement of food allow better absorption of digested food.

(ii) **(a)** Saliva **(b)** Ptyalin

(c) Maltose **(d)** Pepsin

(e) Proteins **(f)** Pancreatic juice

(g) Emulsified fats **(h)** Small intestine

(i) Maltase **(j)** Glucose

20. List the enzymes and their action on the food in the stomach and the intestine.

Ans.

Organ	Enzymes	Function
Stomach	Pepsin	It acts on proteins and convert into peptides.
	Rennin	It acts on milk protein caseing and converts to paracasein.
Small intestine [duodenum]	Amylopsin	It acts on starch and converts into maltose.
	Trypsin	It acts on proteins and peptides and convert into peptides and amino acids.
	Steapsin	It acts on emulsified fats and convert into fatty acids and glycerol.
Small intestine [Illeum]	Erepsin	It converts peptides into amino acids.
	Maltase	It converts maltose to glucose
	Sucrase	It acts on sucrose and converts to glucose and fructose
	Lactase	It acts on lactose and converts into glucose and galactose
	Lipase	It acts on emulsified fats and convert into fatty acids and glycerol.

21. (i) How is thorough chewing of food helpful in digestion?

(ii) Try to swallow the saliva in your mouth and feel your neck with your hand. What happens in the neck?

(iii) Mention two reflexes which occur when a person chews and swallows food.

Ans. (i) Chewing of food helps in breaking the large particles of food into smaller bits. These smaller bits of food provide large surface area for the enzymes to act for better digestion. While chewing more amount of saliva is secreted in our mouth which moistens and lubricates our food so that food particles stick to each other forming a bolus which can be easily swallowed.

(ii) When we swallow the saliva the larynx is pulled upwards so that it comes close to the back of the tongue and a flap like structure called epiglottis closes its opening so that the saliva moves towards oesophagus.

(iii) The two reflexes are:

(1) When a person chews food, salivary glands secrete saliva.

(2) When a person swallows food, the tongue presses upward and back against the palate of the mouth and food bolus is pushed towards pharynx. Epiglottis closes

the opening of trachea and soft palate closes the opening between throat and nasal passage so food bolus passes through oesophagus.

22. Complete the following table by filling in the blanks 1-8.

Organ	Enzyme	Food acted upon	Final product
1	Pepsin	2	3
Mouth	4	5	Disaccharide
6	7	Maltose	8

Ans. 1. Stomach 2. Proteins
3. Polypetides 4. Ptyalin
5. Starch 6. Illeum
7. Maltase 8. Glucose

Chapter 12. Skeleton-Movement and Locomotion

1. Why last two pairs of ribs are called floating ribs?

Ans. The last two pairs of ribs are called floating ribs because they are not attached to sternum.

2. How movement and locomotion help in performing different functions in our body?

Ans. Movement help in change of position of parts of body while locomotion changes place of whole body

3. When do knee joints become stiff?

Ans. With age, joints become stiff due to decrease in lubricating fluid and wearing out of cartilage.

4. Some people in old age complain of stiff joints. What do you think could be a possible region for it?

Ans. Due to decrease in amount of synovial fluid between two bones, some people in old age complain of stiff joints. Synovial joints like shoulder joint and knee joint have lubricating fluid called synovial fluid between the two bones which provides lubrication *i.e.*, it reduces friction between two bones so the movement could be easier. But due to old age this fluid amount decreases causing stiffness in joints and flexibility is also reduced.

5. What are the main functions of skeleton system?

Ans. The main functions of skeleton system are protection, movement, storage of minerals, produce RBC's.

6. What are main functions of skeleton?

Ans. The main functions of skeleton are support, protection, movement, form blood cells, store minerals.

7. How tendons help in movement on bones?

Ans. Muscles are attached to bones with inelastic tendons at one end and fixed at other end. When muscles move, tendon moves the bones.

8. (i) What are antagonistic muscles? Give one example.
(ii) Why antagonistic muscles are required?

Ans. (i) Antagonistic muscles are those which cause opposing movement *i.e.*, a structure moved by one muscle is brought back to its original position by another muscle. For example biceps muscles bends the lower arms over the upper arms and straightening of lower arm is boruth by triceps muscles so these two muscles are antagonistic in nature.

(ii) When muscles contract they become short. They cannot elongate, thus there is need of other muscle to move bone in opposite direction. Such muscles are called antagonistic muscles.

9. What is function of ligaments?

Ans. Ligaments stabilises the bones and prevent dislocation.

10. What are joints and what are its main three types?

Ans. The point where two bones meet is called joint. Joints are of three types – freely movable (synovial), partially moveable (slight bending), immovable (no movement.

11. Complete the missing information in the box

Joint	Location	Main function in our body
Ball and socket	(i)	Circular movement
(ii)	Carpometacarpal joint	(iii) ....
Pivot	Atlas and (iv)	(v) ....
Hinge	(vi)	Movement in one plane
(vii)	Ankle bone	(viii) ....
Immoveable	(ix)	(x) ....

Ans.

Joint	Location	Main function in our body
Ball and socket	Shoulder	Circular movement
Saddle	Carpometacarpal joint	Sideways movement
Pivot	Atlas and Axis	rotation
Hinge	Elbow	movement in one plane
Gliding	Ankle bone	Sliding movement
immoveable	skull	No movement

12. Name the different types of joints? Give one example of each type.

Ans. The different types of joints are:

(i) Immovable joints-Sutures between the bones of skull.

(ii) Partially movable joints-Joint between the vertebrae.

(iii) Freely movable joints-They are of various types.

(1) Gliding joint-Joint between the bones of wrist

(2) Pivot joint-joint between skull and upper end of backbone

(3) Hinge joint-Elbow joint between humerus and ulna

(4) Ball and socket joint-Shoulder joint *i.e.,* between humerus and glenoid cavity of pectoral girdle.

13. (i) What are bones made of? Are the bones living or non-living? Give reason.

(ii) Just as the humerus corresponds to femur, what bones corresponds to tarsals, metacarpals, ulna and radius respectively.

Ans. (i) Bones are made up of organic and inorganic material. Calcium and phosphorus are the main inorganic constituents of bones. It consists of collagen fibres which is a protein-an organic constituent. Bones consists of bone cell called osteocytes. Bones are living as they are richly supplied with blood vessels and in the bone marrow blood cells like RBCs and WBCs are formed.

(ii) Carpals corresponds to tarsals, metatarsals corresponds to metacarpals, fibula correspons to ulna and tibia corresponds to radius respectively.

14. (i) Name the parts of the skeleton where the following are located: transverse process, glenoid cavity, shoulder-blade, acetabulum.

(ii) What are the uses of the skeleton in our body?

Ans. (i) Transverse processes are sideward projections from neural arches of a vertebrae. Shoulder pectorla girdle outer apex has a large cup-shaped cavity the glenoid cavity. Shoulder pectoral girdle contains two flat triangular shoulder blades which lie dorsally on the upper ribs on either side of vertebral column in the thorax region. Acetabulum is a large cup shaped articular cavity present on each side of hip bones.

(ii) The uses of skeleton in our body are:

(1) It provides support or frameword to all the soft parts of our body and gives a definite shape.

(2) It protects all our delicate and important organs of our body.

(3) It provides surface for the attachment of muscles.

(4) Blood cells like RBCs and WBCs are formed in the bone marrow of long bones.

(5) Bones are a storehouse of calcium and phosphorus.

(6) Bones and joints forms levers and increase the speed and distance of movement by a muscle.

Chapter 13. Skin: "The Jack of all Trades"

1. Why we should keep our skin clean?

Ans. Skin is covered with layer of dead cells which is prone to bacterial infections. We must keep it clean to avoid such infections.

2. (i) Cross out the items in the following list which are NOT the functions of the human skin. Sensation, respiration, storage of glycogen, excretion, digestion, temperature regulation, synthesis of Vitamin D, secretion of hormone.

(ii) Name the three sub-layers of the epidermis from outer to inner.

(iii) Name any one modified sweat gland and any one modified sebaceous gland.

(iv) Name the skin glands which when inflamed cause acne.

(v) State any two functions of the mammalian skin other than those concerened with heat regulation.

Ans. (i) Respiration, digestion, secretion of hormone.

(ii) Stratum corneum, Granular layer, Stratum malpighi.

(iii) Mammary gland is modified sweat gland and ceruminous glands are modified sebaceous glands.

(iv) sebaceous glands.

(v) Skin serves as sense organ for touch, pain, pressure, heat, etc. Skin can synthesize vitamin D when exposed to sunlight.

3. (i) Why skin becomes dark in colour if exposed to sun light?

(ii) Why skin becomes reddish when we have fever?

(iii) Why skin becomes blue due to cold?

Ans. (i) Melanin is produced in larger quantity to absorb heat to save underlying layers. Due to more of melanin, skin colour becomes dark.

(ii) Due to fever temperature increases and body tries to lose heat by vasodilation. Due to vasodilation skin becomes red.

(iii) Vasoconstriction reduces blood flow in the vessels which reduces loss of heat and vaporisation of sweat due to that skin becomes blue.

4. (i) Name any two glands found in the numan skin. State their functions.

(ii) Mention two function of the human skin.

Ans. (i) Two glands found in human skin are sebaceous glands and sweat glands. Sebaceous glands produce an oily secretion called sebum which makes hair and outer surface of the skin oily and waterprood, prevents drying of skin, prevents loss of water by evaporation. Sweat glands secrete sweat that helps in regulation of body temperature. It cause cooling effect to the body.

(1) It protects our underlying tissues from mechanical shocks.

(2) It prevents excessive loss of water by evaporation and protects from dehydration.

(3) It prevents the entry of harmful disease causing germs into our body.

(4) It protects the body from entry of UV light which is very harmful.

5. Why skin becomes dry during winters?

Ans. Sebum comes out to skin and keeps it moist. Vasoconstriction constrict the pores through which sebum comes out.

6. How skin protects us?

Ans. By creating barrier to germs ,UV rays, avoid evaporation of water , protect inner delicate tissue from mechanical injury.

7. How does our skin provide protection to our body against the following?
   (i) Entry of germs.
   (ii) Excessive loss of heat in severe cold.
   (iii) Entry of harmful ultra-violet rays.

Ans. It provides protection in the following ways:
   (i) Entry of germs: The skin checks the entry of toxic substances or the agents causing infection into the body.
   (ii) Excessive loss of heat in severe cold: The skin prevents the loss of energy from the body. It preseves the body heat at lower temperatures and promotes heat loss in hot weather.
   (iii) Entry of harmful ultra-violet rays: The skin provides protection against harmful ultra voilet radiations.

8. How animals make their skin waterproof?

Ans. Sebum produced by sebaceous glands is a waxy oily substance which forms a coating on hair to make them waterproof

9. What is goose flesh? Why we experience it?

Ans. Contraction of skin to raise hair causes bumps in skin. We experience it due to fear or due to cold weather.

10. Why old people have wrinkled skin?

Ans. Due to loss of collagen and dryness skin becomes thin and wrinkled.

11. What is peeling of skin? Why does skin peels off?

Ans. Loss of epidermis is called peeling. It is a natural process but due to sunburn, allergy or infection rate of peeling may become enhanced. Cancer and Eczema can also cause peeling of skin.

12. Explain the terms "vasodilation" and "vasoconstriction". How do these processes contribute in the regulation of body temperature?

Ans. Vasodilation is the dilation of blood vessels in the skin that leads to an increase in the blood supply.

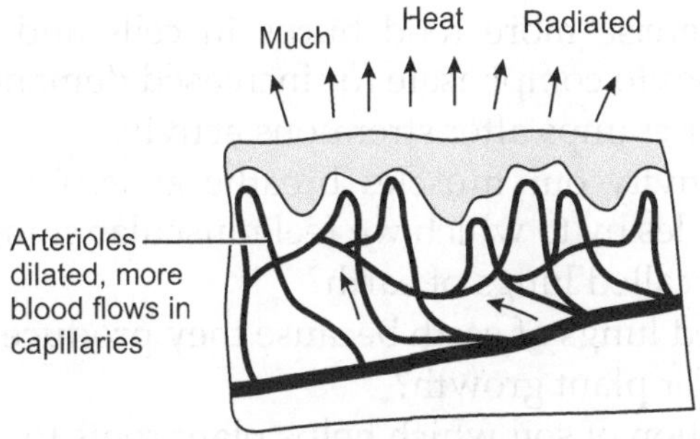

Vasodilation

Vasoconstriction is the narrowing of the blood vessels causing reduction in the blood supply to the skin.

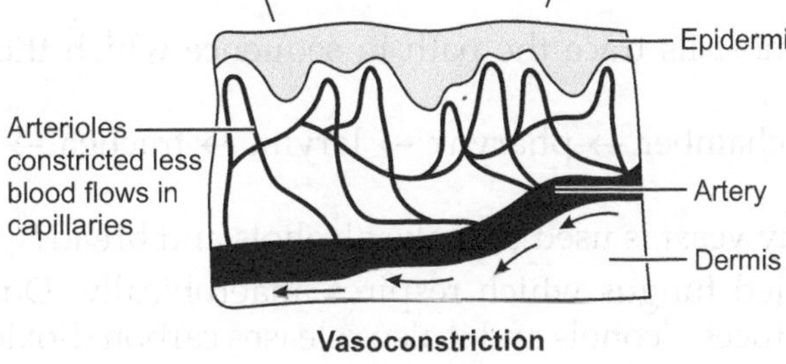

Vasoconstriction

Temperature regulation in different weather are as follows:

In cold weather	In hot weather
The blood vessels are narrowed at lower temperatures reducing blood supply to the skin.	Blood supply to the skin is increases at higher temperatures by dilation of blood vessels in the skin.
Loss of heat is less by conduction, radiation, convection, vaporisation as reduced blood supply lowers the secretion of sweat by sweat glands.	Loss of heat is high by conduction, convection and radiation and vaporisation as more sweat is produces because of rich blood supply to the skin.

13. Complete the missing information in the box :

Part of skin	Location	Main function in our body
Sweat pore	(i)	Produce sweat
Subcutaneous Fat	Hypodermis	(ii) ....
(iii)	Dermis	Produce sebum
Hair shaft	Epidermis	(iv) ....
(v)	Dermis	Sensation

Ans.

Part of skin	Location	Main function in our body
Sweat pore	Dermis	Produce sweat
Subcutaneous Fat	Hypodermis	Storage and temperature regulation
Sebaceous gland	Dermis	Produce sebum
Hair shaft	Epidermis	Heat regulation
Nerve ending	Dermis	Sensation

Chapter 14. The Respiratory System

1. Why breathing rate becomes faster when we run?

Ans. Due to strenuous exercise more food burns in cells and more CO_2 is produced. Thus breathing rate increases to compensate for increased demand of O_2.

2. Why we feel muscular cramps after strenuous activity?

Ans. During strenuous activity our muscles breathe anaerobically. Due to this lactic acid is produced in our muscles by to which we feel muscular cramps.

3. Why large forests are called lungs of earth?

Ans. Large forests are called lungs of earth because they produce a lot of oxygen.

4. Why tilling is useful for plant growth?

Ans. Tilling improves aeration of soil which helps plant roots to respire.

5. Why we feel dizziness at high altitudes?

Ans. We feel dizziness at high altitudes because of lack of oxygen at high altitude.

6. Why you should not sleep under trees during night?

Ans. During night rate of photosynthesis is slow and due to respiration plants also produce carbon dioxide.

7. Starting from the nostrils trace the path in sequence which the inspired air takes until it reaches the air sacs.

Ans. Nostrils → Nasal chamber → pharynx → larynx → trachea → bronchi → bronchioles → Alveoli/air saces.

8. What is Yeast? Why yeast is used to make alcohols and bread?

Ans. Yeast is single celled fungus which respires anaerobically. Due to partial breakdown of glucose ,yeast produces alcohols and it also releases carbon dioxide which raises the dough.

9. What is wrong in the statement "We breathe in oxygen and breathe out carbon dioxide"?

Ans. We breathe in air containing more amount of oxygen and less carbon dioxide and breathe out air containing less oxygen and more carbon dioxide content because the % of oxygen in inspired air is 20.96 and carbon dioxide is 0.04 but in expired air oxygen % is 16.4 and carbon dioxide % is 4.0. So the statement is wrong.

10. Why respiration is reverse of photosynthesis in plants?

Ans. Respiration (CO_2 given out, O_2 used, Energy released, food broken down)

Photosynthesis (O_2 given out, CO_2 used, Energy used, food synthesized.)

11. How would you prove that the air you breathe out is warmer?

Ans. When we breathe out the air on our hand we can feel the warmth of the expired air. This shows that the air we breathe out is warmer. Similarly if we breathe out air in front of a mirror or any cold surface we can see water droplets on it because the warm air condenses on the mirror/cold surface form in droplets of water.

12. How is the respiratory passage kept free of dust particles?

Ans. Respiratory passage *i.e.*, the entire lining of larynx, trachea, trachea, bronchi, bronchioles, nasal chamber is lined with ciliated epithelium. The cilia drives away any foregin particles, dust from the incoming air and keep the passage free from dust particles.

13. Differentiate between burning and respiration.

Ans. Respiration – Series of Cellular biochemical process in which ATP are produced by oxidation of food at body temperature.

Burning – Single step non cellular physico-chemical process in which heat and light are produced by oxidation of fuel at high temperature.

14. Name the two ways by which carbon dioxide is transported in human body?

Ans. (i) Through haemoglobin in RBC'c.

(ii) Through carbonic acid and bicarbonates formed by dissolving O_2 in water in plasma and RBC's

15. Complete the missing information in the box:

Term	Definition	Volume
Tidal volume (TV):	(i)	(ii) .... mL
(iii) .... (IRV)	(iv)	3000 mL.
Expiratory reserve volume (ERV)	(v)	(vi) .... mL.
Vital capacity (VC)	(vii)	(viii) .... mL.
(ix) .... (RV)	Air left in the lungs, even after forcible expiration	(x) .... mL.

Ans.

Term	Definition	Volume
Tidal volume (TV)	(i) Air breathed in and out, in a normal quiet breathing	(ii) 500 mL
(iii) Inspiratory reserve volume (IRV)	(iv) Air that can be drawn in forcibly over and above the tidal air	3000 mL.
Expiratory reserve volume (ERV)	(v) Air that can be forcibly expelled out after a normal expiration	(vi) 1000 mL.
Vital capacity (VC)	(vii) Volume of air that can be taken in and expelled out by maximum inspiration and expiration	(viii) 4500 mL.
(ix) Residual volume (RV)	Air left in the lungs, even after forcible expiration	(x) 1500 mL.

16. The table below the difference between inhaled air and exhaled air:

Componets	% composition of air	
	Inhaled	**Exhaled**
Oxygen	20.7	14.0
Carbon dioxide	0.04	4.0
Water vapour	1.25	3.99
Nitrogen	78.0	78.0

Answer the following questions about exhaled air:

(i) Why has the oxygen content decreased?

(ii) From where has the extra carbon dioxide been producedd?

(iii) From where has the extra water vapour come from?

(iv) Why there is no change in the percentage of nitrogen?

(v) What part does diaphragm play during inspiration?

Ans. (i) Oxygen is used for respiration. So, the oxygen content has decreased.

(ii) Extra carbon dioxide has been produced by the oxidation of glucose in the cells.

(iii) The water vapour has come from the glucose after oxidation.

(iv) Nitrogen is not used for respiration.

(v) During inspiration diapharagm flattens to increase the volume of thoracic cavity.

17. With regard to the respiratory system and the process of respiration in man, answer the following questions.

(i) Name the two muscles that helps in breathing.

(ii) Briefly describe how the above mentioned muscles help in inspiration of air.

(iii) Give the overall chemical equation to represent the process of respiration in humans.

(iv) What is meant by:

(1) Residual air

(2) Dead air space

Ans. (i) Diaphragm and intercostal muscles

(ii) During inspiration, due to contraction of external intercostal muscles ribs move upward and outwards thus the chest cavity gets enlarged. Diaphragm contracts and flattens and moves towards abdominal cavity thus increase the space of the chest cavity lengthwise.

(iii) $C_6H_{12}O_6 + 6O_2 \rightarrow 6CO_2 + 6H_2O + \text{Energy}$

(iv) (1) Residual air is the air which is left in the lungs even after forcibly breathing out. The volume of residual air is 1500 mL.

(2) Dead air space is the tidal air left in respiratory passage like trachea and bronchi where no diffusion of gases can occur. Volume of dead air space is 150 mL.

18. Given below is an overall chemical reaction of a certain process:

$C_6H_{12}O_6 \rightarrow \text{Lactic acid} + 2\text{ATP} + \text{Heat energy}$

(i) Name the process.

(ii) Is this reaction applicable to animals or to plants or to both animals and plants?

(iii) Name one tissue in which this reaction may occur.

Ans. (i) Anaerobic respiration

(ii) It occurs in animals only

(iii) Skeletal Muscles [Muscular tissue]

19. Name the two membranes covering two longs. **[February 2020]**

Ans. Parietal pleura (outer) and visceral pleura [inner] are two membranes covering the lungs.

Chapter 15. Hygiene

1. (i) What is hygiene?
   (ii) What are the major aspects of personal hygiene?

Ans. (i) Hygiene is defined as the science and practice of maintaining good health.
(ii) The major aspects of personal hygiene are as follows:
(1) Cleanliness (2) Physical exercise
(3) Rest and sleep (4) Healthy habits

2. Mention the healthy habits that one should acquire in daily life.

Ans. The healthy habits that one should acquire in daily life are taking food at regular hours, going to bed at night at regular time, etc.

3. (i) Define vectors.
   (ii) Mention some diseases caused by vectors.

Ans. (i) Vector is any agent that acts as an intermediate carrier of the pathogen.
(ii) Some diseases caused by vectors are dysentery, malaria, cholera, etc.

4. (i) Malaria is caused by which organism?
   (ii) Water from the pots, coolers should be removed. Why?
   (iii) Some farmers cultivate *Gambusia* fish. Give reason.

Ans. (i) Malaria is caused by the protozoan *Plasmodium*.
(ii) Water from pots and coolers should be removed as they provide a breeding ground for the mosquitoes.
(iii) Some farmers cultivate *Gambusia* fish as they act as biological control agents and selectively devour mosquito larvae.

5. Write the full form of DDT? Why DDT is harmful?

Ans. DDT: Dichloro Diphenyl Trichloroethane.
DDT is used for pesticides which are most serious water pollutants. Pesticide concentration in humans cause hormonal imbalance, leading to cancer.

6. What was the worst epidemic in history during the earlier part of twentieth century in many parts of the world killing a number of people?

Ans. Plague.

7. (i) What is the meaning of contamination?
   (ii) Write an example of pesticide which is most serious water pollutant?

Ans. (i) Contamination means the unwanted entry of disease germs into drinking water or edible foods.
(ii) DDT

8. (i) What are the type of Hepatitis? How they are transmitted.
   (ii) Write three symptom of Hepatitis.

Ans. (i) There are three types of Hepatitis: Hepatitis A, B and C.
(ii) Three symptoms of Hepatitis are body ace, deep yellow urine, and enlarged liver.

9. (i) How are cockroaches and rats harmful for humans.
   (ii) What are the control measures for preventing diseases caused by them?

Ans. (i) Cockroaches are very common household insects breeding mostly in the connecting manholes and sewers. They eat and spoil food, paper and clothes. They are also suspected to be carriers of certain viral diseases, including possibly a cancer-causing virus. Rats are serious domestic pests. Besides eating grain and other foods they also spread germs of certain diseases. Rat-flea is a parasite which lives on rat's skin. It may carry germs of plague.

(ii) The control measures include elimination of breeding places, spraying DDT or rat poison on stagnant water, and keeping food covered.

10. State the advantages of the following:

(i) Washing hands before taking food

(ii) Keeping hair clean

(iii) Doing proper rest and exercise

(iv) Using clean towel

Ans. (i) Washing hands before taking food helps to remove germs from the hands and fingers.

(ii) Keeping hair clean keeps it healthy and free of parasites.

(iii) Proper rest and exercise keeps the body fit and healthy. Exercise improves blood circulation.

(iv) Using clean towel helps to keep away from eye infections.

11. Mention any three ways by which potable water can be contaminated.

Ans. Contamination of potable water can occur in several ways such as:

(i) Sewage if improperly disposed off leaks into sub soil and may contaminate water supplies.

(ii) People defecating or urinating near lakes, rivers or ponds may contribute to contamination of water.

(iii) Animal wastes and washings from diaries, poultries, etc. may also be drained into water bodies.

12. Give suitable explanations for the following:

(i) One should breathe by nose and never by mouth.

(ii) Hands must be washed before eating food.

(iii) Eating places must be kept free of flies.

Ans. (i) One should breathe by nose and never by mouth because the nose contains mucous secretions that help to entrap harmful particles.

(ii) Hands must be washed before eating food because we are continually handling a variety of things- furniture, books, coins, currency notes, seats and supportingrods in buses, pets and other domestic animals, tools and machineryin workshops, and so on. All these objects may carry germs which may be picked up by our fingers and transferred over other parts of the body or into the mouth through food.

(iii) Eating places must be kept free of flies because they contaminate food particles and spread a number of diseases.

13. Name the two types of dysentery and their causative germs and give one main precaution against them.

Ans. (i) Bacillary dysentery: The causative agent is a bacterium (shigella). Patients are treated with antibiotics. Prevention includes drinking safe (boiled) water and avoiding flies from contaminating food.

(ii) Amoebic dysentery: The causative agent is a protozoan *Entamoeba histolytica*. Prevention includes drinking safe (boiled) water and avoiding flies from contaminating food

Chapter 16. Disease: Cause and Control

1. Define disease. How are the diseases categorised?

Ans. Disease is a departure from normal health through structural and functional disorder of the body.

Disease are categorised into-

(i) Extent of occurrence

(ii) Communicability
(iii) Kinds of pathogen
(iv) Kinds of transmitting agents

2. Based on the extent of occurrence categorise disease and explain them.

An. Categories of diseases based on the extent of occurrence are:

(i) Endemic diseases: When the disease is found in a certain area only attacking a fewer number of people, e.g., yellow fever in African countries.

(ii) Epidemic diseases: When the disease breaks out and spreads from place to place affecting large numbers of people at the same time e.g., plague in India in 1994.

(iii) Pandemic diseases: When the disease is widely distributed worldwide e.g., AIDS.

(iv) Sporadic diseases: When there are scattered individual cases of a disease e.g,. malaria.

3. Mention any two preventive measures of the disease Cholera.

Ans. Two preventive measures of cholera are good sanitation and control of flies.

4. Name the bacterium that causes typhoid. What part of the body does it attack?

Ans. The bacterium that causes typhoid is *Salmonella typhi* and it attacks the intestines.

5. What do you mean by epidemic disease?

Ans. Epidemic disease is the disease that breaks out and spreads from place to place affecting large numbers of people at the same time.

6. What is incubation period? What is the incubation period of Chicken pox?

Ans. Incubation period is the period between the entry of germs and the appearance of the first symptoms of the disease. The incubation period of Chicken pox is14-21 days.

7. Give abbreviations

(i) AIDS (ii) BCG (iii) HIV (iv) STD (v) TB

Ans. (ii) Acquired immunodeficiency syndrome
(ii) BacilleCalmette Guerin
(iii) Human Immunodeficiency Virus
(iv) Sexually transmitted diseases
(v) Tuberculosis

8. What is the role of *Taeniasolium* in the host?

Ans. *Taeniasolium* is a tapeworm which is spread by eating infected pork and beef. The tapeworm remains in the intestine from where it absorbs most of the host's digested food and makes the patient weak. The mature worm passes its eggs into the faeces. When the eggs are accidentally ingested by a pig or cow, they hatch and form larval stages which settle in the muscles of their new host. When man eats infected pork or beef the worm matures in his intestine.

9. Differentiate between communicable and non-communicable diseases.

Ans. In non-communicable diseases, there is no germ of any kind and they cannot spread from a patient to another person by contact or by any other method whereas communicable diseases are caused by germs and they can spread from a patient to another person by contact.

10. What are the diseases caused by bacteria? Explain each.

Ans. The diseases that caused by bacteria are:

(i) Cholera: The disease attacks the intestinal tract, and is caused by a special bacterium called *Vibriocholera*. Incubation period is a few hours to 6 days. The patient suffers from vomiting and loose motion.

(ii) Typhoid: Its major symptoms is a continuous fever which usuallyrises in the afternoon. The causative germ is *Salmonella typhi* which attacks intestines. The incubation period is about 7-12 days.

(iii) Tuberculosis: This is usually a disease of the lungs although other parts including brain, kidney and bones may also be affected. It is caused by *Mycobacterium tuberculosis*. The incubation period is 2-10 weeks.

11. What are the diseases caused by protozoa? Explain each.

Ans. The diseases that caused by protozoa are-

(i) Malaria: The disease is caused by a protozoan *Plasmodium*. The infective stage of the parasite is transmitted through the bite of the female Anopheles mosquito along with its saliva. The incubation period is about three weeks. When the mosquito bites a patient, the parasite is sucked in along with the blood.

(ii) Amoebiasis: It is caused by a kind of amoeba called *Entamoeba histolytica*. It causes destruction of the lining of the large intestine and diarrhoea with griping pains and discharge of mucus and sometimes blood in the stools. The incubation period is about one week.

(iii) Sleeping sickness: It is a disease caused by *Trypanosoma bruceigambiense* and *Trypanosoma bruceirhodesiense*. It gets transmitted through the bites of the Tsetse fly.

12. Give a brief note on chicken pox and hepatitis.

Ans. (i) Chicken pox: The disease is caused by Herpes virus *'Varicella Zoster'*. It is a common contagious disease occurring mainly in children, but can also affect adults. The disease spreads quickly by close contact with an infected person.

(ii) Hepatitis: This disease is caused by 5 strains of viruses namely Hepatitis A virus, Hepatitis B virus, Hepatitis C virus, Hepatitis D virus and Hepatitis E virus. The virus causes inflammation of the liver. It is causes inflammation of the liver.

13. What is vaccine? Name any two viral diseases, write their symptoms and preventive measures.

Ans. Germ or germ-substance introduced into the body to prevent occurrence of an infectious disease is called vaccine.

(i) AIDS is caused by a virus, HIV (Human Immunodeficiency Virus). It attacks the immune system and the patient suffers seriously from even minor infections of other diseases. Most individuals die within 3 years, when AIDS is fully developed, Symptoms during this period may include swollen lymph nodes, fever, night sweats, and weight loss. Preventive measures are disposable syringes should be used, limiting the number of sexual partners.

(ii) Chicken pox is caused by a Herpes virus 'Varicella Zoster'. Its symptoms are highly irritating rashes that appear on the body starting near the chest and back and gradually spread to arms, legs face and head. Preventive measures are bed rest, keeping the rashes clean and dry, do not prick the blister.

14. Complete the following table:

Pathogen	Disease	Target organ	Incubation period
(i)	Hepatitis A	(ii)	(iii) ....
HIV	(iv)	(v)	Up to 12 or more years

Ans.

Pathogen	Disease	Target organ	Incubation period
(i) Hepatitis A Virus	Hepatitis A	(ii) Liver	(iii) 14-45 days
HIV	(iv) AIDS	(v) system	Up to 12 or more years

15. Define the terms infection, pathogen, incubation period and allergen.

Ans. The cause of infectious disease is due to some disease-causing organism called pathogen and its transmission from one person to another is called infection.

Incubation period is the period between the entry of germs and the appearance of the first symptoms of the disease.

An allergen is any substance that has the capable of evoking an allergic reaction.

16. What are the different ways in which infectious diseases can spread?

Ans. Different ways of spreading infectious disease are

(i) Through injection needles
(ii) Mother to child transmission
(iii) Contaminated blood transfusions
(iv) Sexual intercourse
(v) Contaminated food and water
(vi) Through air or dust
(vii) By vectors

17. Name any four non-infectious diseases and their causes.

Ans. Non-infectious diseases are:

(i) Beri-beri which is caused due to nutritional deficiency.
(ii) Diabetes is caused due to metabolic.
(iii) Haemophilia is caused due to genetic disorder.
(iv) Asthma is caused due to allergic reaction.

18. Why is it important to know how the germs leave the body of a patient?

Ans. It is important for a person to know how the germs leave the body of a patient because it will help to decide the time and type of a medicine that should be give to the patient. It will also help to know its precautions and preventions of the disease so that the disease that the sick one is suffering must not be transmitted to a healthy one.

19. Write very briefly about the following:
(i) BCG (ii) Incubation period (iii) Chicken pox (iv) Hepatitis A

Ans. (i) BCG vaccination has been found useful in developing immunity for tuberculosis.

(ii) Incubation period is the period between the entry of germs and the appearance of the first symptoms of the disease. Infectious disease usually does not appear immediately after the infection but it may take some time varying from a few hours to a few days, this period is the incubation period.

(iii) Chicken pox is caused by a Herpes virus 'Varicella Zoster'. Its symptoms are highly irritating rashes that appear on the body starting near the chest and back and gradually spread to arms, legs face and head. Preventive measures are bed rest, keeping the rashes clean and dry, do not prick the blister.

(iv) Hepatitis A virus has an incubation period of 14 to 45 days and is most common in children and young adults, mainly transmitted by contaminated food and water. Hepatitis causes inflammation of the liver.

20. What are the causes and symptoms of malaria, Chicken pox and tuberculosis? How can these diseases be prevented?

Ans. (i) Malaria is caused by a protozoan Plasmodium. The infective stage of the parasite is transmitted through the bite of the female Anopheles mosquito along with its saliva. The incubation period is about three weeks. When the mosquito bites a patient, the parasite is sucked in along with the blood. Chills and high fever are the symptoms. Control measures are destruction of the mosquito at all stages and to avoid mosquito-bite by using mosquito nets or repellants.

(ii) Chicken pox is caused by a Herpes virus 'Varicella Zoster'. Its symptoms are highly irritating rashes that appear on the body starting near the chest and back and gradually spread to arms, legs face and head. Preventive measures are bed rest, keeping the rashes clean and dry, do not prick the blister.

(iii) Tuberculosis is usually a disease of the lungs although other parts such as brain, kidney, and bones may also be affected. It is caused by a bacterium called Mycobacterium tuberculosis, and spread by the sputum from the infected persons. The germs may be carried by air or dust. In the body the surrounding tissues form a kind of wall or tubercle round the germs. The germs are widespread and almost every one receives them. BCG vaccine is useful in preventing.

21. Classify the following diseases into communicable and non-communicable diseases: Cholera, beri-beri, colour blindness, diabetes, malaria, plague, heart-attack.
    (i) Communicable
    (ii) Non-communicable

Ans. (i) Communicable – Cholera, malaria, plague
(ii) Non-communicable- beri-beri, colour blindness, diabetes, heart-attack

22. Complete the following table:

Disease	Mode of transmission	Symptoms	Incubation period
Tuberculosis	Through air, dust	(i)	(ii) ....
Cholera	(iii)	(iv)	Up to 12 or more years
Typhoid	(v)	High fever, red nose, acute headache, diarrhoea	(vi) ....

Ans.

Disease	Mode of transmission	Symptoms	Incubation period
Tuberculosis	Through air, dust	(i) Persistent cough, bloody mucus, loss of weight	(ii) 2 -10 weeks
Cholera	(iii) Contaminated water, spread by flies	(iv) Vomiting, abdominal pain, dehydration	Up to 12 or more years
Typhoid	(v) Contaminated water, milk, through flies	High fever, red nose, acute headache, diarrhoea	(vi) 7 -21 days

Chapter 17. Aids to Health

1. Define immunity. What are the two levels of defence system? Explain.

Ans. Immunity means body's defence against disease.The two levels of defence system are: local defence system and immune system.

2. What do you mean by local defence system? Mention the merits of it.

Ans. Local defence systems prevent the entry of germs. It tackles the germs at their possible entry points. Merits of local defence system:
(i) It start working instantaneously.
(ii) It is not dependent on previous exposure to infection.
(iii) It is effective against a wide range of potentially infectious agents.

3. Choose any vaccine from the living weakened germs category, name the disease for which it provide immunity. Write a short not on the vaccine.

Ans. BCG is vaccine that comes under the category living weakened germs. It is stands for Bacilli Calmette Guerin used against tuberculosis. This vaccine is given to the children who have high risk of contracting TB. The BCG vaccine helps the child's immune system fight the germs that cause TB and help stop them from getting serious TB disease.

4. Write the full form of the following:
(i) DTP (ii) BCG (iii) STD (iv) HIV
(v) AIDS (vi) DT (vii) DDT (viii) WHO

Ans. (i) DTP- Deptheria, Tetanus, and Pertusis
(ii) BCG- Bacillus of Calmettes and Guerin.
(iii) STD- Sexually transmitted disease.

Short Answer Type Questions

(iv) HIV- Human immunodeficiency virus.
(v) AIDS- Acquired immunodeficiency syndrome
(vi) DT- Diphtheria and Tetanus.
(vii) DDT- Dichlorodiphenyltrichloroethane
(viii) WHO – World health organisation

5. Complete the following table:

Vaccine	Disease	The nature of vaccine
TAB	(i)	**(ii)**.....................
Salk's vaccine	(iii)	(iv) ....
BCG	(v)	Living weakened germs
Vaccine for measles	Measles	(vi) ....
Cowpox virus	(vii)	(viii) ....
Toxoids	(ix)	Extracts of toxins
	(x)	Secreted by bacteria

Ans.

Vaccine	Disease	The nature of vaccine
TAB	(i) Typhoid	(ii) Killed germs
Salk's vaccine	(iii) Poliomyelitis	(iv) Killed germs
BCG	(v) Tuberculosis	Living weakened germs
Vaccine for measles	Measles	(vi) Living fully poisonous germs
Cowpox virus	(vii) Smallpox	(viii) Living fully poisonous germs
Toxoids	(ix) Diphtheria	Extracts of toxins
	(x) Tetanus	Secreted by bacteria

6. Given below is a scheme of classifying immunity against human diseases. Fill up the types of immunity in the blanks.

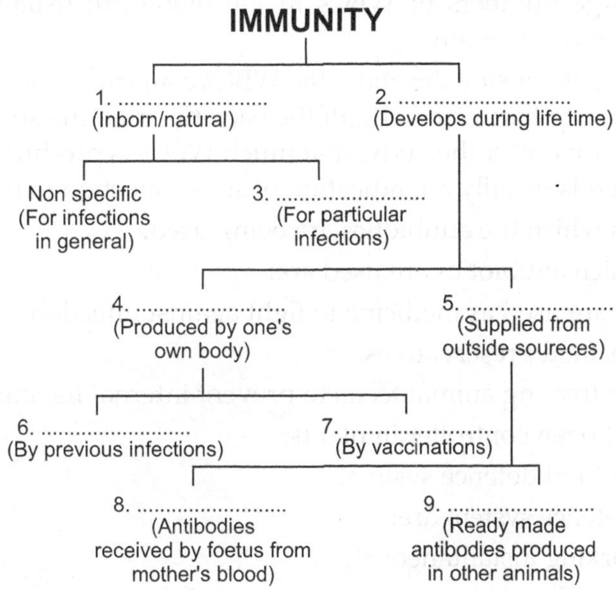

Ans. (i) Innate immunity
(ii) Acquired immunity
(iii) Specific innate immunity
(iv) Actively acquired immunity
(v) Passively acquired immunity
(vi) Naturally acquired active immunity
(vii) Artificially acquired active immunity

(viii) Naturally acquired passive immunity

(ix) Artificially acquired passive immunity

7. Explain the local defence mechanism.

Ans. The skin, nails, and hairs form the local defence system that prevent dirt and pathogen from entering inside the body. We also have internal defence system which includes white blood cells and antibodies they produce, otherwise known as immune system.

8. Write a summary of our body's defence against disease.

Ans. The body's defence system is also called immunity. The body is invaded by various harmful substances like germs. The germs may enter inside our body through skin, mucus membrane, eyes, nose, urinary tract, through water or food, etc. The entry of the foreign substances is first prevented by the physical barrier and if somehow they enter the body, then it fights with them without causing any harm to the body.

9. Which body process work by throw out method, if any foreign substance enters the body?

Ans. The throw out process includes coughing, vomiting, and sneezing that throw out germs and pathogen out of the body and avoid their entry inside the respiratory and digestive system.

10. Write a short note on the lymphocytes.

Ans. Lymphocytes are white blood cells that are also one of the body's main types of immune cells.They are made in the bone marrow and found in the blood and lymph tissue. These cells work together to defence the body against foreign substances.

11. Write about the source of penicillin production and its role in our body.

Ans. The source of Penicillin production is *Penicillium notatum.* The penicillin works by bursting the cell wall of the bacteria.

12. The principle of vaccination is to produce immunity against a disease. Explain.

Ans. Vaccination is a process in which the germs or the germ substances are introduced inside the body to gain the resistance against the particular disease caused by the germ and the substance introduced inside the body is termed as vaccine.

13. "Abnormally, large numbers of WBCs in the blood are usually an indication of some infection in our body". Explain.

Ans. Whenever germs enter inside the body, the WBCs come in action. It gets squeezed out of the walls of the blood capillaries and engulf the bacteria or germs and destroy them. Therefore, as much as the germs enter the body, that much WBCs is produced. Hence large number of WBCs in the blood is usually an indication of some infection in our body.

14. List four ways in which the antibiotics are being used.

Ans. Four ways in which antibiotics are used are:

(i) Antibiotics are used as medicine to fight against infection.

(ii) It is used as food preservatives.

(iii) It is used in treating animal feeds to prevent internal infection.

(iv) Usedas pathogen controller in plants.

15. List the merits of local defence system.

Ans. Merits of local defence system are:

(i) It starts working instantaneously.

(ii) It is not dependent on previous exposure to infection.

(iii) It is effective against a wide range of potentially infectious agents.

16. Suppose a person develops the disease diphtheria. Comment upon the principle of the treatment he should receive.

Ans. Vaccination is a process in which the germs or the germ substances are introduced inside the body to gain the resistance against the particular disease caused by the germ and diphtheria is a disease that causes infection of the nose and throat and it is prevented by vaccine. Thus the person should be treated with pre-prepared antibodies in which the disease is produced in a mild form.

17. What first aid steps would you take in the following cases of accidents?

(i) Bleeding from cut in the skin

(ii) A fracture darm

(iii) Stoppage of breathing due to electric shock

(iv) Burn injuries

Ans. (i) Apply antiseptic

(ii) Apply ice to avoid swelling

(iii) Apply pressure on the chest to restore circulation.

(iv) Apply ice over the burnt area to cool.

18. Given below are the groups of certain substances of particular categories. Mention the category of each group and identify the wrong example given reason.

(i) Lysol, benzoic acid, DDT, mercurochrome

(ii) Formallin, iodine, lysol, phenol.

(iii) BCG, DTP, ATP.

(iv) Tears, skin, nasal secretion, HCl (in stomach).

Ans. (i) **Lysol, benzoic acid, DDT, mercurochrome:** Antiseptics DDT is a wrong example for this category as it is a disinfectant which is not good for human skin.

(ii) **Formalin, iodine, lysol, phenol:** Disinfectants iodine is a wrong example as it is an antiseptic.

(iii) **BCG, DTP, ATP:** Vaccines ATP is a wrong example as it is an energy carrier in the cells of all known organisms.

(iv) **Tears skin, nasal secretion, HCl (in stomach):** Germ-killing secretions skin is a wrong example as it is a protective mechanical barrier and prevents the entry of germs in our body.

Chapter 18. Health Organisation

1. Mention any two common health problems in India.

Ans. Two common health problems in India are:

(i) Food and water borne diseases: Typhoid, dysentery are very common in areas deficient in proper water supply.

(ii) Insect and air borne diseases: The public in general, and specially the uneducated village folk, are not conscious of the flies which alight on exposed food stuff and contaminate them.

2. Write the full form of the abbreviations:

(i) WHO (ii) UNO (iii) NICD

Ans. (i) World Health Organisation.

(ii) United Nations Organisation.

(iii) National Institute of Communicable Diseases

3. Define Health Organisations. Name any two health organisations.

Ans. Health organisation is the health care system and resources that deliver health care services to meet the health needs of target population.

The two main health organisations are Red Cross and WHO.

4. What is the main function of Red Cross Society?

Ans. The main function of Red Cross Society is to perform activities which should prevent or remove human sufferings in peace time as well as at the time of war.

5. State any two activities of Red Cross societies.

Ans. The major activities of Red Cross Societies are as follows:

(i) To procure and supply blood for the needy victims of war or other calamities.

(ii) To extend all possible first-aid in any accident.

6. List any two main activities of WHO. **[February, 2020]**

Ans. The two main activities of WHO:

(i) To collect and supply information about the occurrence of diseases of epidemic nature.

(ii) To promote and support projects for research on diseases.

7. Write a short note on the responsibilities of city municipalities in caring for people's health.

Ans. (i) It makes arrangements for the cleanliness of the city.

(ii) It makes arrangements for safe drinking water.

(iii) It runs dispensaries and hospitals for the sick people, thereby controls diseases.

8. Differentiate on the basis of what is given in bracket.

Plasma and serum (Composition).

Ans. Plasma: Fibrinogen is present.

Serum: Fibrinogen is absent.

9. Categorise the following activities as per the functions of the Red cross society and the WHO.

(i) To suggest quarantine measures to prevent spread of disease.

(ii) Humanitarian services to vitims of war.

(iii) To edudcate people in accident prevention.

(iv) To promote projects for reaserch on disease.

Ans. WHO

(i) To suggest quarantine measures to prevent spread of disease.

(ii) To promote projects for research on disease.

Red Cross society

(ii) Humanitarian services to victims of war.

(iii) To educate people in accident prevention.

10. Mention two activities of Red cross. **[November, 2019]**

Ans. (i) To help the victims of the calamities.

(ii) To provide first aid in any accident.

11. Mention two functions of the World Health organisation.

Ans. (i) Assists countries to improve health and sanitation services.

(ii) Promotes medical education and training.

12. List any three health problems in India.

Ans. Health problems in India:

(i) **Food and water-borne diseases:** Contaminated food and water cause several diseases.

(ii) **Insect-and air-borne diseases:** Lack of cleanliness leads to breeding of houseflies and mosquitoes which are carriers of certain diseases.

(iii) **Lack of medical facilities:** Lack of medical facilities especially in rural areas leads to unavoidable deaths and damage to health. Lack of knowledge and superstitious beliefs also delay timely treatment which may result in serious consequences.

13. Describe the reasons for the formation of WHO.

Ans. The world Health Organisation (WHO) is a specialised agency of the United Nations (UN) which is concerned with international public health. It was established on 7 April 1948, with headquarters in Geneva, Switzerland, and is a member of the United Nations Development Group. Member countries of the UNO focused on the need for creating an international body to look after the health problems of the people of the world. This need was particularly felt in the field of research on the causes and cures of diseases. Combined efforts in this direction were to give better and faster results. Poor and developing countries were to benefit quickly.

Chapter 19. Waste Generation and Management

1. Why do we need to manage waste?

Ans. Ever increasing population is producing a lot of waste, so it is important to keep environment clean.

2. What will happen if waste is not disposed?

Ans. Piles of garbage will collect and spoil the environment.

3. How farmers should dispose agricultural waste?

Ans. By making compost from wastes.

4. Why plastic bags have been banned?

Ans. Plastic bags have been banned because plastic is not degraded easily.

5. What can you do to keep environment clean?

Ans. Recycle, reuse and reduce.

6. Why is it useful to separate waste into biodegradable and non biodegradable?

Ans. Separation of wastes into biodegradable and non-biodegradable is useful for disposing-off wastes by different methods having concernabout less pollution.

7. What can we do to reduce waste production?

Ans. Recycle –Reuse – Reduce

8. Why hospitals and laboratories should dispose their waste carefully? Explain.

Ans. Biomedical waste may be contaminated with germs

9. Effluents from factories should not be directly discharged into rivers. Give a suitable reason.

Ans. Effluent may be containing poisonous chemicals which should be removed before disposal

10. State two methods by which waste from industries can be treated to make it less harmful.

Ans. Electrostatic precipitators and scrubbers.

11. Name the following.

(i) The waste water from different sources that run off into the drains.

(ii) The wastes which are degradable by microbial action.

Ans. (i) Sewage, (ii) Biodegradable wastes.

12. Incineration, the disposal method of solid wastes leads to pollution. Comment.

Ans. Yes, the incineration method involves burning of waste materials which causes air pollution.

13. Open landfill sites are potential centres to harm the environment. How?

Ans. Open landfill sites are potential centres to harm environment as leaching of chemicals and toxins may contaminate environment.

14. Define e-wastes.

Ans. Electronic wastes generated due to discarded electronic equipments e.g. compurters, mobiles, etc.

15. Give the name of any two ways to reduce particulate air pollutants.

Ans. Air scrubbers and electrostatic precipitator.

16. Mention the function of catalytic converter in an automobile.

Ans. Catalytic converter reduces the emission of poisonous gases by converting unburnt hydrocarbons to CO_2 and waste.

17. Differentiate between biodegradable and non-biodegradable and non-biodegradable wastes.

Ans. Differences between biodegradable wastes and non-biodegradable wastes are as follows:

Biodegrable wastes	Non-biodegradable wastes
They can be broken down into simple substances by decomposers.	They cannot be broken down by decomposers.
They become non-toxic after decomposition.	They are recycled and reused.
These are environement friendly.	These are not environment friendly.

18. What are the sources of radioactive wastes?

Ans. The sources of radioactive wastes are as follows:

(i) Accidental leakage of nuclear material or radiations from nuclear power plants.

(ii) Improper disposal of nuclear or radioactive wastes.

19. Give the methods used to dispose the solid wastes.

Ans. The methods used to dispose the solid wastes are:

(i) Segregation (ii) Dumping

(iii) Composting (iv) Incineration

(v) Landfills

20. Broken glass materials are considered as non-biodegradable. Why?

Ans. Glass is non-biodegradable because it cannot be decomposed into simpler forms. Natural means of decomposition are unable to degrade it and they persist in the environment for a very long time.

21. It is observed that the landfills are forming fast near larges cities. Give reason.

Ans. Increased amount of waste is being generated in large cities. All the garbage is collected and dumped in landfills. Due to increased waste production, the sites are being filled rapidly. New landfill sites are needed fast. Hence, sites near to large cities are used. This is the reason why, landfills are forming near large cities.

22. Explain the treatment of liquid wastes.

Ans. Treatment of liquid wastes is carried out in three stages:

(i) Primary treatment: The liquid is passed through coarse screen to remove large objects; like cans, plastic packets, rags and sticks. After that, liquid is passedd through a grit chamber to remove grit and sand. Then the liquid is sent to a flocculation tank sedimentation and precipitation are carried out to remove solid impurities from the liquid.

(ii) Secondary treatment: First of all, the waste water treated with anaerobic bacteria to change sludge into harmless substances.

Then, liquid is aerated so that aerobic bacteria can remove other wastes. After that, liquid is subjected to chlorination to kill harmful germs in water.

(iii) Tertiary treatment: This is an advanced stage of water treatement. It removes nitrogenous wastes from water.

23. It is necessary to separate the constituents of municipal sewage before any treatment disposal. Give reason.

Ans. The municipal sewage waste is first segregated into two categories of biodegradable and non-biodegradabale wastes. This seregation is important for selecting a proper disposal method.

The biodegradable wastes are isolated and converted into compost which can be used by farmers and gardeners as manure. Since, non-biodegradable wastes cannot be decomposed they are treated further to decrease their toxicity. Then, they are dumped in landfill sites in proper manner and covered.

24. How does an electrostatic precipitaror work to remove particulate pollutants released from the thermal power plants?

Ans. Electrostatic precipitator removes the particulate matter with 99% efficiency. It has electrode wires maintained at high voltage, which produces a corona that releases electrons.

These electrons get attached to the dust particles and give them negative charge. A positively chargedd collecting plate is also there, which attracts the negativly charged dust particles. Thus, separating them from the air and collecting them on itself. Afterwards, the air passes through the collecting plate and gets cleaned.

25. Discuss the various methods of solid wastes disposal:

Ans. Various methods of solid wastes disposal are as follows

(i) Segregation: It is the first step in solid wastes disposal. Waste is segregated into biodegradable and non-bidegradable; before being disposed. For this, municipal authorities put garbage bins of two colours.

The green one is for biodegradable wastes and the blue one is for non-biodegradable wastes.

(ii) Dumping: solid waste is usually dumped in landfill sites. For toxic waste, landfill sites are covered with soil layer after dumping the waste.

(iii) Composting: biodegradable wastes can be converted into nanure by this process. Sometimes, earthworms are used for hastening the process known as vermicomposting.

(iv) Drainage: liquid waste is usually removed through a system of drainage. Sewage is sent to waste water treatment plant.

(v) Incineration: solid wastes can be burnt in incinerators. Incineration reduces the volume and mass of waste and thus, reduces the need for area of landfill. Incineration also converts toxic substances into less toxic or non-toxic substances.

26. Identify A, B, C and D in the figure given below.

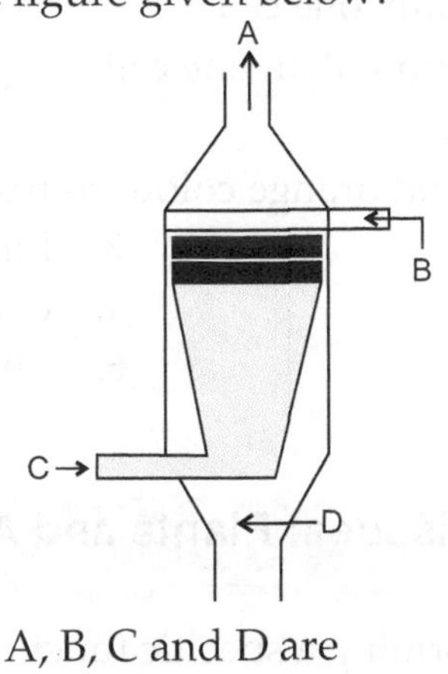

Ans. The correct identification for A, B, C and D are

(A) Clear air (B) Water/lime spray

(C) Dirty air (D) Particulate matter

27. Describe the usefulness of incineration of wastes, and also mention the precautions required in it.

Ans. **Usefulness of incineration of wastes:**

It reduces the weight of the waste.

It reduces the volume of the waste.

It renders toxic wastes into less toxic or even non-toxic wastes.

Precautions required for incineration:

It should be carried out at very high temperatures.

It should be equipped with pollution control devices.

It should be installed away from residential areas.

Give Technical Term

Set 15

Give techanical terms ofr the follwing:

Chapter 1. Introducing Biology

1. It helps us to aware of environmental hazards caused by man's activity, like cutting forests, increased use of automobiles, etc.
2. The study of origin and descent of organisms.
3. The science of grouping or classifying the organisms.
4. The study of improvement of human race.

Ans. 1. Biology 2. Evolution 3. Taxonomy 4. Eugenic

Chapter 2. Cell: The Unit of Life

1. The powerhouse of the cell.
2. Small granules either scattered in the cytoplasm or attached to the endoplasmic reticulum.
3. The flattened sac-like structure contain vesicles, vacuoles and tubules.
4. The organisms consisting of only one cell.
5. The organisms consisting of more than one cell.
6. An organelle that stores ATP. **[November, 2019]**
7. The plastid which gives red and orange colour to fruits and flowers. **[February, 2020]**

Ans. 1. Mitochondria 2. Ribosomes
3. Golgi body 4. Unicellular organisms
5. Multicellular organisms 6. Mitrochondria
7. Chromoplasts

Chapter 3. Tissues: Plants and Animal tissue

1. A group of similar cells performing a specific function.
2. Cells least specialized in the plants.
3. Cells responsible for increase in diameter of the stem and root of dicot plants.
4. A tissue enables the plant body to withstand various strains by giving strength, rigidity.
5. Muscles rhythmically contracts and relaxes to pump and distribute blood to the various parts of body.
6. The fluid connective tissue which help in transportation of oxygen and carbon dioxide.
7. The soft matrix covered by a fibrous membrane, perichondrium.
8. Study of tissues.
9. A group of organs that perform a particular function, e.g. digestive system.
10. The tissue helping in transportation of water, minerals and food in the plant.
11. The tissue that conducts food plants. **[November, 2019]**
12. The type of tissue that lines the respiratory tracts. **[February, 2020]**
13. The involuntary muscles found in head. **[February, 2020]**

Ans. 1. Tissue 2. Parenchyma 3. Cambium 4. Sclerenchyma
5. Cardiac muscles 6. Blood 7. Cartilage 8. Histology
9. Organ system 10. Conductive tissue 11. Phloem
12. Ciliated columnar epithilium 13. Cardiac muscles

Chapter 4. The flower

1. Floral part which attracts insects for pollination.
2. Floral part which performs the process of photosynthesis when they are green.
3. When male and female flowers grow on the same plant.
4. The mode of arrangement of the flowers on the axis of the plant.
5. The structure when tip of the flower stalk expands to form a cup shaped structure.
6. A condition in which the petals are united. **[November, 2019]**

Ans. 1. Petals 2. Sepals
3. Monoecious plants 4. Inflorescence
5. Receptacle or Thalamus 6. Gamopetalous

Chapter 5. Pollination and Fertilisation

1. A condition (called a type of self-pollination) in which pollen grains are transferred to the stigma of another flower of the same plant.
2. The male gamete of flowering plant which plays an important role in pollination.
3. Two wall layers of pollen grain.
4. Different timings for maturation of gynoecium and androecium.
5. A flower containing both male and female parts.
6. When stigma and anthers do not grow up to same height, which favours only cross-pollination.
7. When maturation time of a flower reach the stigma of the same flower.
8. Pollination of flowers by insects.
9. Arrangement of flowers on a twig / stem.
10. When stigma and anthers do not grow up to same height which favours only cross – pollination.
11. The arrangement of ovules inside the ovary of a flower. **[November, 2019]**

Ans. 1. Geitonogamy 2. Pollen grain 3. Exine and intine 4. Dichogamy
5. Bisexual flowers 6. Heterostyly 7. Dichogamy 8. Entomophily
9. Influorescence 10. Heterostyly 11. Placentation

Chapter 6. Seeds-Structure and Germination

1. The seeds which lack endosperm containing albumin protein.
2. The seeds with two embryonic leaves.
3. The process of formation of a seedling developed from the embryo.
4. The suitable temperature which favours seed germination.
5. The type of germination in which cotyledons are pushed above the ground.
6. The type of germination in which cotyledons remain underground.
7. The opening through which water enters a seed. **[November, 2019]**

Ans. 1. Exalbuminous seeds 2. Dicotyledonous seeds

3. Germination
4. Optimum temperature
5. Epigeal germination
6. Hypogeal germination
7. Micropyle

Chapter 7. Respiration in Plants

1. Respiration which occurs only in presence of excess oxygen.
2. Respiration which occurs in absence of oxygen.
3. A destructive or catabolic process occurring in all living cells.
4. A physio-chemical process where all energy is liberated in form of heat and light.
5. An anaerobic phase of respiration which occurs in cytoplasm.
6. The aerobic phase of respiration which occurs in mitochondria.
7. The organic substances undergoing oxidative break down during respiration.
8. The incomplete break down of sugar into ethanol and carbon dioxide.

Ans. 1. Aerobic respiration
2. Anaerobic respiration
3. Respiration
4. Combustion
5. Glycolysis
6. Kreb's cycle
7. Respiratory substrates
8. Alcoholic fermentation

Chapter 8. Five Kingdom Classification

1. A group of organisms which can interbreed themselves to produce fertile offsprings.
2. Animals whose body temperature remain constant.
3. A muscular structure separating thorax from abdomen.
4. Animals whose body temperature fluctuates directly with the temperature of the environment.
5. Animals which live partly on land and partly in water.
6. A rod-like structure present in the mid-dorsal axis of the body.
7. Chordates which possess a backbone.
8. Body can be divided into two similar halves only along one plane.
9. Method of naming organisms.
10. Largest division in classification of animals and plants.
11. Category of organism which can inter breed among themselves. **[November, 2019]**

Ans. 1. Species
2. Homeothermal
3. Diaphragm
4. Poikilothermal
5. Amphibians
6. Notochord
7. Vertebrates
8. Bilateral symmetry
9. Binomial nomenclature
10. Phylum
11. Species

Chapter 9. Economic Importance of Bacteria and Fungi

1. The process of boiling water at high temperature(about 110°C) to kill microbes in it.
2. Heating milk at 60°C and then suddenly cooling it.
3. The nutrition in which organism feed on dead and decayed organisms.

4. Introduction of vaccine into the body to develop immunity to a particular disease.
5. A substance having antitoxin of a particular disease-causing microbes.
6. Complete breakdown of organic matter by bacteria without emission of foul smell.

Ans. 1. Sterilisation 2. Pasteurisation 3. Saprotrophic nutrition
4. Vaccination 5. Serum 6. Decay

Chapter 10. Nutrition

1. Indigestible part of food.
2. Micro nutrient which we get from salts.
3. Macronutrient required for growth and development.
4. A diet that is popular for a time.
5. Part which closes glottis during swallowing.
6. Deficiency disease commonly called as bleeding gums.
7. The disease due to which a person is not able to see in dim light.
8. Breakdown of food into to smaller parts.
9. Group of sour fruits which are rich source of vitamin C.
10. A gap present between incisors and premolars due to absence of canines in herbivore like rabbit.
11. A diet having all the nutrients in the right proportion. **[November, 2019]**

Ans. 1. Roughage 2. Minerals 3. Proteins 4. Food fads
5. Epiglottis 6. Scurvy 7. Night blindness 8. Digestion
9. Citrus 10. Diastema 11. Balanced diet

Chapter 11. Digestive System

1. Converting insoluble complex food to soluble ,simpler and smaller molecules.
2. Soaking of useful materials from membrane.
3. Gulping in of food.
4. Making body material from absorbed molecules.
5. Removal of waste food material.
6. Killing of germs.
7. Rhythmic waves of contractions and relaxation which makes food move.
8. Finger-like projections on inner side of small intestine.
9. Type of epithelium present in wall of stomach which produces digestive juices.
10. Tissue which stores fats.
11. A protein found in milk that forms curd, it ends to be difficult for infants to digest.
12. A lymphatic vessel in an intertinal villus, it aids in the absorption of fats.
13. Sphere formed by bile salts that surround lipids and make them temporatily water soluble.
14. Fine microscopic, finger-like projections lining the small intestine that serve to increase the absorprive surface area.
15. The muscular structure that manipulates food in the mouth. **[February, 2020]**

Ans. 1. Digestion 2. Absorption 3. Ingestion 4. Assimilation
5. Egestion 6. Disinfection 7. Peristalsis 8. Villi
9. Glandular epithelium 10. Adipose 11. Casein
12. Lacteal 13. Micelle 14. Villi 15. Tongue

Chapter 12. Skeleton: Movement and Locomotion

1. Other name for vertebral column.
2. Hard frame work of body.
3. Bones of arms and legs.
4. Tissue which hold bones together.
5. Brain box.
6. Only bone which is not attached to any other bone.
7. Four vertebrae of tail region.
8. Dagger shaped central bone of ribcage.
9. Muscles of heart.
10. Muscles which work under command.

Ans.

1. Backbone	2. Skeleton
3. Apendicular skeleton	4. Ligament
5. Cranium	6. Hyoid
7. Caudal	8. Sternum
9. Cardiac	10. Voluntary

Chapter 13. Skin: "The Jack of all Trades"

1. Layer of skin made of dead cells having keratin.
2. Layer of skin which protects us from infections due to bacteria.
3. Layer of skin which has melanin in it.
4. Glands which produce sebum.
5. Summer sleep.
6. Winter sleep.
7. Disease characterised by total loss of pigmentation.
8. Disease characterised by loss of pigmentation in patches.
9. Modified sebaceous glands.
10. Layer of fats under the skin.
11. The process of dilation of blood vessels in the skin due to high temperature. **[November, 2019]**
12. The principal heat regulating centre in the brain. **[February, 2020]**

Ans.

1. Stratum corneum	2. Stratum corneum
3. Malpighian layer	4. Sebaceous glands
5. Aestivism	6. Hibernation
7. Albinism	8. Leucoderma
9. Ceruminous gland	10. Subcutaneous fats
11. Vasodilation	12. Hypothalamus.

Chapter 14. The Respiratory system

1. Respiratory organ of Insects
2. Method of respiration in unicellular organisms
3. Process of intake of air
4. Ratio of rate of CO_2 output to O_2 uptake
5. Innermost part of nasal chamber having hairs
6. Film covering alveoli which lowers surface tension
7. It closes glottis during swallowing of food

8. A condition of painful breathing
9. Organ of sound production
10. Body cavity in which lungs are located
11. The flap that guards the entrance of the trachea **[November, 2019]**
12. The end product of glycolysis **[November, 2019]**
13. Blind sacs in lungs where exchange of respiratory gases takes place **[February, 2020]**

Ans. 1. Spiracles and tracheae 2. Diffusion
3. Inhalation 4. Respiratory Quotient
5. Vestibule 6. Lecitin
7. Epiglottis 8. Dyspnea.
9. Larynx 10. Thoracic cavity
11. Epiglottis 12. Pyruvate
13. Alvoeli

Chapter 15. Hygiene

1. An agent which is carrier of pathogens.
2. It is a disease caused by *Entamoeba histolytica*.
3. The science and practice of maintaining good health.
4. The condition in which the normal functioning of the body is disturbed.
5. The unicellular organisms showing metabolism of their own and cannot be crystallised.
6. The disease spread from an infected person to a healthy person.

Ans. 1. Vector 2. Amoebiasis 3. Hygiene 4. Disease
5. Bacteria 6. Infectious disease

Chapter 16. Disease: Cause and Control

1. The period between the entry of germs and the first symptom of theappearance of disease.
2. The condition in which the normal functioning of the body is disturbed.
3. The organism which is a disease-causing micro-organism of malaria *i.e.,* Anopheles mosquito.
4. Disease in which there is no germ and cannot spread from one person to another **[February, 2020]**

Ans. 1. Incubation period 2. Disease
3. Vector 4. Non-communicable disease

Chapter 17. Aids to Health

1. The process by which a person is protected against diseases by inoculating the person with vaccine.
2. Mild chemical applied on the skin to kill germs.
3. Treatment by the use of chemicals.
4. Artificially introducing weakened germs or germ substances into the body foe the developing resistance to a particular disease.
5. The process by which white blood cells engulf bacteria.
6. The substances produced in the body when an antigen enters into the body.
7. The hypersensitivity of the body to a particular substance.
8. The chemical substances produced by microorganism, which inhibit or kill other microorganisms. **[November, 2019]**

9. Resistance developed in the body due to previous infection. **[February, 2020]**

Ans. 1. Immunation 2. Antiseptic 3. Chemotharapy 4. Vaccination
5. Phagocytosis 6. Antibody 7. Allergy 8. Antibotics
9. Actively acquired immunity

Chapter 18. Health Organisation

1. Any foreign molecule or microorganism that induces the formation of an antibodies in the body.
2. Immunisation of the infants and others to protect them from several infectious diseases.
3. A national as well as international agency.
4. Removal and proper disposal of garbage and other waste, elimination of breeding places of flies, mosquitoes, etc.
5. The disease found in a certain area only attacking a fewer number of people.
6. The disease breaks out and spread from place to place affecting large number of people at the same time.
7. Introduction of weakened germs into the body to develop resistance against a disease. **[February, 2020]**

Ans. 1. Antigen 2. Vaccination 3. Red Cross 4. Sanitation
5. Endemic disease 6. Epidemic disease 7. Vaccination

Chapter 19. Waste Generation and Management

1. Waste derived from non living things and not decomposed by bacteria.
2. The type of waste generated in house kitchens.
3. Waste which is harmless to humans.
4. The process of separating waste in to different categories.
5. The waste water coming out of a factory.
6. The disposal of garbage by burning it.
7. The treatment plants where sewage waste is treated.
8. The areas where waste material is thrown to let it decay.
9. The chemical method of removing toxic gases from industrial chimneys.
10. The process of making fertiliser from garbage.

Ans. 1. Non biodegradable 2. Biodegradable
3. Non-toxic 4. Segregation
5. Affluent 6. Incineration
7. Sewage treatment plants 8. landfills
9. Electrostatic precipitation 10. Composting

Explain the Terms

Set 16

Chapter 1. Introducing Biology

1. Palaeontology

Ans. Study of fossils of plants and animals is called palaeontology.

2. Bioinformatics

Ans. Management and analysis of biological information stored in databases is bioinformatics.

3. Cloning

Ans. Producing offsprings(clones) from a single parent. They show similar characteristics.

4. Biotechnology

Ans. The branch of biology which deals with the use of cells or micro-organisms in technology and industry.

5. Pisciculture

Ans. The technique of rearing and cultivation of fish is known as pisciculture.

Chapter 2. Cell: The Unit of Life

1. Cytology

Ans. The branch of biology which deals with the study of cells and their parts.

2. Nerve cells

Ans. The cells which are long to conduct impulses from the distant parts of the body to the brain and vice versa.

3. Cell membrane

Ans. Cell membrane is the outer boundary of a cell. It is a thin elastic and living membrane which allow some selective materials to pass through it.

4. Chloroplast

Ans. Chloroplast is the green colour plastid which contain chlorophyll which helps in photosynthesis.

5. Vacuole

Ans. A hollow space bound by the membrane called tonoplast and enclosing a fluid called cell sap.

Chapter 3. Tissues: Plants and Animal Tissue

1. Ligament

Ans. A fibrous tissue that connects bone to bone.

2. Tendons

Ans. Tendons are the tissues that connects muscle to bone.

3. Lymph

Ans. Lymph is the fluid connective tissue containing filtered blood plasma, found in lymphatic vessels.

4. Adipose tissue

Ans. Adipose tissue is connective tissue having specialized cells which store fat. This tissue forms padding under the skin and around kidneys, eye ball.

5. Striated muscles

Ans. Striated muscles are under the control of the will which are found in the arms, legs, etc.

6. Unstriated muscles

Ans. Unstriated muscles are not under the control of the will that are found in the walls of the intestine helping in the passage of food.

7. Tissue **[February, 2020]**

Ans. Tissue is a group of similiar cells performing a specific function.

Chapter 4. The Flower

1. Complete flower

Ans. A flower which contains all the four whorls is called a complete flower, e.g. rose.

2. Bisexual flower

Ans. A flower which contains both male and female reproductive parts, e.g. rose.

3. Unisexual

Ans. A flower which contains either male or female reproductive part is called unisexual, e.g. ray florets and disc florets or Chrysanthemum.

4. Pedicel

Ans. The stalk of a flower.

5. Perianth

Ans. The condition in which calyx and corolla connot be distinguished.

6. Placenta

Ans. A tissue present in the wall of the ovary where ovules are attached.

7. Polypetalous

Ans. The condition where all the petals are free.

8. Pistillate flower

Ans. A flower which contains only female reproductive part, e.g. ray floret of chrysanthemum.

9. Neuter flower

Ans. A flower in which stamens and pistil are not present.

10. Monadelphous stamens

Ans. When the stamens are united in one group, it is called monadelphous.

11. Diadelphous stamesn

Ans. When the stamens are arranged in two bundles it is called diadelphous.

12. Polyadelphous

Ans. When the stamens are arranged in several groups, it is called polyadelphous.

13. Staminate flower

Ans. A unisexual flower which contains only the stamen.

Chapter 5. Pollination and Fertilisation

1. Ornithophily

Ans. The process of pollination affected by birds is known as ornithophily.

2. Elephophily

Ans. The process of pollination affected by elephants is known as elephophily.

3. Artificial pollination

Ans. The process of pollination affected by men through artificial modes is referred to as artificial pollination.

4. Triple fusion

Ans. The fusion of polar nuclei and male gametes.

5. Fertilization

Ans. The fusion of male gamete and egg nucleus.

6. Homogamy

Ans. The condition in which the anther and the stigma mature at the same time is known as homogamy.

7. Double fertilisation

Ans. The fusion of one sperm nucleus with egg nucleus and the other sperm fuses with two polar nuclei together is termed as double fertilisation.

Chapter 6. Seeds: Structure and Germination

1. Grain

Ans. Grain is a fruit in which the fruit wall and seed coat are fused together to form a protective layer.

2. Aleurone layer

Ans. Aleurone layer is the outermost protein rich layer of endosperm.

3. Embryo

Ans. Embryo is the tiny plant inside the seed which contains radicle and plumule.

4. Testa

Ans. Testa is the outermost hard covering of seed which protects the inner layers of seed from attack of microbes as well as mechanical injury.

5. Tegmen

Ans. Tegmen is the thin protective layer lying inner to testa of seed.

6. Epicotyl

Ans. Epicotyl is the region of the axis between the point of attachment of cotyledons and plumule.

7. Hypocotyl

Ans. Hypocotyl is the region of axis below the cotyledons.

8. Scutellum

Ans. Embryo containing a single cotyledon called scutellum.

9. Pyrogallic acid

Ans. Pyrogallic acid is a chemical which absorbs oxygen.

10. Endospermic seed

Ans. The seeds where food is stored in endosperm and cotyledons are thin and membranous are called endospermic seed.

Chapter 7. Respiration in Plants

1. Anoxybiotics

Ans. Anoxybiotic involes the incomplete breakdown of glucose in absence of oxygen to release ethanol [plants] and carbon dioxide.

2. Oxybiotic

Ans. Oxybiotic involves the complete breakdown of glucose in presence of oxygen to release carbon dioxide and water.

3. Metabolism

Ans. Metabolism is the sum of all chemical reactions occurring in our body.

4. Catabolism

Ans. Catabolism is the process of breakdown of large molecules into simpler ones with release of energy.

5. Anabolism

Ans. Anabolism is the process of build up of complex molecules from simpler ones by using energy.

6. ATP

Ans. ATP is a form of chemical energy released during respiration process which is stored in form of molecules in mitochondria.

7. Respiration

Ans. Respiration is a catabolic process catalysed by enzymes where glucose is oxidised in a series of steps to release energy.

8. Control set-up

Ans. Control set-up is the set-up used in experiments where the condition under study is missing.

Chapter 8. Five Kingdom Classification

1. Diploblastic

Ans. Diploblastic is an animal that develops from blastula with two germ layers ectoderm and endoderm.

2. Triploblastic

Ans. Triploblastic is an animal that develops from blastula with three germ layers ectoderm, mesoderm and endoderm.

3. Radially symmetrical

Ans. If an animal is cut in any plane through the central axis, the body is divided into two similar parts and termed as radially symmetrical.

4. Acoelomate

Ans. Acoelomate are the animals without a body cavity.

5. Coelomate

Ans. Coelomate are the animals with a body cavity.

6. Invertebrates

Ans. Invertebrates are those animals that are without a backbone.

7. Tentacles

Ans. Tentacles are organs to catch prey/food organisms.

8. Prokaryotic

Ans. The condition of the cell where nuclear material lie free in cytoplasm without being enclosed in a nuclear membrane.

9. Eukaryotic

Ans. The condition of the cell where a well defined nucleus is present *i.e.,* nuclear material is enclosed within nuclear envelope.

10. Hermaphrodite

Ans. Hermaphrodite is an individual organism which possess both sex.

11. Viviparous

Ans. Viviparous animals give birth to young ones.

12. Nocturnal

Ans. Nocturnal are those animals that are active at night.

13. Cold-blooded animals

Ans. Cold-blooded animals are animals whose body temperature changes according to the surrounding temperature.

14. Warm-blooded animals

Ans. Warm-blooded animals are animals whose body temperature remains constant irrespective of the surrounding temperature.

Chapter 9. Economic Importance of Bacteria and Fungi

1. Nitrification

Ans. Nitrification is the process by which ammonia is converted into nitrites by nitrifying bacteria.

2. Antibiotics

Ans. Antibiotics are the chemical compounds produced by microbes to kill or inhibit the growth of other microbes.

3. Putrefaction

Ans. Incomplete breakdown of organic matter by bacteria with emission of foul smell is called putrefaction.

4. Serum

Ans. Serum is blood without corpuscles and fibrinogen.

5. Botulism

Ans. Botulism is a very serious food poisoning caused due to a special bacteria sometimes found in tinned and sealed foods.

6. Fermentation

Ans. Fermentation is the process by which microbes breakdown carbohydrates into ethanol or lactic acid in absence of oxygen.

7. Anthrax

Ans. Anthrax is a bacterial disease caused commonly in cattle leading to swelling of body and reduced milk yield. This bacteria is also used as a bioweapon.

8. Penicillin

Ans. Penicillin is the first antibiotic produced from a fungus *Penicillium notatum* by Alexander Fleming in 1929.

9. Saprophyte

Ans. Saprophyte are organisms derived food from dead and decayed organisms.

10. Vaccine

Ans. Vaccine is a preparation of dead or weakened germs.

Chapter 10. Nutrition

1. Malnutrition

Ans. Malnutrition is the condition in which a person suffers due to lack or deficiency of one or more essential elements of food.

2. Balanced diet

Ans. A balanced diet is one which contains all the principal constituents of food in proper quantity.

3. Roughage

Ans. Roughage is the undigested portion of the diet. The fruits, vegetables and the outer coats of the seeds and grains are particularly rich in roughage.

4. Kwashiorkor

Ans. Kwashiorkor is a disease mostly present in young children which means the sickness in baby when mother stops breast-feeding to their babies.

5. Deficiency disease

Ans. The diseases caused when sufficient nutrients are not taken in diet are known as deficiency diseases.

6. Scurvy

Ans. A disease caused due to deficiency of vitamin C. Its symptoms are spongy and bleeding gums, internal haemorrhage, etc.

Chapter 11. Digestive System

1. Digestion **[February, 2020]**

Ans. Digestion is a catabolic process in which complex and large components of food are broken down into respective simper and smaller forms with the help of various enzymes.

2. Teeth

Ans. Teeth are hard structures present on the bones of both lower and upper jaw.

3. Peristaltic movement

Ans. The necessary action of digestive process is essential for moving food in a regular manner along digestive tube. The inner lining of alimentary canal are muscles which contract rhythmically in order to push the food forward is called peristaltic movement.

4. Dental formula

Ans. The kinds of teeth are represented by equation which is constant for each species. The two halves of each jaw are identical, the teeth of only one side are recorded in the formula.

5. Deglutition

Ans. The process of passing bolus or mass of food in oesophagus from buccal cavity's deglutition.

6. Assimilation

Ans. Assimilation is the utilisation of absorbed food for various purposes.

7. Enzyme

Ans. Biological catalysts which alter the rate of a biochemical reaction without unddergoing any change are called as enzymes.

8. Carnivores

Ans. The animals which feed on the flesh of other animals are known as carnivores.

9. Peristalsis **[February, 2020]**

Ans. Peristalsis is a special movement of food which occurs through all regions of the gut and is a wave of constriction caused by the circular muscles of the gut which pushes the food along the gut.

10. Omnivore

Ans. Omnivore feeds on both plants and animals.

11. Pylorus

Ans. Pylorus is the opening of the stomach into the small intestine.

12. Kilocalorie

Ans. Kilocalorie represent the energy value of foods which is 1000 times of the unit calorie.

13. Basal metabolic rate

Ans. Basal metabolic rate is the amount of energy per unit time that a person needs to keep the body functioning at rest.

Chapter 12. Skeleton-Movement and Locomotion

1. Joint

Ans. The junction between two articulating bones is known as joint.

2. Exoskeleton

Ans. The skeleton present on the body of an organism is exoskeleton, e.g., scales of fish, lizard and horns of animals.

3. Appendicular skeleton

Ans. Appendicular skeleton is the part of endoskeleton which consists of bones of pectoral and pelvic girdles, arms and legs.

4. True ribs

Ans. The ribs which are directly connected to the sternum through the cartilage are called true ribs. These are seven pairs.

5. Glinding joints

Ans. Glinding joints occurs between the bones of wrist and also between the bones of ankle as well as in between vertebrae.

6. Hinge joints

Ans. Hinge joint moves like a hinge in one point example elbow joints between humerus and ulna.

Chapter 13. Skin-"The Jack of all Trades"

1. Dermatology

Ans. Dermatology is the branch of science that deals with diagnosis and treatment of skin disease.

2. Skin

Ans. Skin is the largest organ, consisting of tissues to perform specific functions.

3. Vasodilation

Ans. The blood versels get dilated due to an increased in temperature of body is vasodilation.

4. Vasoconstriction

Ans. Due to low temperature, blood vessels get constricted and less blood flows is vasoconstriction.

5. Endothermal

Ans. Animals with an internal heat-producing mecanism are endothermal animals.

6. Ectothermal

Ans. Animals have body temperature that rise and fall according to outside temperature one ectothermal animals.

7. Heat stroke **[February, 2020]**

Ans. A condition in which sweat production is unable to keep pace with in evaporaion is very hot winds that result in rise of body's temperature which may prove fatal.

Chapter 14. The Respiratory System

1. Respiration

Ans. It is biochemical reaction in which glucose is oxidised by stepwise reactions with the help of enzymes.

2. Tidal volume

Ans. It is the volume of air inspired or expired during normal breathing.

3. Inspiration

Ans. It is the inhalation of oxygen into lungs.

4. Asphyxia

Ans. The condition in which oxygen supply to the respiratory system is cut off.

5. Ventilation of lungs.

Ans. The breating of air in and out of the lungs is called ventilation of lungs.

Chapter 15. Hygiene

1. Health

Ans. Health is a condition when the body is functioning normally.

2. Personal hygiene

Ans. Personal hygiene includes how we take care of our body such as bathing, washing hands before and after meals, avoid breathing by mouth, etc.

3. Contamination

Ans. Contamination means unwanted entry of disease germs into drinking water or edible foods.

4. Bacillary dysentery

Ans. Bacillary dysentery is a disease caused by a bacterium which is present in contaminated water and food. It have symptoms like loose motions, abdominal pain.

5. Amoebic dysentery

Ans. Amoebic dysentery is a disease caused by a protozoa which is spread by drinking contaminated water.

Chapter 16. Disease: Cause and Control

1. Virus

Ans. Viruses are highly specific. They attack only one kind of host and only certain tissues. Virus is like a gene with no cell of its own. It dictates the host cell to produce more of its own kind of virus particles only. The host cell ultimately dies and the liberated virus particles attack new host cells.

2. Endemic

Ans. Endemic diseases- when the disease is found in a certain area only attacking a fewer number of people, e.g., yellow fever in African countries.

3. Epidemic

Ans. Epidemic diseases- when the disease breaks out and spreads from place to place affecting large numbers of people at the same time e.g., plague in India in 1994.

4. Non infectious disease

Ans. In non-communicable diseases, there is no germ of any kind and they cannot spread from a patient to another person by contact or by any other method.

5. *Wuchereria bancrofti*

Ans. Wuchereria bancrofti causes filariasis. It transmitted through the bite of the Culex mosquito. These tiny worms lie in the lymphatic system and connective tissues of the human body mainly the limbs

6. BCG

Ans. BCG is a vaccine which is effective against the bacterial disease tuberculosis (TB). It develops immunity to TB. BCG. stands for Bacillus Calmete Guerin.

7. Incubation period

Ans. Incubation period is the period between the entry of germs and the appearance of the first symptoms of the disease. Example: Incubatioin period of pneumonia is 1-3 days.

8. Chicken pox

Ans. It is a viral disease caused by the Herpes Varcella zoster virus. It spreads rapidly by close contact with an infected person. A live attenuated vaccine Varicella is administered to children of 12-18 months for active immunisation.

9. Hepatitis A

Ans. It is a viral disease caused by Hepatitis A virus which result in inflammation of the liver. It has an incubation period of 14-45 days. It is mainly transmitted through contaminated food and water.

Chapter 17. Aids of Health

1. Immunity

Ans. Whenever our body is attacked by any foreign substances which are harmfull for us our body releases defense mechanism against it which is known as immunity. This helps us to fight against the germs.

2. Antigen

Ans. The foreign substance which can harm the body is known as antigen. For example:bacteria, fungus, etc.

3. Antibiotics

Ans. Antibiotics are chemical substances or in other words protein that is found in the blood which works against the germs.

4. Toxin

Ans. Toxin is the term,used for any poisonous substance produced by any plant or animals.

5. Deodorant

Ans. Deodorants are mask for bad smell, its neither antiseptics nor disinfectants.

6. Antiseptics **[February, 2020]**

Ans. Antiseptics are chemicals that apply on the body to kill germs. Lyson, carbolic acid, iodine, benzoic acid, mercurochrome, boric acid, etc., are dilute and work as good antiseptics.

7. Disinfectants

Ans. Disinfectants are strong chemical substances used to clean the floor to kill the germs. Cresol, phenol, Lysol, 40% formalin, lime, etc are common disinfectants.

8. Vaccines

Ans. The vaccine is introduced into the body usually by injection and sometimes orally e.g., polio drops, it involve in the stimulation of WBCs to produce antibodies against the particular disease.

Chapter 18. Health Organisation

1. Local bodies

Ans. All cities, towns and even the larger villages have some kind of local body to look after the health and sanitation of their people. These include municipal corporations, municipalities, etc.

2. National bodies

Ans. Several agencies, institutes and research centres are actively engaged in studing the health problems of the country. For each disease, they undertake a survey, identification of the transmitting agent and its breeding places, arranging immunation and some steps to be taken to fight against respective disease.

3. WHO

Ans. This WHO (World Health Organisation) is an agency of United Nations Organisation, established in 1948 to collect and supply information about occurrence of diseases, to promote and support research projects on diseases, to organise campaigns for the control of epidemic and endemic diseases.

4. Red Cross Society

Ans. Red Cross is the international agency whose function is to perform activities which should prevent or remove human sufferings. Red Cross Society was founded in 1864

Chapter 19. Waste Generation and Management

1. Incineration

Ans. Reducing the weight and volume of solid waste by burning it at high temperature.

2. Sewage

Ans. The liquid waste that run off into the drains from the house is called sewage.

3. Biodegradable waste

Ans. The waste that can be decomposed by micro-organisms is called biodegradable waste.

4. Compost

Ans. Compost is the manure obtained by decomposing biodegradable waste by mictro-organisams.

5. e-waste

Ans. e-waste is the waste, generated from the discarded electronic devices.

6. Segregation

Ans. Segregation means separating the wastes as reusable, degradable and non-degradable.

7. Drainage

Ans. Removal of waste or sewage (surface or sub surface) by drains from given area.

❑